D0787877

Agricultural Recycling of Sewage Sludge and the Environment

Agricultural Recycling of Sewage Sludge and the Environment

S.R. Smith

WRc
Marlow
Buckinghamshire
UK

CAB INTERNATIONAL

CAB INTERNATIONAL
Wallingford
Oxon OX10 8DE
UK

Tel: +44 (0)1491 832111
Fax: +44 (0)1491 833508
E-mail: cabi@cabi.org
Telex: 847964 (COMAGG G)

A catalogue record for this book is available from the British Library.

ISBN 0 85198 980 2

Printed and bound in the UK by Biddles Ltd, Guildford

Contents

Preface

The centralized collection and treatment of wastewater produces a residual sludge requiring safe and economic disposal. Application to agricultural land is the principal way of deriving a beneficial use for the residual sewage sludge by recycling plant nutrients and organic matter to soil for crop production. Agricultural utilization also provides a cost-effective method of sludge disposal, but it is essential that sludge recycling in agriculture is controlled to minimize potential environmental problems. The regulatory system for the agricultural use of sewage sludge in the UK is based on research which has defined soil concentration limits for potentially toxic elements (PTEs) to safeguard human health and crop yields, and sludge treatment and management practices to minimize risks of infection from pathogenic organisms which may be present in sludge. Despite the regulatory controls, however, some concerns remain about the practice of applying sewage sludge to agricultural land. In particular, the soil PTE limits have been criticized because they do not appear to consider potential effects on long-term soil fertility. Furthermore, the implications for human health arising from the contamination of sludge with organic contaminants may result in public anxiety from agricultural recycling. Conversely, regulations which are too precautionary adversely affect sludge recycling and may have significant economic implications. However, research can inform decisions about potential risks and provide a sound basis for framing regulations and codes of practice which minimize potentially detrimental environmental effects whilst maintaining agriculture as a recognized outlet of value for the recycling of sewage sludge. The scientific research involved is multi-disciplinary and interpretation of the results is complex, not least because of the range of soil types to be considered and the multi-factorial relations involved in field situations. The main purpose of this book is to provide a comprehensive source of information about the environmental impact of sludge application to agricultural land, particularly in relation to other accepted agricultural practices, to put the potential environmental affects into perspective and distinguish between perceived and real risk.

Acknowledgements

The preparation of this book was made possible by the generous support of the Foundation for Water Research (FWR) whose staff also provided guidance and encouragement throughout. An overview of a subject as complex as sewage sludge recycling in agriculture inevitably required a multi-disciplinary approach and I am especially indebted to several specialists in the different fields of research for their constructive and critical comments on the manuscript. In particular, I would like to thank Barbara Stark (Consultant) for commenting on the animal ingestion sections, Ken Giller (Wye College, University of London) for reading the chapter on soil microbial effects, and many colleagues at WRc for help with various sections, notably the grassland contamination model, organic contaminants, animal pathogens and sludge regulations, and for typing the manuscript. Finally my wife, Toni, deserves a very special thank you for her patience, support and encouragement during the writing of this book.

Chapter one:

Introduction

The need to treat sewage effluents to maintain high quality discharges into receiving water courses ultimately produces a sewage sludge which requires safe and economical disposal. In the UK, it is estimated that 1.1 million tonnes of sludge (dry solids) are produced annually (CES, 1993) and within the European Union (EU) the total amount is approximately 6.5 million tonnes (dry solids) (Hall and Dalimier, 1994, see Fig. 1.1). The options currently available for dealing with the sludge include application to agricultural land, incineration, land reclamation, landfill, forestry, sea disposal and dedicated sacrificial land. The relative importance of these outlets is shown in Fig. 1.2 for both the UK, and for the EU as a whole.

Agricultural use represents the largest outlet for sludge in the UK accounting for 44% of the sludge currently produced and is also the second most important outlet when sludge disposal within the EU is considered accounting for 37% of total EU sludge production (CES, 1993; Hall and Dalimier, 1994). The land area utilized for sludge application in the UK represents less than 1.0% of the total area of agricultural land (MAFF, 1990a; CES, 1993). Although sea disposal represents a relatively small outlet for sludge within the EU, this outlet currently takes 30% of the UK's sludge. However, the practice of sea disposal ends in 1998 due to the Council of the European Communities (CEC) Directive concerning urban waste water treatment (CEC, 1991a). This Directive also requires that sewage currently discharged through sea outfalls is treated, the extent of treatment depending on the size of the contributing population and the sensitivity of the receiving waters. Consequently, by the year 2006 it is anticipated that the amount of sludge to be disposed of in the UK could increase by approximately 50-60% compared with production in 1990 (Hall and Dalimier, 1994). Sludge production within the EU is also likely to increase by more than 50%.

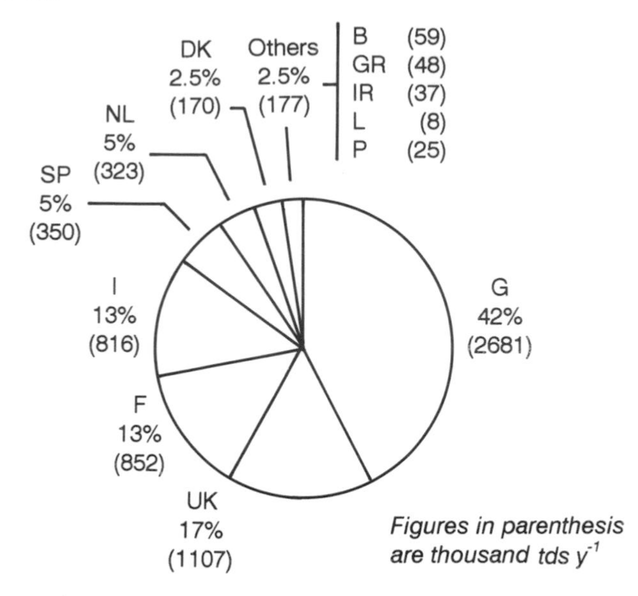

Fig. 1.1. Sludge production in the European Union (Hall and Dalimier, 1994).

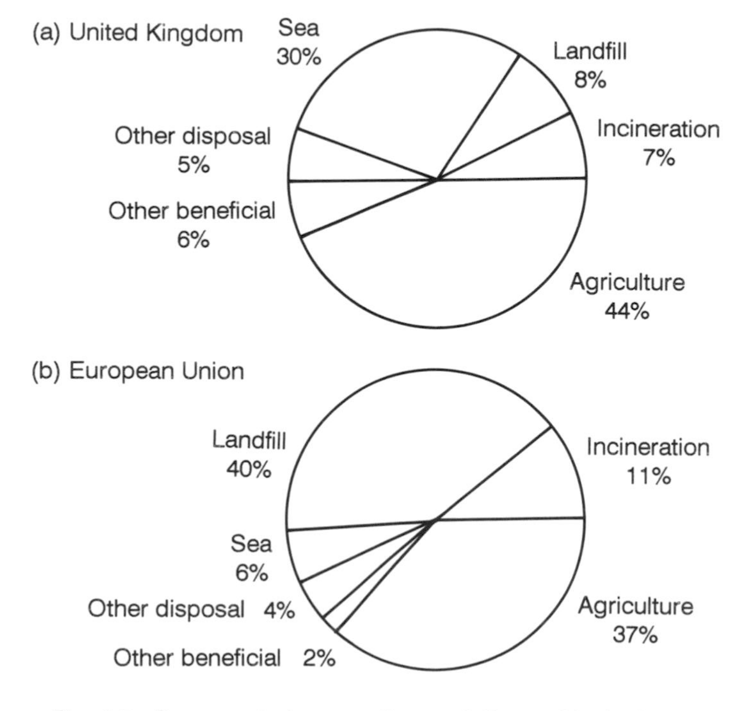

Fig. 1.2. Sewage sludge recycling and disposal in the United Kingdom and the European Union (Hall and Dalimier, 1994).

Sludge may be produced by one or more sewage treatment processes. Raw or primary sludge is produced by gravity settlement after initial screening of litter and grit removal. Subsequent biological treatment of the settled sewage, by the activated sludge process or percolating filters, produces a secondary sludge largely composed of bacteria and is usually co-settled with the primary sludge. This sludge can be applied to land or otherwise disposed of without further treatment, but most sludges are treated to reduce bulk and to avoid potential problems from odour and pathogens. Such treatment processes affect the properties of the sludge products making them more amenable for reuse or disposal to particular outlets, and can influence their agronomic value. The main sludge processing options are summarized in Fig. 1.3. UK sludge treatment and disposal costs are in excess of £250 million per year (Hall, 1992), accounting for about half of the total costs of sewage treatment, and these costs are likely to be much higher in the future due to EU legislation.

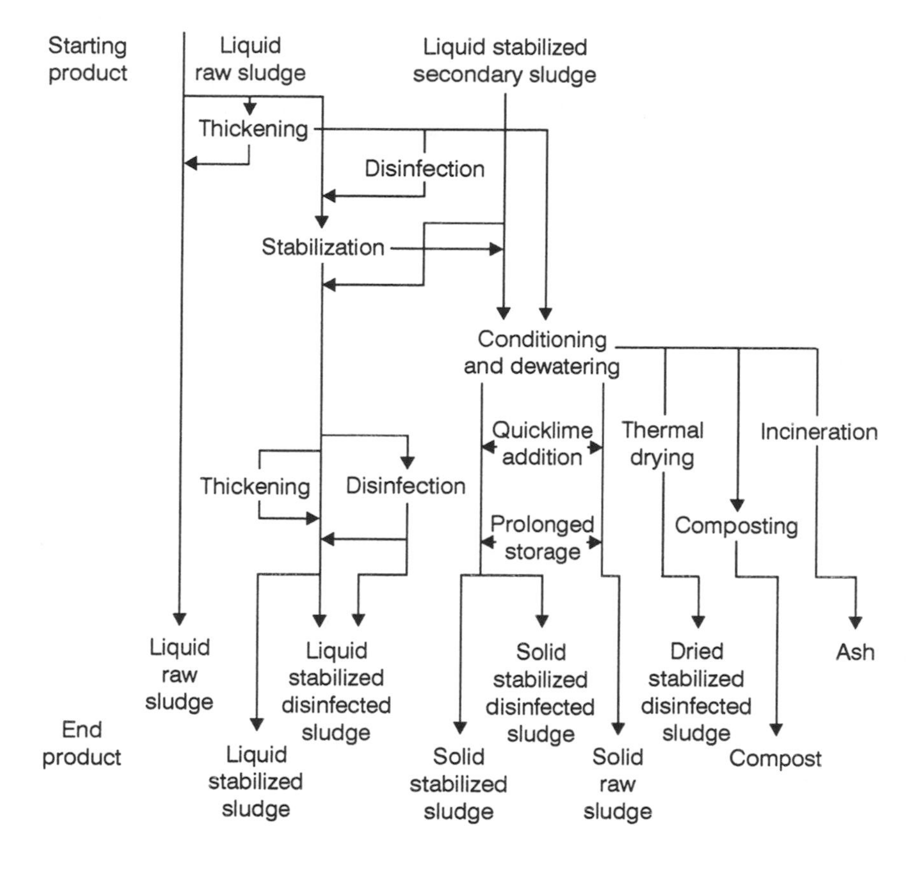

Fig. 1.3. Flow chart of sludge processing options for production of suitable end-products for utilization or disposal.

The application of sewage sludge to farmland is generally the most economical outlet for sludge and provides an opportunity to recycle beneficial plant nutrients and organic matter to soil for crop production. For example, the N and P fertilizer replacement value of sewage sludges has been reported frequently (e.g. Coker and Carlton-Smith, 1986; Coker *et al.*, 1987a,b) and fertilizer recommendations for sludge have been published (WRc, 1985; MAFF, 1994a). Hall (1992) estimated that the potential savings in fertilizer provided by sludge spread on farmland each year were in excess of £15 million. Crop productivity can also be increased by improving soil physical properties through the application to soil of organic matter contained in sludge (Pagliai *et al.*, 1981; Smith *et al.*, 1992a).

With increasing quantities of sludge being produced within the EU, the water industry will seek to recycle increasing quantities of sewage sludge to agriculture. In order to allay public and political concern about health and safety, this method of disposal needs to be well researched so that regulatory controls can be seen to be soundly based. Some of the potential concerns were highlighted by Kelley *et al.* (1984) from a survey of technical personnel within the waste disposal industry on research and education needs for land application of sewage sludge. Educating the public and accumulation of heavy metals in soils were seen as priority areas. Other important concerns included the entry of heavy metals into the food chain, protection of water resources and methods of applying sewage sludge to agricultural land. In general, the surveyed population targeted the health aspects of land application, public acceptance of sludge and protection of the soil and water environment as the key issues for additional emphasis and research.

To ensure the safe and beneficial use of sewage sludge in agriculture the Council of the European Communities has adopted a Directive designed to protect the agricultural environment where sludge is used (CEC, 1986a). This Directive has been implemented in the UK through the *Sludge (Use in Agriculture) Regulations 1989* (SI, 1989a) which are complemented by a *Code of Practice for Agricultural Use of Sewage Sludge* (DoE, 1989a) providing further non-statutory measures to protect the environment when sludge is spread on farmland. Furthermore, regulations designed to reduce nitrate leaching losses from agricultural soils to protect potable water supplies (CEC, 1991b) will also constrain the spreading of sludge on farmland. National and European legislation and codes of practice relevant to the agricultural use of sewage sludge are described in the Appendix.

The treatment and use of sewage sludge has been the subject of intensive nationally funded research for the past 25 years in the EU, which has been coordinated by a COST (Cooperation in Science and Technology) programme, 68 and 681, under the auspices of the Commission of the European Communities (DGXII). Hall *et al.* (1992) reviewed the work and extensive output of COST 68/681 from 1972 until 1990 when the programme ended. The UK Government recently commissioned a scientific review of the potential environmental implications of the rules on sewage sludge recycling to agricultural land. In particular, the scientific basis to the regulations on potentially toxic elements (PTEs) were carefully scrutinized in relation to the possible effects on soil fertility (MAFF/DoE, 1993a) and food safety (MAFF/DoE, 1993b).

Considerable research effort on the environmental effects of applying sewage sludge to agricultural soil has also taken place in the US (Joint Conference on Recycling Municipal Sludges and Effluents on Land, 1973; Page *et al.*, 1983; Page *et al.*, 1987; Page and Logan, 1989; Chaney, 1990a,b). Here, the US Environmental Protection Agency (US EPA) has promulgated a new regulation, the *Standards for the Use or Disposal of Sewage Sludge (40 CFR Part 503)* (US EPA, 1993), which encourages the beneficial recycling of sludge on agricultural land (see Appendix). The controls stipulated in the Final Part 503 Rule on sludge contaminants were developed from a rigorous and arguably conservative risk assessment of critical environmental pathways for potential exposure of soils, plants, animals and humans. The US approach takes advantage of the soil's capacity to assimilate and detoxify pollutants. Consequently the US regulations permit significantly larger PTE additions to agricultural land in sludge and appear considerably more pragmatic to sludge recycling compared with the highly precautionary approach adopted in some European countries. Nevertheless, the procedures and limit values laid down in Directive 86/278/EEC offer a rational basis for recycling sewage sludge in agriculture.

For the foreseeable future, sewage treatment works will continue to function as 'sludge factories' with increasing and unstoppable output. Indeed, the Urban Waste Water Treatment Directive (CEC, 1991a) will lead to a substantial increase in the amount of sludge requiring disposal in the UK and other EU Member States. However, the disposal outlets are decreasing, yet economic pressures still require low-cost solutions to sludge disposal problems. For example, some Member States have adopted approaches to regulating the use of sludge in agriculture more precautionary than the well-founded provisions of Directive 86/278/EEC for agricultural utilization (CEC, 1986a). This is demonstrated by the limit values for heavy metals in sludges and sludged soils listed in the Appendix (Tables A4 and A5). Consequently, increased demands are being placed upon sludge disposal routes when regulations controlling the traditional outlets for sludge are being tightened.

The ultimate endpoint of the precautionary principle is the balancing of PTE inputs with their offtakes in crops thereby maintaining a *status quo* and avoiding the long-term accumulation of PTEs in soil. It is argued that this may be the only approach which will guarantee maintaining the multifunctionality of soils for any form of land use activity in the future. Whilst this is a laudable goal it is probably unachievable in practice due to the diverse range of PTE inputs to soil including those from other fertilizer materials, essential for agronomic productivity, as well as atmospheric deposition. Furthermore, the approach is not without potential environmental impact itself because the sludge must be disposed of by some means and normally by either landfilling or by incineration (Fig. 1.2). Consequently the precautionary principle also brings with it a significant economic penalty which may be an unnecessary burden in achieving an environmentally safe method of sewage sludge disposal.

It also follows that the very precautionary approach taken by some EU Member States, in setting highly stringent limits on PTEs and trace organic contaminants in sludge for agricultural use, could generate undue alarm among farmers. There is growing concern that the sensitivity of food industries and farmers' cooperatives to alarmist attitudes which prevail in some countries, could close off this outlet for sludge (e.g. Hellström and Dahlberg, 1994; Tritt, 1994). Sludge quality restrictions are such that in some countries they will in any case ensure the cessation of agricultural use of sludge. Strict limitations on permissible soil concentrations and annual rates of addition of PTEs in other countries may reduce the allowable rate of application of sludge so far as to make the disruption caused by spreading operations no longer acceptable to the farmer since the nutrient benefits may be too small.

Unnecessarily restrictive limits on sludge quality effectively precluding agricultural utilization may even be counter-productive to environmental protection in general. Without a beneficial route for reuse, sludge is effectively a waste material requiring disposal. The impetus to 'clean up' by reducing point and diffuse sources of contamination may be diminished when standards are set for recycling which cannot be reasonably or practically achieved.

Despite the adoption of different philosophical approaches to regulating the agricultural use of sludge in Europe and the US, ironically, the quality of municipal sewage sludge (measured by concentrations of PTEs) is very similar in all of the developed nations. In other words, the potential problems associated with pollutants occurring in sewage sludge are more or less the same in all developed countries: it is the perceived level of environmental risk from the contaminants in sludge which differs. Evidence from recent surveys in the UK (CES, 1993; Hall and Dalimier, 1994) suggests that there may not be much scope for further improvements in sludge quality although reductions may be possible in the future as new legislation is phased in which restricts discharges of PTEs to sewers. This issue is discussed further in Chapter two.

In view of these developments, there is a need for a more cohesive policy within the EU regulating sludge application to agricultural land so that sludge disposal authorities are able to implement secure and cost-effective disposal strategies for sludge. This process will not be assisted by the developing regulations on hazardous waste, although it is highly unlikely that sewage sludge would be designated as hazardous (see Appendix). However, it could be difficult, for example, to explain to the public, and to the agricultural industry, that even though certain contaminants present in sewage sludge may be designated as hazardous, sludges themselves can still be used beneficially on land. Such developments could have a significant negative impact on public and farmer perceptions of the use of sludge on agricultural land, irrespective of sludge quality. This emphasizes the strategic importance of the scientific basis to agricultural recycling of sewage sludge ensuring that sludge is used beneficially without detriment to the environment.

In the UK, agricultural utilization of sewage sludge is a well established practice which has developed successfully over the last 30 years (Davis, 1989). Since agricultural use of sludge is a cost-effective method of disposal, and is the only outlet associated with obvious environmental benefits from recycling of plant nutrients and organic matter, spreading sludge on farmland would appear to be a natural and sensible solution. However, to maintain or develop this outlet for sludge, anxieties must be allayed about potential long-term harmful effects on the environment. The purpose of this book, therefore, is to assess the environmental impact of sludge recycling to agricultural land based on a critical and extensive review of the scientific literature.

In an overview of soil contamination problems, Berrow (1986) noted that an enormous amount of data had accumulated over the past 20 years on the benefits and hazards of sewage sludge application to soils. Many reviews have also been published on nearly all the principal topic areas which might be considered of concern. The views expressed by those authors have been considered here, and integrated with recently published information in an attempt to gain an overall consensus of the importance of possible environmental impacts resulting from spreading sewage sludge on farmland.

The scientific literature can be conveniently divided according to the potential environmental effects of the four principal groups of sludge components which include: (1) PTEs, (2) major plant nutrients (nitrogen and phosphorus), (3) organic contaminants, and (4) pathogenic agents. Effects of the main sludge contaminant groups on soils, crops, grazing animal and human health and impacts on water quality and natural ecosystems have been considered and the impacts of sludge nutrients on water and air quality have also been examined. In Chapter twelve, an environmental assessment of sludge recycling to agricultural land is made of each of the potentially impacted areas which include (1) human health, (2) crop yields, (3) grazing animal health, (4) groundwater quality, (5) surface water quality, (6) air quality, (7) soil fertility, and (8) natural ecosystems.

Chapter two:

Potentially Toxic Elements (PTEs) in Sewage Sludge and Soil

Inputs of PTEs to the Sewer System

The occurrence of potentially toxic elements (PTEs) in sewage sludge arises principally through domestic, road run-off and industrial inputs to the combined sewerage system in the UK. Industrial inputs now represent a relatively small proportion of total metal discharge to sewers generally and domestic sources account for the largest amounts of certain important elements, particularly of Cu and Zn (Table 2.1). For example, Critchley and Agg (1986) reported that 62% of the Cu entering the sewer system was from domestic sources compared with 5% from road run-off and only 3% from industry. The Cu is derived principally from Cu plumbing systems; Coppoolse (1992) estimated that more than 80% of the Cu discharged from households originated from corrosion of Cu tubing.

Table 2.1. Summary of estimated metal inputs to UK sewer system.

Source	Metal input (t y^{-1})						
	Cd	Pb	Hg	Cu	Zn	Cr	Ni
Domestic	10.7	193	0.5	634	816	24	71
Road run-off	1.7	219	0.05	64	305	19	38
Industry	32.4	11	7.9	30	150	403	110
Total identifiable inputs	44.8	423	8.45	728	1271	446	219
Estimated total load discharged to sewers	53.1	758	>13.7	1024	2945	1136	414
% unaccounted for	16	44	>38	29	57	61	47

Source: Critchley and Agg (1986)

Critchley and Agg (1986) also demonstrated a similar pattern of discharge for Zn with domestic, road and industrial sources representing 64%, 24% and 12% of the total identifiable inputs of Zn. The presence of Zn in household discharges is explained through corrosion of both galvanized iron domestic plumbing systems (Kumper, 1985) and of cast iron water mains which form 80% of the water distribution network in the UK (Hedgecott and Rogers, 1991). The concentrations of Cu and Zn in sewage sludges found by Page (1974) from a review of about 300 sewage treatment works serving only residential areas in the UK, USA, Canada and Sweden were over 500 mg Cu kg^{-1} and 1000 mg Zn kg^{-1} (dry solids) and were therefore attributed to inputs from domestic plumbing.

Not all the Cu and Zn entering the sewer system, based on measured concentrations in sewage sludges, could be accounted for by Critchley and Agg (1986). These unaccounted for fractions represented 29% and 57% of the estimated total loads of Cu and Zn, respectively. However, this discrepancy could be explained due to insufficient information being available on actual diffuse sources of heavy metals entering the sewer system.

More recently, Comber and Gunn (1994) conducted a survey of diffuse sources of heavy metals going to sewer within a representative catchment at Bracknell in the UK with a fairly typical mix of domestic, run-off, commercial and light industrial sources of metals. The study confirmed the importance of Cu inputs from tap water, and particularly from the domestic hot supply. At Bracknell, the Cu leached from plumbing contributed 55% of the total Cu entering the sewage treatment works (STW) and 66% of this came from the hot water tank alone. Contributions from other plumbing sources, such as those found in commercial and light industrial premises, increased the Cu input to approximately 75% of the total load. Newer plumbing was shown to leach greater quantities of Cu than older fitments. Significant differences in tap water Cu concentrations were also identified in different areas of the UK depending on the hardness of the mains supply. Hard water is particularly aggressive to Cu plumbing such that Cu inputs to the sewer system in tap water were an order of magnitude higher in regions supplied with hard water compared with soft water areas. This also had a pronounced effect on Cu concentrations in sewage sludges which were directly proportional to increasing hardness of the mains supply.

The domestic inputs of Zn and Ni also dominated the total load entering the STW. In contrast to Cu, plumbing contributions only amounted to between 25 and 30% of the total loads of these metals. The majority of the input was attributed to domestic activities including clothes and dish washing, and from faeces. For Zn, washing machines were shown to contribute almost 20% of the total domestic load and faeces approximately 50%. Dishwashing supplied 19% of the Ni going to sewer whereas faeces provided 40% of the total Ni load. Comber and Gunn (1994) noted that Zn may also be used as an active ingredient in a diverse range of body care products although the inputs of Zn to the sewer system from these sources are difficult to estimate. However, medicated shampoos in particular can contain up to 0.5% of Zn as an active ingredient to combat dandruff and these products represent a significant proportion (26%) of total shampoo sales in the UK. Comber and Gunn (1994) estimated that the use of these products could be equivalent to approximately half of the Zn which is estimated to be added through bathing.

As would be expected, industry is the major source of Pb, Cd, Cr and Ni (Critchley and Agg, 1986; Comber and Gunn, 1994). Effluent from a light industrial estate in the Bracknell catchment examined by Comber and Gunn (1994) had similar concentrations of Cu, Zn and Hg to those derived from domestic sources. For other determinands, however, the concentrations in the industrial effluent were elevated above those measured from the purely domestic source. Lead showed the greatest enhancement of over five times domestic levels and Cd was approximately four times domestic values. Chromium and Ni concentrations were 2.3 and 1.4 times the domestic level. The predominance of automobile-related activity (bodyshops and engine repairs) was considered the principal cause of the elevated concentrations. In terms of total load to the STW only Pb was considered of major importance (34%) with the contribution from the other metals being less than 15%.

Run-off was the most important diffuse source of Pb entering the sewer system in catchments without Pb in the domestic plumbing, forming 44% of the total load to Bracknell STW. This high loading reflected contributions of Pb from car exhausts. Comber and Gunn (1994) considered that, with the gradual phasing out of leaded fuel, inputs from this source should become less significant in the future. In areas with appreciable runs of Pb piping, the domestic Pb input to the sewer may account for up to 60% of the total load to the STW. Run-off accounted for less than 14% of the total load for the other metals.

In general, the concentrations in sewage sludge of the contaminants derived predominantly from industry are small relative to those of Cu and Zn (Tables 2.2 and 2.3) and rarely limit the recycling of sludge to agriculture given the current regulations in the UK (Table A2; SI, 1989a). However, Hg has the potential to become a limiting element at the present concentrations in sewage sludge if Cu and Zn concentrations in sludge continue to decline but Hg concentration remains static (Tables 2.2 and 2.3). Comber and Gunn (1994) could not account for a large proportion of the Hg going to sewer in their catchment model. In the absence of industrial inputs, the short-fall was attributed to inputs from dental practices. Hutton and Symon (1986) estimated that 5.6 t of Hg were discharged to the sewer system each year from dentists, accounting for 74% of the total sewer load. Comber and Gunn (1994) recommended that an in-depth survey is undertaken within the UK, to assess the measures being taken to reduce Hg contamination of dental waste and to quantify these inputs to sewers.

Table 2.2. PTE content in sewage sludges (mg kg^{-1} dry solids) utilized on agricultural land according to DoE/WRc 1982/83 UK sludge survey.

Element	Min	Mean	Max	Median	90th percentile
Zinc (Zn)	**279**	**1144**	**27600**	**1205**	**2058**
Copper (Cu)	**69**	**589**	**6140**	**625**	**1087**
Nickel (Ni)	**9**	**61**	**932**	**59**	**303**
Cadmium (Cd)	**<2**	**9**	**152**	**9**	**33**
Lead (Pb)	**43**	**398**	**2644**	**418**	**761**
Mercury (Hg)	**<2**	**4**	**140**	**3**	**7**
Chromium (Cr)	**4**	**197**	**23195**	**124**	**696**
Molybdenum (Mo)	**<2**	**5**	**154**	**5**	**17**
Selenium (Se)	**<2**	**3**	**15**	**3**	**5**
Arsenic (As)	**<2**	**6**	**123**	**5**	**11**
Fluoride (F)[1]	**60**		**40000**	**250**[2]	
Beryllium (Be)[1]	1		30		
Boron (B)[1]	15		1000	30[2]	
Titanium (Ti)	355	1677	11629	1795	2579
Vanadium (V)	7	29	660	26	45
Manganese (Mn)	55	376	13902	318	761
Iron (Fe)	2480	16299	106812	12479	31312
Cobalt (Co)	<2	10	617	8	34
Gallium (Ga)	<2	3	15	3	7
Germanium (Ge)	<2	<2	9	<2	2
Bromine (Br)	4	38	1049	29	68
Rubidium (Rb)	<2	23	232	16	37
Strontium (Sr)	45	158	1335	174	302
Yttrium (Y)	<2	8	34	7	14
Zirconium (Zr)	14	91	2500	70	148
Niobium (Nb)	<2	5	41	5	8
Silver (Ag)	<2	25	1252	25	109
Tin (Sn)	19	90	683	101	209
Antimony (Sb)	<2	8	572	7	15
Tellurium (Te)	<2	<2	<2	<2	<2
Barium (Ba)	23	323	3104	363	622
Tungsten (W)	<2	7	1418	4	11
Thallium (Tl)	<2	<2	5	<2	<2
Bismuth (Bi)	<2	10	557	8	15
Uranium (U)	<2	2	18	2	5

[1] Davis (1980)
[2] Represents 'common value'
Elements in bold controlled under UK regulations (SI, 1989a) and Code of Practice (DoE, 1989a)
Source: Sleeman (1984)

Table 2.3. PTE concentrations (mg kg^{-1} dry solids) in sewage sludges spread on agricultural land in 1990/91.

Element	Percentile[(1)] 10		50		90	
Zn	454	(643)	889	(1205)	1471	(2058)
Cu	215	(261)	473	(625)	974	(1087)
Ni	15	(21)	37	(59)	225	(303)
Cd	1.5	(4)	3.2	(9)	12	(33)
Pb	70	(164)	217	(418)	585	(761)
Hg	1.1	(<2)	3.2	(3)	6.1	(7)
Cr	27	(25)	86	(124)	489	(696)
Mo	0.4	(2)	1.0	(5)	6.2	(17)
Se	0.05	(2)	0.28	(3)	0.65	(5)
As	1.0	(2)	3.2	(5)	6.1	(11)
F	15		100		240	

(1) Figures in brackets denote 1982/83 values from Sleeman (1984)
Source: CES (1993)

PTE Concentrations in Sewage Sludge

The sludge trace metal content data used by Critchley and Agg (1986) in estimating the loadings to sewers was obtained from the survey of sewage sludge composition and disposal practice in 1980/81, carried out by the DoE/NWC Standing Committee on the Disposal of Sewage Sludge (DoE/NWC, 1983). Over the past 30 years, there has been a dramatic reduction in the inputs of metals to sewers resulting from: (a) improved trade effluent control imposed by the water undertakings; (b) changes in the nature of traditional manufacturing industries; and (c) adoption of cleaner manufacturing technologies. These trends have been associated with a concomitant decrease in the metal concentrations in sewage sludge. For example, Rowlands (1992) recently reported a 98% reduction in the concentration of Cd in sewage sludge from the Nottingham STW between 1962 and 1992 (Fig. 2.1a). Zinc content was reduced by approximately 80% at Nottingham (Fig. 2.1b) and at Coventry STW the concentration of Cr decreased by about 70% (Fig. 2.2) over the same period. These remarkable reductions in PTE concentrations have occurred despite the increased use and improved efficiency of anaerobic digestion which increases concentrations of heavy metals due to the loss of volatile solids from sludge (Rowlands, 1992). Further reductions in the concentrations of PTEs in sludge may be anticipated in the future with the phased implementation in the

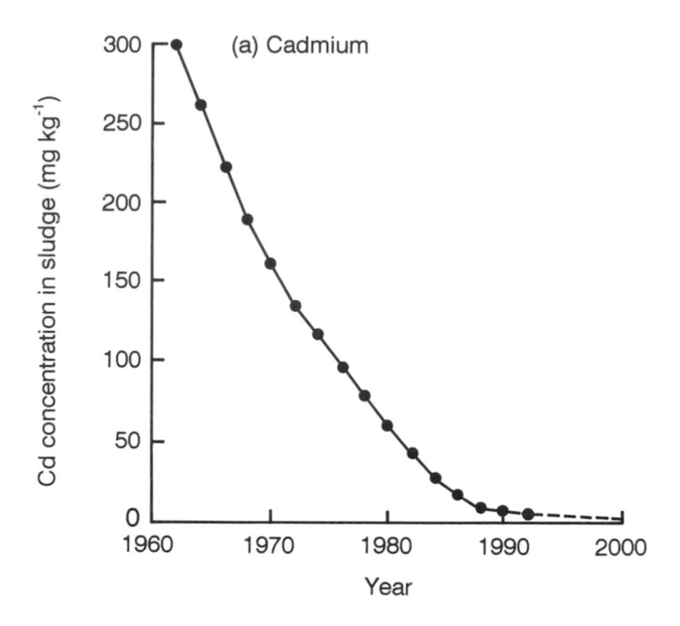

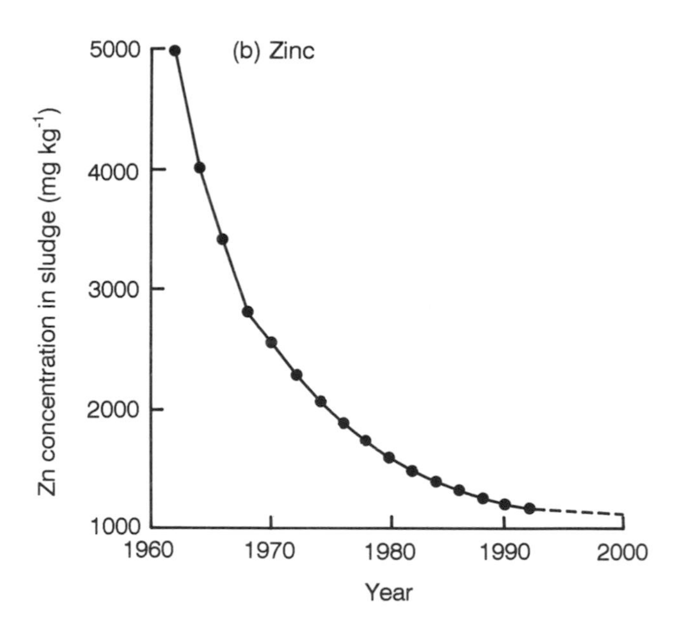

Fig. 2.1. Trends in (a) cadmium and (b) zinc concentrations in sewage sludge from Nottingham STW in the UK (adapted from Rowlands, 1992).

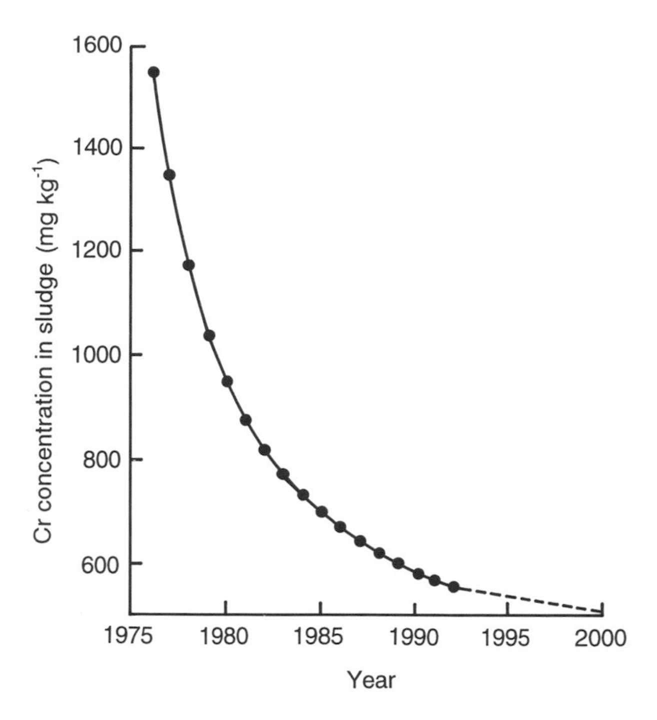

Fig. 2.2. Trend in chromium concentrations in sewage sludge from Coventry STW in the UK (adapted from Rowlands, 1992).

UK of the Dangerous Substances Directives (CEC, 1976, 1986b, 1988 and 1990a). However, this will become progressively more difficult as the contributions from diffuse sources gradually predominate the total heavy metal load in sewage and sewage sludge.

The improvement in sludge quality which has occurred nationally in recent years can be demonstrated by comparing the 1982/83 UK sludge survey (Sleeman, 1984) with the 1990/91 survey results (CES, 1993) in Tables 2.2 and 2.3, respectively. The median concentrations of Zn, Cu and Cr have decreased by 20-30% in the last eight years; Ni content has declined by nearly 40% and Pb by about 50%. By contrast, there has been no change in the Hg content of sludges which remains at a low level, but is significant in relation to the soil limits for PTEs in sludge-treated agricultural land. However, the most dramatic reduction in sludge contamination has been achieved for Cd. Sludges utilized on agricultural land now contain approximately 30% of the Cd compared with the early 1980s. Indeed the median sludge concentration value of 3.2 mg Cd kg^{-1} (dry solids) is approaching the background concentration of Cd present in human faeces of 2 mg kg^{-1} (dry solids) (Davis and Coker, 1980). On this basis the Dutch and Danish regulations (Table A4) for the agricultural use of sludge, which stipulate maximum permitted concentrations of Cd of 1.25 and 0.8 mg kg^{-1} (dry solids), respectively (SO, 1989; Dirkzwager, 1991), appear highly restrictive and may effectively preclude agricultural utilization altogether (Chapter one).

Other Sources of PTEs in Agricultural Soil

There are a wide range of organic wastes spread on land as manures, such as livestock wastes and various industrial sludges (e.g. food, vegetable, meat and drink processing wastes, paper mill sludges, etc.). These wastes contain nutrients and organic matter and thus can potentially be recycled beneficially on land. However, there are currently no specific legislation or quality standards to control their use, but in comparison with sewage sludge, the PTE content of livestock wastes for instance can be appreciable. For example, average and maximum concentrations of Zn in pig slurry of 919 and 1686 mg kg^{-1} (dry solids), respectively, and in cattle slurry average and maximum values of 580 and 840 mg kg^{-1} (dry solids), respectively, have been reported (Fleming, 1993). Copper is another potentially major contaminant of animal wastes (MAFF, 1985). The manure from fattening pigs may contain 300-2000 mg Cu kg^{-1} (mean 870 mg Cu kg^{-1}) in the dry solids and 200-1500 mg Zn kg^{-1} (mean 600 mg Zn kg^{-1}) (MAFF, 1985). However, there are situations when the concentrations of Zn in farm wastes may even exceed these levels. For example, it is common veterinary practice to supplement the diets of fattening pigs with 2500 mg kg^{-1} of Zn (as ZnO) added to the concentrate feed as a growth stimulant and antibacterial agent, reducing the reliance upon antibiotic drugs (Mounsey, 1994). The animals may be fed a 100% diet of the supplemented concentrate and this regime may continue for an indefinite period and until 28 days before slaughter. On the basis that only 5% of the mineral is absorbed by the gut (Stark, 1988) and the digestibility of these feed products is high (50-80%), the Zn content of the waste could be in excess of 5000 mg kg^{-1} (dry solids). Poultry manures may also have elevated Zn with reported average and maximum concentrations of Zn in the range 226-341 mg kg^{-1} (dry solids) and 669-937 mg kg^{-1} (dry solids), respectively (Sims and Wolf, 1994). The scale of the potential problem of soil contamination with PTEs from livestock wastes, compared with the comparatively small inputs from sewage sludge within the EU, is shown in Table 2.4. Even by taking low estimates of Zn and Cu in livestock wastes (500 and 100 mg kg^{-1} dry solids (ds), respectively), the amounts of these metals added to agricultural land are likely to be considerable, and many times higher than for sewage sludge (at least 30 times more in the case of Zn). Ironically, sewage sludge application to agricultural land is regulated in respect of its metal content, but there are no controls for metals on the disposal of animal wastes to farmland.

Table 2.4. Comparison of amounts of Zn and Cu applied to agricultural land annually from sewage sludge and livestock wastes in the European Union.

	PTE	Concentration ($mg\ kg^{-1}$ ds)	Metal addition to EU agricultural land ($t\ y^{-1}$)
Sewage sludge	Zn	1000[(1)]	2400
(2.4 M t ds y^{-1})	Cu	380[(1)]	910
Livestock waste	Zn[(2)]	1000	154000
(154 M t ds y^{-1})	Cu[(2)]	380	58500
	Zn[(3)]	500	77000
	Cu[(3)]	100	15400

(1) Mean PTE concentrations in sludge used in agriculture in the EU (Hall and Dalimier, 1994)
(2) Assuming equivalent average concentrations to that in sewage sludge
(3) Low estimate of current quality
Source: Davis and Dalimier (1994)

In contrast with the amount of data held on PTEs applied to agricultural land in sewage sludge, information on the extent of metal applications in the wastes of farm animals is very limited. Therefore, it is difficult to substantiate the view that the problems of Zn and Cu pollution from livestock manures has been virtually eliminated (Webb and Archer, 1994) on the basis of the published, albeit circumstantial, evidence of metal levels in different types of animal waste. Indeed, the Independent Scientific Committee reviewing soil fertility aspects of PTEs in sludge-treated soil (MAFF/DoE, 1993a) recommended that the contribution from livestock wastes should be considered with respect to regulating heavy metal concentrations in agricultural land.

In addition to livestock wastes, however, other important inputs of heavy metals to agricultural land are from the application of fertilizer materials and atmospheric deposition (Davis and Coker, 1980; Bergback *et al.*, 1994; van der Voet *et al.*, 1994). Cadmium is potentially a major contaminant of rock phosphates used in P fertilizer manufacture. Phosphatic fertilizers are considered the principal cause of Cd accumulations in agricultural soils (Bergback *et al.*, 1994; van der Voet *et al.*, 1994) and this realization has resulted in various restrictive measures being introduced by some countries, such as prohibiting the importation of rock phosphate from high Cd sources, and also by imposing Cd:P ratio limits on fertilizers. Sludges generally have Cd:P ratios much lower than the current fertilizer limits, where these are imposed.

Within the EU there is an estimated input of 334 t Cd y^{-1} to agricultural soils from phosphate fertilizers, which equates approximately to 50% of the total annual addition (van der Voet *et al.*, 1994). In comparison, sewage sludge contributes only 3% of the total Cd load. Approximately 14% of the Cd input to agricultural land is from atmospheric deposition and livestock wastes contribute 18% of the total annual addition. However, it could be argued that sludge applications represent a localized problem of soil contamination with Cd compared with normal phosphate fertilizer application practice. Whereas this may have been true in the past, the improvements in sludge quality in recent years have markedly reduced the inputs of Cd to soil from agricultural recycling of sludge. Furthermore, sewage sludges are rich in P and replace the need for mineral P fertilization (Chapter nine). Therefore, the inclusion of sewage sludge in the nutrient programme for agricultural crops is unlikely to raise soil Cd excessively or increase it much above the level already occurring with conventional sources of inorganic P and from other diffuse inputs of Cd (Cabrera *et al.*, 1994).

In addition to atmospheric and fertilizer inputs of pollutants, there are other perhaps less obvious sources of PTEs which can result in significant accumulations of heavy metals in agricultural soils. For example, Mellor and McCartney (1994) estimated that there was an annual input to soil of at least 6000 t of lead shot in the UK from clay pigeon shooting. Total Pb concentrations at shooting ranges can commonly exceed 5000 mg kg^{-1} which is more than 16 times higher than the UK maximum permissible soil limit for Pb in sludge-treated agricultural land (Table A2).

PTE Accumulations in Sludge-treated Agricultural Land

Although sludge quality has improved markedly in recent years, the concentrations of PTEs in sludge remain larger than those in the soil (Tables 2.5 and 2.6). Once applied to soil in sewage sludge, PTEs are retained indefinitely in the cultivated layers (McGrath and Lane, 1989) such that repeated applications of sludge will gradually increase the trace element content of soil. To place in context the effect of applying sludge to farmland on soil PTE concentrations, the time taken to increase the soil levels up to the maximum permissible limit values for each of the regulated elements (SI, 1989a) was calculated assuming that a single application of sludge was made each year (Table 2.7). This effectively equates to the minimum number of years to reach the maximum soil limit for the most limiting PTE. In practice, there may be several years between applications.

This analysis showed that Zn and Cu are the principal elements limiting sludge recycling to agricultural land in the long-term (Table 2.7). Both elements limit sludge applications to a similar extent given the current soil regulations and sludge quality, reaching their respective maximum soil concentration values in approximately 70-80 years when sludge is applied annually at a rate supplying 170 kg N ha^{-1} (Table 2.7). Interestingly, the Zn/Cu ratio of median sludges has remained relatively constant at about 1.9 over the last decade indicating no major shift has occurred in the relative inputs of Zn and Cu to sewers for moderately contaminated sludges (Table 2.3). However, there is evidence suggesting that inputs of Zn from industrial sources have declined to a greater extent relatively compared with Cu inputs. For example, in 1990/91 the 90th-percentile Zn/Cu ratio was approximately 1.5 whereas the ratio was nearer 2.0 in the 1982/83 survey. This is explained because of the difficulty in controlling inputs of Cu to sewers due to the domestic loading discussed earlier. Consequently, in the long-term Cu could become the main PTE nationally restricting the application of sludge to farmland. Interestingly, PTEs arising from industrial contamination sources were of much less importance compared with Zn and Cu in limiting sludge applications. For example, sludge with median Cr content can be applied for over 600 years before the provisional soil limit set in the UK for Cr of 400 mg kg^{-1} (DoE, 1989a) is reached when sludge applications are governed on the basis of 170 kg N ha^{-1} y^{-1} (Table 2.7). After Zn and Cu, Hg is the next element most likely to limit the recycling of sewage sludge to agricultural land in the long-term.

Table 2.5. Concentrations of heavy metals in soils of England and Wales from the National Soil Inventory (mg kg^{-1} dry soil).

Element	Min.	Max.	Normal range[1]	Mean	Geometric mean	Median	90th percentile
Zn	5.0	3648.0	24.1-260.9	97.1	79.2	82.0	147.0
Cu	1.2	1507.7	5.1-63.5	23.1	18.0	18.1	36.5
Ni	0.8	439.5	4.5-85.8	24.5	19.6	22.6	41.5
Cd	<0.2	40.9		0.8		0.7	1.3
Pb	3.0	16338.0	9.4-220.1	74.0	45.6	40.0	130.0
Cr	0.2	837.8	8.6-135.4	41.2	34.2	39.3	64.2

[1] Geometric mean ± 2 standard deviations
Sources: Gaunt *et al.* (1990); McGrath and Loveland (1992)

Table 2.6. Range and mean values of PTEs in soil (mg kg^{-1} dry soil).

Element	Symbol	Ranges	Mean
Zinc	**Zn**	**1.5 - 2000**	**59.8**
Copper	**Cu**	**<1 - 390**	**25.8**
Nickel	**Ni**	**0.1 - 1520**	**33.7**
Cadmium	**Cd**	**<0.005 - 8.1**	**0.62**
Lead	**Pb**	**<1 - 888**	**29.2**
Mercury	**Hg**	**0.004 - 4.6**	**0.098**
Chromium	**Cr**	**0.9 - 1500**	**84**
Molybdenum	**Mo**	**0.07 - 27.5**	**1.92**
Selenium	**Se**	**0.03 - 2**	**0.4**
Arsenic	**As**	**0.1 - 194**	**11.3**
Fluoride	**F**	**6 - 7070**	**270**
Beryllium	Be	0.5 - 30	1.5[1]
Boron	B	0.9 - 1000	38.3
Titanium	Ti	<60 - 34000	5100
Vanadium	V	0.8 - 1000	108
Manganese	Mn	<1 - 18300	760
Iron	Fe	100 - 210000	32000
Cobalt	Co	0.3 - 200	12
Gallium	Ga	2 - 200	21.1
Germanium	Ge	0.1 - 50	3.0[1]
Bromine	Br	0.27 - 850	42.6
Rubidium	Rb	1.5 - 1800	120
Strontium	Sr	<3 - 3500	278
Yttrium	Y	5 - 213	27.7
Zirconium	Zr	<10 - 3000	345
Niobium	Nb	<6 - 300	14[1]
Silver	Ag	0.01 - 8	0.4[1]
Tin	Sn	0.1 - 40	5.8
Antimony	Sb	0.29 - 8.6	1.7
Barium	Ba	<1 - 10000	568
Tungsten	W	0.5 - 3	1.1[1]
Thallium	Tl	0.1 - 0.8	0.25[1]
Bismuth	Bi	0.1 - 13	0.5[1]
Uranium	U	0.76 - 14	2.18

[1] Tentative value

Elements in bold controlled under UK regulations (SI, 1989a) and Code of Practice (DoE, 1989a)

Source: Ure and Berrow (1982)

Table 2.7. Number of applications of sludge necessary to reach the UK maximum permissible concentrations of PTEs in sludge-amended agricultural land.

Element	Mean agric.[1] sludge conc. (mg kg^{-1} ds)	Total annual metal[2] load to agric. (t y^{-1})		Background[3] soil conc. (mg kg^{-1} ds)	Statutory[4] soil limit conc. (mg kg^{-1} ds)	Application rate limits[6] (kg metal ha^{-1} y^{-1})			Min. no. applications to reach soil limits[7]		
		Now	Future			Metal basis[4] statutory	Nitrogen basis[5] 250 kg ha^{-1}	Nitrogen basis[5] 170 kg ha^{-1}	Metal basis[4] statutory	Nitrogen basis[5] 250 kg ha^{-1}	Nitrogen basis[5] 170 kg ha^{-1}
Zn	922	429	644	80	300	15	7.65	5.26	29 (44)	57 (86)	83 (125)
Cu	574	267	400	20	135	7.5	4.76	3.27	31 (46)	48 (72)	71 (106)
Ni	65	30	45	20	75	3	0.54	0.37	37 (55)	204 (306)	297 (446)
Cd	5	2.3	3.5	0.8	3	0.15	0.04	0.03	29 (44)	110 (165)	147 (220)
Pb	201	93	140	50	300	15	1.67	1.15	33 (50)	302 (453)	438 (657)
Hg	3.5	1.6	2.4	0.1	1	0.1	0.03	0.02	18 (27)	60 (90)	90 (135)
Cr	208	97	145	35	400	15	1.73	1.19	49 (73)	422 (633)	613 (920)

(1) CES (1993)
(2) Sludge production x sludge concentration; UK sludge production utilized in agriculture now: 465000 t ds y^{-1} (CES, 1993); future: 697000 t ds y^{-1} (Hall and Dalimier, 1994)
(3) Geometric mean values from the National Soil Inventory except for Hg
(4) SI (1989a); DoE (1989a)
(5) Annual nitrogen application limits: 250 kg ha^{-1} (MAFF, 1991a); 170 kg ha^{-1} (CEC, 1991b)
(6) t ds ha^{-1} x sludge concentration (quantity of sludge: 8.3 t ds ha^{-1} at 250 kg N ha^{-1} and 5.7 t ds ha^{-1} at 170 kg N ha^{-1})
(7) $\frac{\text{(soil limit - background)} \times 2}{\text{Annual rate limit}}$

Figures in brackets denote values using 3 as the multiplier on the basis that the cultivation depth of soil is 23 cm and soil density is 1.3 (S.P. McGrath, 1993, personal communication). The water undertakings in the UK generally use a value of 2, assuming a cultivation depth of 20 cm and density of 1.0, which may be preferred since it provides a more precautionary estimate of the concentrations of PTEs in soil arising from sludge application.

Chapter three:

Effects of PTEs on the Yield of Agricultural Crops

Metal Uptake

The application of PTEs to soil in sewage sludge may increase the concentrations in crop tissues of certain elements, but has little or no effect on others depending on the chemistry and behaviour of particular elements in soil and subsequent partitioning within the plant. The overview by Vigerust and Selmer-Olsen (1986) of more than 100 crop trials treated with sludge provides a useful illustration of how the uptake of PTEs compares relatively and to untreated controls (Table 3.1). This study indicated that the application of sludge increased PTE concentrations in crops in the decreasing order Zn>Cd>Ni>Cu>Pb=Hg=Cr. The levels of Zn and Cd in crops were more or less doubled relative to controls (117% and 84% increase, respectively) and for Ni and Cu crop contents increased by about a half (65% and 56% increase, respectively). There was no effect overall of sludge applied Pb, Hg or Cr on the concentrations of these metals in tissues of crop plants.

Davis *et al.* (1978) determined the upper critical concentrations of PTEs supplied in nutrient solutions to young barley plants which resulted in a reduction in dry matter yield due to their toxic effects (Table 3.2). These and other data (Davis and Beckett, 1978) indicate such values are probably widely applicable to different plants as it appears plant species, and even cultivars, vary more in their ability to assimilate elements from soil than in their response to given concentrations of elements in their tissues (Davis and Carlton-Smith, 1980). Whilst this may be true in principal, there may be a discrepancy between the actual tissue concentrations which are associated with a reduction in yield of plants grown in solution culture or in soil possibly due to differences in the way chelated Fe is absorbed from nutrient solutions compared with soil (Logan and Chaney, 1983). Påhlsson (1989) concluded from a review of literature on heavy metal toxicity to vascular plants that the upper critical leaf concentrations of Zn

Table 3.1. Average content of metals in crops for treatment without sludge (O) (mg kg^{-1} dry matter) and relative concentrations for treatment with sludge (S), without sludge = 100. Numbers of experiments referred in brackets.

		Cd		Pb		Hg		Ni		Zn		Cu		Cr		Sludge application (t ha^{-1} dry solids)
		O	S	O	S	O	S	O	S	O	S	O	S	O	S	
Barley	grain	0.10 (10)	130	0.90 (6)	100	0.024 (8)	50	0.3 (5)	133	58 (13)	109	4.9 (13)	173	0.33 (4)	106	57
	straw	0.17 (9)	158	300 (9)	113	0.041 (7)	95	1.3 (7)	92	36 (9)	208	4.9 (11)	116	0.90 (4)	89	57
Oat	grain	0.11 (6)	163	0.42 (4)	90	0.010 (2)	100	1.2 (2)	446	35 (6)	140	3.6 (6)	131	0.42 (2)	95	70
	straw	0.12 (6)	158	1.70 (6)	135	0.042 (6)	93	1.7 (6)	100	54 (6)	117	6.1 (6)	125	0.38 (6)	126	50
Wheat	grain	0.07 (22)	214	0.44 (13)	61	0.030 (2)	67	5.4 (10)	259	36 (17)	228	4.3 (17)	221	0.30 (6)	93	114
	straw	0.19 (11)	200	1.10 (11)	91	0.148 (3)	84	0.6 (7)	150	23 (12)	700	3.3 (9)	239	429 (6)	80	139
Rye	grain	0.15 (6)	186	0.93 (2)	108			0.9 (6)	189	25 (6)	240	4.6 (6)	172	1.10 (4)	127	60
Grass		0.21 (8)	105	2.40 (8)	100	0.035 (4)	106	2.4 (8)	129	58 (9)	121	6.5 (9)	131	0.85 (2)	88	40
Clover		0.13 (6)	223	131 (6)	100	0.041 (2)	132	2.8 (6)	204	73 (6)	223	7.4 (6)	142	0.75 (2)	61	73
Rape		0.32 (7)	213	0.78 (7)	85	0.014 (1)	150	1.2 (7)	225	50 (7)	610	5.4 (7)	361	0.33 (7)	148	103
Carrot	root	0.77 (7)	187	1.83 (3)	86			9.0 (4)	108	46 (7)	135	6.5 (6)	122	1.10 (2)	91	140
	tops	1.53 (4)	105	5.15 (2)	107	0.020 (1)	160	4.3 (4)	121	63 (4)	135	8.5 (4)	109	1.60 (2)	119	55

Table 3.1 continued

		Cd		Pb		Hg		Ni		Zn		Cu		Cr		Sludge application (t ha^{-1} dry solids)
		O	S	O	S	O	S	O	S	O	S	O	S	O	S	
Potato	tuber	0.23 (4)	113	1.50 (2)	133			1.0 (2)	170	21 (12)	129	5.1 (12)	143	1.20 (1)	100	90
Red beet	root	0.35 (4)	287	15.0 (2)	89			8.8 (2)	202	82 (4)	122	13.9 (2)	89			29
	tops	1.20 (4)	292	15.0 (2)	67			8.2 (2)	100	450 (2)	125	27.6 (2)	97			29
Beans	seed	0.18 (10)	194	1.57 (3)	51			4.6 (3)	161	36 (10)	150	4.4 (8)	189			54
	veg. part	0.50 (2)	160	5.10 (2)	92			5.0 (2)	52	34 (8)	259	6.4 (6)	173			56
Tomato	fruit	0.46 (8)	235	5.60 (6)	132					24 (8)	146	8.3 (7)	133			180
	veg. part	0.80 (3)	75	5.60 (3)	98					30 (3)	140	6.7 (3)	128			90
Salad		0.81 (19)	285	4.6 (11)	117	4.42 (2)	100	4.2 (6)	119	43 (11)	300	10.0 (13)	120			79
Mean rel. numbers			184		98		104		164		217		156		102	78

Source: Vigerust and Selmer-Olsen (1986)

Table 3.2. Upper critical concentrations of PTEs for barley and background levels in plant material.

Element	Background concentrations (mg kg^{-1})	Critical concentrations for barley[1] In tissue (mg kg^{-1})	In solution (mg l^{-1})	Visual symptoms of toxicity
Zn	40	290 (160-320)	9 as Zn^{2+}	Yellow leaves, brown patches and pale green stripes on leaves
Cu	8	20 (18-21)	4 as Cu^{2+}	Bluish leaves
Ni	2	26 (4-26)	1.5 as Ni^{2+}	Longitudinal white stripes and brown patches on leaves
Cd	<0.5	15 (14-16)	0.5 as Cd^{2+}	Red-brown patches on leaves, stunted stems
Pb	3	35 (20-35)	25 as Pb^{2+}	
Hg	0.05	3 (2-5)	4 as Hg^{2+}	Yellow leaves, red stems
Cr	<1	10 (5-20)	8 as Cr^{3+}	Yellow leaves, pale green longitudinal stripes on leaves
Mo	1	135 (130-140)	70 as MoO_3^{2-}	
Se	0.2	30 (7-90)	5 as Se_3^{2-}	Red stems
As	<1	20 (11-26)	4 as $HAsO_4^{2-}$	Yellow leaves, red stems
Be	<0.1	0.6	0.6 as Be^{2+}	Yellow leaves, stunted stems
B	30	80 (40-130)	2 as BO_3^{3-}	Brown specks on leaves, yellow leaves
V	1	2 (1-2)	1 as V^{3+}	Pale green longitudinal stripes on leaves
Co	0.5	6 (3-9)	1 as Co^{2+}	Pale green leaves with pale longitudinal stripes
Zr	-	15 (5-18)	40 as Zr^{4+}	Red stems
Ag	0.06	4 (4-5)	0.5 as Ag^{+}	Red-brown patches on leaves, red stems
Sn	<0.3	63	110 as Sn^{2+}	Red stems
Ba	-	500 (400-800)	160 as Ba^{2+}	
Tl	<1	20 (11-45)	0.5 as Tl^{+}	Yellow leaves

[1] Corresponding to a 10% reduction in dry matter yield
Elements in bold controlled under UK regulations (SI, 1989a) and Code of Practice (DoE, 1989a)
Sources: Davis *et al.* (1978); Davis (1980)

and Cu affecting growth in most species were in the range 200-300 mg Zn kg^{-1} and 15-20 mg Cu kg^{-1} (dry matter). Macnicol and Beckett (1985) proposed more conservative critical levels above which a reduction in growth may occur of 100 mg kg^{-1} for Zn and 10 mg kg^{-1} for Cu and Ni in plant tissues based on an analysis of 1000 references on the subject. However, neither of these reviews differentiated between nutrient solution or soil culture in estimating the ranges of critical plant tissue concentrations of potentially phytotoxic elements. In contrast, Logan and Chaney (1983) argued that the nutrient solution approach underestimated the concentrations of foliar metals associated with a reduction in yield compared with soil grown plants. Under field conditions yield reductions (25%) may be observed at leaf tissue concentrations of 500 mg Zn kg^{-1}, 20-40 mg Cu kg^{-1} and 50-100 mg Ni kg^{-1} (dry matter) (Logan and Chaney, 1983). As a general guide, the ranges of normal and potentially toxic concentrations of PTEs in mature leaf tissue for various plant species are listed in Table 3.3 from a comprehensive review by Kabata-Pendias and Pendias (1992).

Table 3.3. Typical concentration ranges of PTEs in plant leaves (mg kg^{-1} dry matter).

Element	Deficiency	Sufficient or normal	Excessive or toxic	Tolerable in agricultural crops
Zn	10 - 20	27 - 150	100 - 400	300
Cu	2 - 5	5 - 30	20 - 100	50
Ni		0.1 - 5	10 - 100	50
Cd		0.05 - 0.2	5 - 30	3
Pb		5 - 10	30 - 300	10
Hg			1 - 3	
Cr		0.1 - 0.5	5 - 30	2
Mo	0.1 - 0.3	0.2 - 5	10 - 50	
Se		0.01 - 2	5 - 30	
As		1 - 1.7	5 - 20	
F		5 - 30	50 - 500	
Be		<1 - 7	10 - 50	
B	5 - 30	10 - 100	50 - 200	100
Ti			50 - 200	
V		0.2 - 1.5	5 - 10	
Mn	10 - 30	30 - 300	400 - 1000	300
Co		0.02 - 1	15 - 50	5
Zr			15	
Ag		0.5	5 - 10	
Sn			60	
Sb		7 - 50	150	
Ba			500	
Tl			20	

Elements in bold controlled under UK regulations (SI, 1989a) and Code of Practice (DoE, 1989a)
Source: Kabata-Pendias and Pendias (1992)

Certain PTEs clearly accumulate in crop plants grown in soil more readily than other sludge contaminants and in particular Zn, Cu and Ni are readily absorbed to potentially phytotoxic levels. Therefore, the regulations in the UK (SI, 1989a) have been developed to avoid damage to crop yields from elevated concentrations of these elements in soils receiving sewage sludge. The US EPA pathway risk analysis (US EPA, 1992a) also identified phytotoxicity to crops (Pathway 8; Table A6) as the most limiting route of environmental exposure to Zn, Cu and Ni in sludge-treated soil (Table A7). The US approach to developing statutory controls on sludge utilization based on risk assessment and environmental pathway analysis is described in the Appendix.

Zinc, Copper and Nickel

Recommendations for the regulation of Zn, Cu and Ni in sludge-amended agricultural soils were first established in the UK by Chumbley in 1971. These recommendations implied that the relative phytotoxic responses of Zn/Cu/Ni were additive and that the amount of toxic metal in sludge could be expressed as a single figure, known as the Zn equivalent, obtained by adding together the Zn content, 2 x the Cu content and 8 x the Ni content. This approach was later criticized because subsequent experimental work suggested that the toxicity to metals was not additive but acted independently for each element below the critical plant tissue concentration values (Beckett and Davis, 1982). Thus the amounts of sludge-borne metals which could be safely applied to soils were greatly underestimated by the Zn equivalent approach. Furthermore, the Zn equivalent model did not apply uniformly over a broad range of plant species (CAST, 1976). However, plants respond to the presence of contaminants in soils in a more or less uniform manner as demonstrated by the league tables of 39 crops grown on two soils of contrasting PTE content by Davis and Carlton-Smith (1980). More recently, Kim *et al.* (1988) also showed that the relative concentrations of Cd and Zn in food plants grown in several sludge-treated soils differed significantly, but the relative concentrations in plants in one soil have the same statistical characteristics as those in another. Consequently, the application limits for trace metals in soil should be calculated on the basis that they allow even the most vulnerable crop to be grown safely (Webber, 1981).

The interactive effects of metals on crops were studied further by Davis and Carlton-Smith (1984) investigating the relative phytotoxicities of Zn, Cu and Ni to ryegrass grown in a sandy loam soil (pH 7.0) in large plant growth containers and using sewage sludges of controlled metal content. Upper critical total concentrations of the trace elements in soil were determined as 319 mg Zn kg^{-1}, 105 mg Cu kg^{-1} and 221 mg Ni kg^{-1} which corresponded to critical tissue concentrations in ryegrass of 140 mg Zn kg^{-1}, 22 mg Cu kg^{-1} and 90 mg Ni kg^{-1}. The relationships between ryegrass yield and total concentrations of Zn, Cu and

Ni in soil are shown in Fig. 3.1. It was concluded that although the metals always occur together in practice in operationally sludge-treated soil, the three elements could be dealt with separately in setting limits for sludge utilization in agricultural land. This was because phytotoxic effects of mixtures of the three elements were independent at subcritical concentrations in the soil. Additivity occurred only when one or more elements greatly exceeded their critical soil concentration (for example at 707 mg Zn kg^{-1}). Mitchell *et al.* (1978a) came to a similar conclusion about the interactive effects of Zn, Cu and Ni on the yield of other crops including wheat and lettuce.

Sludges of controlled metal addition were also used in a pot trial by Sanders *et al.* (1987) to assess effects of sludge-borne Zn, Cu and Ni on the growth of clover, barley and red beet in a range of soil types. The soils used included two sandy loams (pH 6.5 and 7.1), a heavy clay (pH 6.3) and a calcareous clay (pH 7.8). No phytotoxic effects of Cu and Ni on crop yields were measured at the maximum soil concentrations used of 95 mg Cu kg^{-1} and 77 mg Ni kg^{-1} for a sandy loam of pH 6.5 and with the lowest cation exchange capacity (CEC) (10.1 meq 100 g^{-1}) of the soil types examined. The total concentration of Zn in the same sensitive soil likely to give rise to critical tissue concentrations was estimated as 250 mg Zn kg^{-1}. Reductions in yield occurred when leaf Zn concentration was greater than 900-1100 mg kg^{-1} in red beet and 400-500 mg kg^{-1} in clover. Comparison of yields for the 'low Zn' soil treatments with those for Zn-Cu and Zn-Ni combinations (all with the same amounts of applied Zn) showed that yield was not affected by the presence of either Cu or Ni. Unfortunately this study did not derive critical soil concentrations for Cu or Ni, but the authors indicated the results were consistent with the earlier work of Davis and Carlton-Smith (1984). Sanders *et al.* (1987) noted, however, that total Zn concentrations in soil which are necessary to give phytotoxicity will vary between soils, because of differences in their pH and cation exchange capacities. Therefore, toxic metal limits should also protect crops grown on the most vulnerable soils. However, the lower critical soil concentration for Zn reported by Sanders *et al.* (1987) compared with Davis and Carlton-Smith (1984) could have arisen out of an artefact of this experimental approach in studying the effects of toxic elements on plant growth. Furthermore, the concentrations of PTEs in the tissues of crop species grown in sludge-amended soil are normally smaller than those measured in the soil itself whereas the opposite occurred in the pot experiment described by Sanders *et al.* (1987). This observation would further suggest that these data may be atypical of the behaviour of sludge-treated soil systems in practice.

Logan and Chaney (1983) described the potential limitations of using greenhouse pot experiments in assessing the environmental effects of toxic metals on plants which arise because accumulations of PTEs in plant tissues can be increased 1.5-5-fold compared with field studies with the same soil, sludge and crop. Greater accumulations occur for a number of reasons including: (1) use of NH_4-N fertilizers which lowers soil pH more in pots than in the field; (2)

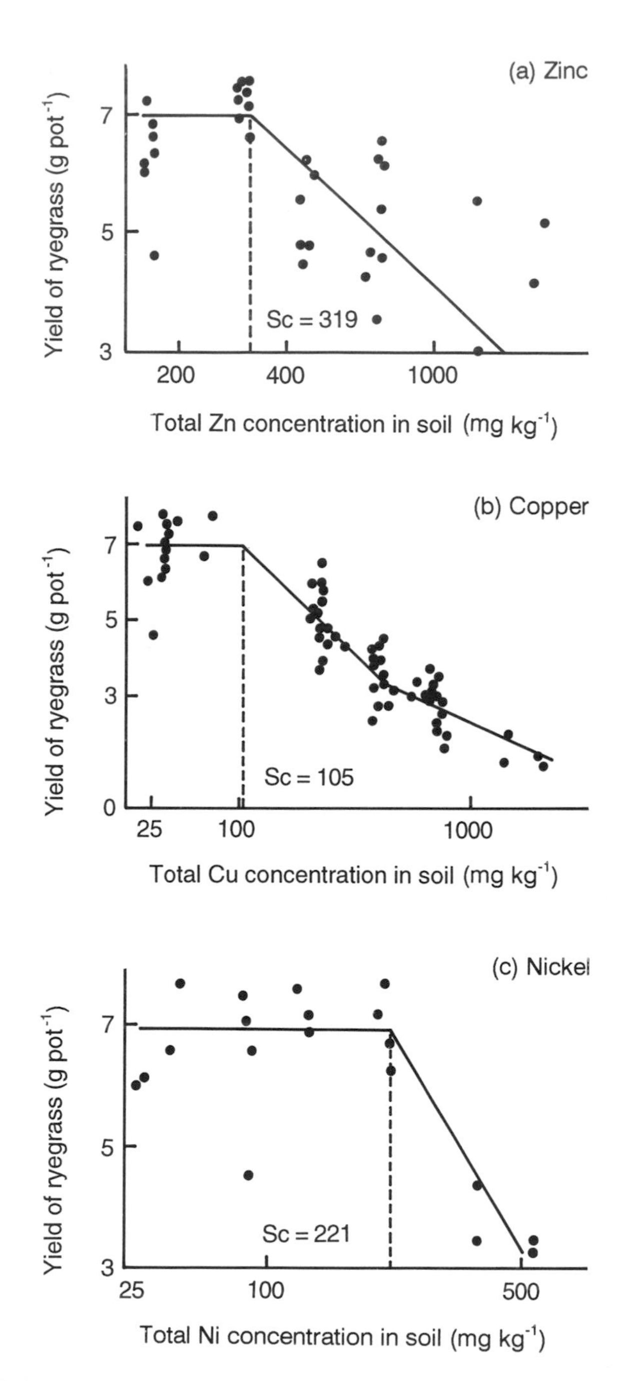

Fig. 3.1. Relation between ryegrass yield and total concentrations (log_{10} scale) of (a) zinc, (b) copper and (c) nickel in soil. Sc is the critical soil concentration for each element (Davis and Carlton-Smith, 1984).

higher soluble salt levels in pots than in the field due to the smaller soil volume required for fertilizer nutrients; (3) confinement of plant roots to the small volume of treated soil in pots; and (4) abnormal watering pattern and relative humidity in greenhouse pot studies. The smaller the pot, the greater the expected error. The pots used by Sanders *et al.* (1987) were particularly small holding only 1 kg of soil; those used by Davis and Carlton-Smith (1984) were significantly larger holding 7 kg. Chaney *et al.* (1987) considered that large pots of similar capacity were a prerequisite for trace element risk assessment using pot culture techniques.

In spite of these criticisms, pot studies in controlled environments enable close control over experimental variables needed to characterize soil-plant interactions of metals. However, Logan and Chaney (1983) argued that regulations for the utilization of sludge should be based upon field research because of the problems encountered with pot trial studies. In the UK, certain pot trials using controlled metal sludges have provided a highly precautionary, but nevertheless toxicological basis, for soil regulations for Zn, Cu and Ni protective against phytotoxicity in sludge-treated agricultural soils.

The first field trials in the UK designed specifically to test the effects of metal contaminated sewage sludges on crops were conducted by the Agricultural Development and Advisory Service (ADAS) at the Luddington and Lee Valley Experimental Horticulture Stations commencing in 1968 (Marks *et al.*, 1980). At the Lee Valley site the soil was a stony silt loam with pH in the range 6.2-7.0. At Luddington the soil was a sandy loam with a typical pH of 6.2, but maintaining pH was difficult at this site presumably due to the coarse texture of the soil and lime was applied twice during the course of the experiment to maintain the soil pH value. There was some evidence that soil pH declined during the trial with a mean value of pH 6.0 being measured at the end. The crops grown were red beet (at both centres) and either celery (Lee Valley) or lettuce (Luddington) for four years at each site.

The experiments were designed to study the effects of sewage sludge contaminated to varying degrees with different potentially toxic metals (Zn, Cu, Ni and Cr). However, the results were only of limited value because treatments (31.25 t ha^{-1} of dry solids annually for four years or 125 t ha^{-1} of dry solids in the first year) generally either gave a normal yield or a complete crop failure. For example, the high Zn sludge contained 48 000 mg Zn kg^{-1} (dry solids) which is about 54 times the amount of Zn found currently in sludges of median quality used in agriculture (CES, 1993). This was mixed with a 'relatively uncontaminated' sludge, itself containing up to 3000 mg Zn kg^{-1}, to provide a sludge with up to 16 000 mg kg^{-1} of Zn (dry solids) for application to the field plots. Other sludges used at Luddington and Lee Valley contained up to 8000 mg Cu kg^{-1} or 4000 mg Ni kg^{-1} (dry solids). Marks *et al.* (1980) concluded, therefore, that it was impossible to derive metal limits for sludge-treated soils from these field experiments. However, the results indicated that reduced tolerance to metals at Luddington could be explained by the coarser texture and lower pH of soil at this site compared with Lee Valley (see Chapter five). Large applications of sludge in field experiments are necessary to raise soil concentrations of PTEs within the relatively short time scales that trials are normally conducted over. However, the addition of large rates of high metal concentration sludges to soil can grossly overestimate metal availability and toxicity compared with repeated dressings of 'normal' low level sludges (Logan and Chaney, 1983). The effects of sludge properties and application rates on metal availability are discussed in Chapter five.

In another series of field trials described by Carlton-Smith (1987), single applications of sludge were applied to a sandy loam, clay and calcareous loam soils and a range of crops grown for five years including wheat, potato, lettuce, red beet, cabbage and ryegrass to assess human dietary exposure to sludge-borne Cd (see Chapter four). In contrast to Marks *et al.* (1980) these sludges contained more or less normal amounts of Zn, Cu and Ni and were applied at approximately 20, 40, 70 and 150 t ha^{-1} (dry solids). No phytotoxic effects on crop yield were measured even though soil metal content was raised to 79 mg Ni kg^{-1}, 237 mg Cu kg^{-1} and 474 mg Zn kg^{-1}. These field trials serve to demonstrate the large margins of safety which are built into precautionary soil limits set in the UK for Zn, Cu and Ni against phytotoxicity (Table A2).

The Woburn Market-Garden experiment was started in 1942 to investigate the soil conditioning properties of bulky organic manures including farmyard manure and sewage sludge applied annually to an impoverished sandy loam soil (Le Riche, 1968). The use of sewage sludge was discontinued after 1961 and no further applications have been made. Le Riche (1968) first noted that significant accumulations of heavy metals in soil treated with sludge had occurred and that concentrations of certain elements (particularly Zn, Cu and Ni) in tissues of crop plants were also elevated compared with untreated controls. A full description of the metal contents of the manures and soils from the field experiment has been given by McGrath (1984). At the largest rate of addition (25 applications of approximately 40 t ha^{-1} of sludge dry solids between 1942-61) the soil concentrations of Zn, Cu and Ni increased to 635.4 mg Zn kg^{-1}, 239.4 mg Cu kg^{-1} and 42.3 mg Ni kg^{-1} in 1960. However, no reductions in the yields of crops at these levels of soil contamination following sludge application were noted by Mann and Patterson (1963) in a summary of the results from 1944 until 1960 which described only beneficial effects of sludge on crop yields. The soil pH was 6.2 in 1944 and the site was regularly limed after that date. The mean pH in 1960 was 6.7. In later studies using soils from the same trial, McGrath *et al.* (1988) reported no detrimental effect on clover yield due to phytotoxicity at soil concentrations of 469 mg Zn kg^{-1}, 163 mg Cu kg^{-1} and 35 mg Ni kg^{-1}. Effects of metals in sludge-treated soils on the clover-*Rhizobium* symbiosis are discussed later in Chapter eight in relation to soil fertility.

For the past 15 years researchers at WRc have used sites with a long history of sludge application, of at least 10 years or often for very much longer

(100 years +), to assess the long-term environmental impact of utilizing sewage sludge on agricultural land (Coker and Davis, 1979). Carlton-Smith and Stark (1987) established 40 trial plots with ryegrass at 12 such sites and measured yields and metal concentrations over two growing seasons. As would be expected, crop yields varied considerably between the sites, but there was no evidence that yields were reduced on the high metal soils compared with untreated controls at any one site. Total concentrations of Zn, Cu and Ni measured in the sludge-amended soils reached 2200 mg Zn kg^{-1}, 850 mg Cu kg^{-1} and 510 mg Ni kg^{-1}. Predicted tissue metal concentrations in relation to soil content were well below values likely to cause phytotoxicity in ryegrass at the maximum permissible concentrations of these PTEs in sludge-treated agricultural land (Davis and Carlton-Smith, 1984). Interestingly, maximum concentrations of Zn and Cu measured in ryegrass (517 mg Zn kg^{-1} and 34 mg Cu kg^{-1} (dry matter)) exceeded the considered critical phytotoxic levels of these elements at certain sites although no significant reduction in yield was detected.

Complementary glasshouse pot experiments (Carlton-Smith *et al*., 1987) with eight historically sludged soils, mixed in various proportions with uncontaminated soil to provide a range of soil concentration levels, showed reductions in crop yields of leaf beet, barley, lettuce, ryegrass and clover only when soil metal content exceeded or greatly exceeded the CEC Directive upper limits for Zn, Cu and Ni (CEC, 1986a). The soils studied had not received any sludge for about 10 years prior to the trial and the metals contained in them had probably attained a state of equilibrium in relation to their availability for crop uptake (see Chapter five). As would be expected, the yield response depended to a large extent on the sensitivity of different crops to sludge-derived metals. Barley was the least sensitive crop grown and yield reductions were measured only in one sludged soil where the soil metal concentrations were three times the maximum permissible amount (CEC, 1986a). On the other hand, leaf beet readily accumulates metals in its tissues and is therefore particularly susceptible to soil contamination effects on plant yield (Davis and Carlton-Smith, 1980). The yield response of leaf beet is shown in Fig. 3.2. Interestingly, leaf beet yield was reduced to a very much larger extent for soil from the Luddington experimental site (indicated on Fig. 3.2 as 'L') which was also included in the pot trial for comparison with operationally managed soils. This indicates further that there were fundamental problems with the approach to the ADAS trials described earlier in that they did not reflect likely effects on crops obtained from low rates of application of sludge over long periods of time under operational practice in the field.

On balance, phytotoxicity has generally only been observed in glasshouse pot trials and when prepared sludges have been applied to soil or when highly metal contaminated sludged have been applied in large dressings in the field. These experimental approaches to estimating appropriate soil limits for PTEs would appear to aggravate the effects of potentially phytotoxic heavy metals. It

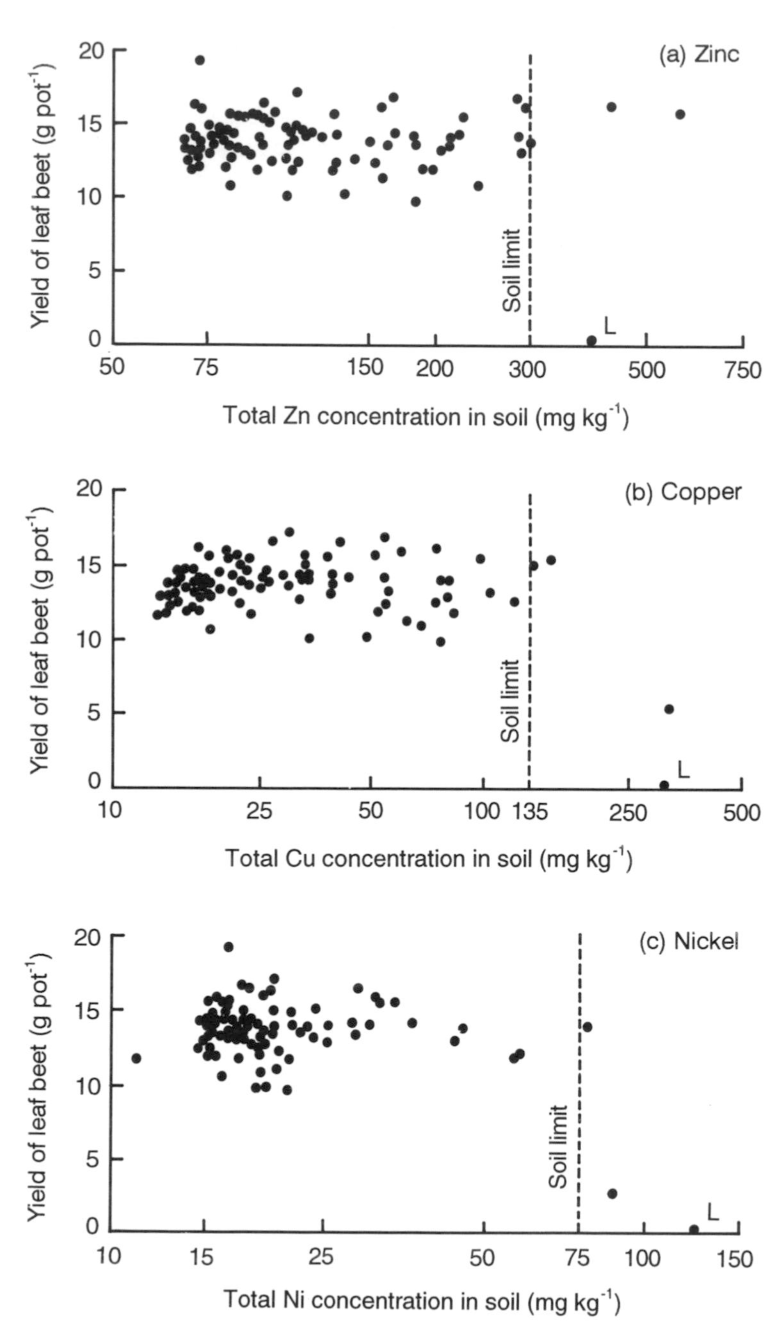

Fig. 3.2. Relation between leaf beet yield and total concentrations (log_{10} scale) of (a) zinc, (b) copper and (c) nickel in soil. L indicates Luddington soil (Carlton-Smith *et al.*, 1987).

is emphasized that they can therefore provide a highly conservative estimate of the effects of potentially toxic metals on crop yield. For example, the concentrations of each of the major potentially phytotoxic metals in individual prepared mono-metallic sludges used in experimental work in the UK (Davis and Carlton-Smith, 1981) may be as high as 42 000 mg Zn kg^{-1}, 15 000 mg Cu kg^{-1} or 2500 mg Ni kg^{-1} (dry solids). Chaney and Ryan (1993) argued that it is a combination of glasshouse pot trials and sludges of high metal content which caused phytotoxicity in these studies. The effect of the high metal content is explained in terms of filling of specific metal binding sites in the sludge and because the strength of the binding sites vary such that as the metal content of the sludge increases, the least strongly bound metal is more phytoavailable (Chapter five). Whether these studies should be used as the basis for establishing appropriate and acceptable controls to protect crop yields from agricultural recycling of sludge is another matter. Under field conditions and with normal sludges, however, phytotoxicity is rarely seen and then only when soil metal concentrations greatly exceed the UK statutory limits for PTEs in sludge-treated agricultural land. Normally under these conditions only an increase in yield response is observed resulting from the fertiliser value of sewage sludge (Carlton-Smith, 1987). For example, Johnston *et al.* (1983) observed little or no evidence of Zn or Cu toxicity in a field trial from single applications of up to 1610 kg Zn kg^{-1} and 213 kg Cu ha^{-1} in sludge. A demonstration of the apparent conservatism of the UK regulations which protect agricultural crops from phytotoxicity is possible by examining the US exposure analysis for effects on crop yield (Pathway 8; Table A6).

Chang *et al.* (1992) describe a methodology of evaluating phytotoxicity criteria for metals, when sewage sludges are applied on agricultural land. This is based on an analysis of the relationships between metal concentrations in plant tissue and growth retardation obtained from laboratory experiments, where plants were grown in media treated with single elements. Data from field experiments with sewage sludge were used to estimate appropriate soil metal loading rates corresponding to the phytotoxicity threshold concentration in plant tissues. The probability of plants, grown on soils with a given range of metal loadings, exceeding the phytotoxicity threshold was calculated and an appropriate loading range selected based on the risk of exceeding the phytotoxicity threshold that is considered acceptable. Chang *et al.* (1992) arbitrarily selected 50% yield reduction as the threshold for testing the approach for corn (*Zea mays* L.), lettuce (*Lactuca sativa* L.) and bush bean (*Phaseolus vulgaris* L.) although adequate field data were only available for corn. Other toxicity thresholds could be applied in the methodology depending on the required basis for regulation development. The leaf tissue concentration of Zn corresponding with a 50% growth retardation was 1975, 475 and 375 mg kg^{-1} for corn, lettuce and bush bean, respectively. The probability level (level of tolerable risk) for plant tissue metal concentrations to exceed the 'acceptable' phytotoxic threshold was set at 0.01 although this could also be adjusted. This value is five times more stringent than the minimum accepted probability level of significance in experimental biology.

Based on the 50% yield reduction threshold, Chang *et al.* (1992) estimated that at least 3500 kg ha^{-1} of Zn may be added through sludge application without significantly exceeding the selected leaf tissue concentration threshold in corn and therefore reducing yield significantly below the 'acceptable' 50% level. However, in practice any detrimental impacts of sludge application on crop yield performance would be seen to undermine the fundamental reason for recycling sewage sludge to agricultural land in the first place. Indeed, sludge application should be regarded beneficially and intended to increase crop yield through the application to soil of plant nutrients and organic matter. At the 8% yield reduction threshold (the lowest examined by Chang *et al.*, 1992), the probability that leaf Zn would exceed the level causing 8% yield retardation was 0.024 at a loading rate for Zn in sludge <200 kg ha^{-1}. The probability of causing yield reduction decreased to 0.002 at <100 kg ha^{-1} of Zn. According to this analysis, exceeding 200 kg ha^{-1} of Zn may cause a significant reduction in corn yield from the 8% tolerance threshold (probability = 0.142).

The estimated loading rates of Zn resulting in minimal yield reduction of corn appear relatively low compared with the concentrations of Zn measured in sludge-treated soils associated with phytotoxic effects of this element (Davis and Carlton-Smith, 1984; Carlton-Smith, 1987; Carlton-Smith and Stark, 1987). One possible explanation for this may be due to the frequency distribution characteristics of the data and that metal concentrations in plant tissues in particular appear to be log_{10}-normally distributed (Beckett and Davis, 1977). Whilst Chang *et al.* (1992) explained that a comparison of normal and log_{10}-normal data produced similar loading rates for the selected phytotoxicity threshold (50%) and risk level (0.01), the effect of log_{10} transformation of the data may become increasingly important in estimating loading rates at the lower phytotoxicity threshold levels. This is because phytotoxic threshold levels approaching the upper critical concentration in plant tissues may be difficult to quantify on the linear scale since large changes in crop yield may be associated with only small changes in leaf Zn concentration. In contrast, higher 'acceptable' phytotoxicity thresholds occur in the range of leaf Zn concentrations where proportionally smaller effects on yield are caused by large changes in the Zn content of plant tissues.

The probability method (Chang *et al.*, 1992) was one of two approaches used by US EPA in evaluating the potential risk of phytotoxicity from sludge-borne contaminants (Environmental Pathway 8; Table A6) for the Final Part 503 Rule (US EPA, 1992a; US EPA, 1993). In the second approach (Approach 2), plant tissue concentrations associated with the lowest observed adverse effect level (LOAEL) on yield were obtained for sensitive crops, which is a more sensitive indicator than yield decrease. This was prompted because the probability method (Approach 1) was based on data for corn, which is relatively insensitive to sludge-borne metals. Therefore, it was considered that applying loading rates for metals in sludge based on the criterion that a 50% reduction in the yield of corn was 'acceptable', may not protect sensitive species such as lettuce, bush beans and swiss chard. Consequently, the LOAEL analysis on Zn was followed for lettuce, one of the most sensitive crops, using a foliar tissue concentration of 400 mg Zn kg^{-1} (dry matter), to indicate the first detectable yield suppression and using the geometric mean of the crop uptake slopes for Zn in leafy vegetables to calculate the appropriate loading rate for Zn. Interestingly, the selected critical plant tissue concentration in lettuce exceeded the phytotoxicity tolerance threshold of 250 mg Zn kg^{-1} (dry matter) associated with an 8% yield reduction of the more tolerant corn crop (Chang *et al.*, 1992) highlighting the inconsistencies apparent in the data sets on phytotoxicity. A value of 400 mg Zn kg^{-1} in plant leaves is at the upper end of the range of Zn concentrations considered excessive or potentially toxic in plant tissues (Table 3.3). Cumulative Pollutant Loading Rates in the Final Rule were estimated by Approach 2 for Zn (Table A7).

Both approaches were also applied to Cu and Ni although loading rates (Table A7) were estimated on the basis of Approach 1 for these elements using the maximum loading rates available from experimental studies as the upper safety boundary due to the absence of available data on larger rates of addition. Indeed, loading rates calculated by Approach 2 were 2500 kg Cu ha^{-1} and 2400 kg Ni ha^{-1} using threshold phytotoxic concentrations of 40 mg Cu kg^{-1} (dry matter) in snap beans, and 40 mg Ni kg^{-1} (dry matter) in corn. The same value for Cu was used for corn in estimating the loading rate for this element by the probability approach. However, a more conservative tolerance threshold of 3.0 mg kg^{-1} in corn plants was used in Approach 1 for Ni. These leaf tissue concentrations correspond or are likely to correspond with no observed adverse effect levels (NOAELs), compared with the Zn concentration in corn plants associated with a 50% yield reduction considered 'acceptable' for assessing Zn by this approach. Indeed, the probability of Cu in corn grown in sludge-treated soils exceeding the phytotoxic concentration was <0.0001 for cumulative loading rates of 1550 kg Cu ha^{-1} at the upper range of the available data (Chang *et al.*, 1992). By the same analysis, using maximum loadings for Ni as the upper boundary, 425 kg Ni ha^{-1} can be safely applied without affecting yields (probability of exceedance 0.0045) (Chang *et al.*, 1992). Consequently, it could be argued that the limits set for Cu and Ni, as well as Zn, in the 503 Regulation (USEPA, 1993) are highly protective against potential phytotoxic effects on crop yields.

Chromium

Chaney (1990b) reported that no adverse effects on crop yields have been demonstrated for any level of Cr applied to soil in sewage sludge. The results of Carlton-Smith and Davis (1983) and more recently of Smith *et al.* (1992b) also show that Cr does not affect plant growth unless concentrations are very large. Forage crops, radish and carrots grown in these studies showed little uptake of Cr and no effect on yield for soil containing up to approximately 7000 mg Cr kg^{-1}. Root growth and root crops may be particularly sensitive to the toxic effects of Cr (Yamaguchi and Aso, 1977; Williams, 1988). Vigerust and Selmer-Olsen (1986) similarly showed no uptake of Cr for crops grown in sludge-amended soil (Table 3.1). Carlton-Smith and Stark (1987) measured a maximum concentration of Cr in historically sludge-treated trial plots of 15 300 mg Cr kg^{-1} (dry soil). However, the concentrations of Cr in ryegrass were commonly below the limit of analytical detection of 0.2 mg kg^{-1} (dry matter). Titanium analysis of the plant tissue indicated that the increased amounts of Cr in ryegrass above the detection limit were probably caused by the possible soil contamination of the plant material. A significant uptake of Cr and detrimental effect on yield was measured by Smith *et al.* (1992b) only for one soil which contained 1.4% Cr representing a seriously contaminated soil.

Given the extreme conditions of the sludge treatments already described for Luddington and Lee Valley, it is interesting that the Cr sludges (containing up to 8800 mg Cr kg^{-1}) had no adverse effects on vegetable crop yields (Marks *et al.*, 1980). Indeed, the uncontaminated and Cr sludge treatments frequently gave yields higher than for the controls in response to the addition of plant nutrients and organic matter to the soil in sludge. Berrow and Burridge (1980) indicated that the yields of later grass crops at Luddington were similarly unaffected by Cr at a soil concentration of 600 mg Cr kg^{-1}. Therefore, it seems somewhat ironical that based largely on a re-analysis of these data Williams (1988) later concluded that the limit concentration for Cr which could be safely tolerated in soil was in the range 150-250 mg Cr kg^{-1}.

Williams (1988) indicated that the yield of red beet was reduced significantly in one year out of four between 1969-1972 at a soil Cr concentration approximating to 400 mg Cr kg^{-1}. The yields of celery and lettuce, however, were not affected by soil concentrations of 440 and 350 mg Cr kg^{-1} respectively. Several crops were grown on the plots each year and because there were only two replicates of each treatment the yield data were averaged for each year in an attempt to overcome problems associated with high variability in the yield of individual crops as well as seasonal differences (Marks *et al.*, 1980). Crop failures for reasons other than phytotoxicity also contributed to difficulties in obtaining a meaningful statistical treatment of the data. Consequently, the significance of the detrimental responses reported by Williams (1988) should be viewed cautiously as the effects could have occurred due to normal random variation given the number of crops grown over the duration of the trial and because treatments were compared only at the 5% significance level. The apparent conflicts with the earlier reports (Marks *et al.*, 1980; Berrow and Burridge, 1980) also cast doubt on their validity. It is paradoxical, however, that the European Commission had considered an even lower maximum permissible concentration range for Cr (CEC, 1990b) than that recommended originally by

Williams (1988) which is a conservative range for sludge-treated soil (McGrath, 1995). The proposal on Cr has been dropped by the European Commission for the time being (CEC, 1993), but it is highly probable that the issue of regulating Cr in sludge will be reviewed in the future.

One reason for the concern is because Cr(VI) can be highly toxic to plants (Williams, 1988), but sewage sludges and sludge-treated soils contain only the Cr(III) form which is non-toxic being almost insoluble under most soil conditions (James and Bartlett, 1984). In contrast to the large amount of information indicating the non-activity of Cr(III) in soil, the evidence for Cr(VI) toxicity is very limited and only comes from pot studies with inorganic chromate and dichromate salts (Williams, 1988) and there is no evidence of Cr(VI) formation in sludge-treated soils. Furthermore, absorbed Cr(III) is retained in plant roots and is not translocated to plant foliage in toxic amounts even when soils are heavily contaminated. This process has been described as the 'soil-plant barrier' (Chaney, 1980) which protects the environment from the potential transfer and toxicity of certain sludge-borne contaminants. Few if any studies demonstrate crop uptake to reported critical tissue concentrations for Cr (barley:10 mg Cr kg^{-1}, see Table 3.2; ryegrass 1.3 mg Cr kg^{-1} see Cottenie *et al.*, 1976). For example, Berrow and Burridge (1981) detected no relationship between soil Cr concentration and herbage content of pasture crops grown at Luddington and Lee Valley; amounts in leaves were small and below the detection limit of the analytical apparatus used of 0.1-0.3 mg Cr kg^{-1}. The virtual insolubility of Cr in sludge-amended soil was also demonstrated recently by McGrath and Cegarra (1992). Chromium in soil treated with sewage sludge from the Woburn experiment remained predominantly in the residual aqua-regia soluble fraction, with no change occurring 20 years after sludge application to the field had ceased.

The redundant proposed Directive limit for Cr in sludge-amended soil of 100-150 mg Cr kg^{-1} (CEC, 1990b) reflects concern, which is apparently not justified by the available scientific evidence, that this element may be potentially phytotoxic to plants grown in sludged soil (Williams, 1988). By contrast, in their evaluation of the US EPA Proposed Rule on Standards for the Disposal of Sewage Sludge (US EPA, 1989a) the Peer Review Committee (Page and Logan, 1989) concluded that there was no scientific basis to limit application of sludge-Cr based on phytotoxicity (or any other sensitive environmental pathway) and suggested a cumulative application of >2000 kg Cr ha^{-1} which equates approximately to a soil limit of >1000 mg Cr kg^{-1} in soil (Chaney, 1990b). In developing the Final Part 503 Regulation (US EPA, 1993) the US EPA identified the sludge → soil → plant pathway (Pathway 8) as the most critical route of exposure to Cr (Table A6) and set a cumulative loading for Cr in sludge which equates to a soil concentration limit of approximately 1540 mg Cr kg^{-1} (Table A7). In fact, the only reasonable basis which could be found to regulate Cr was to set the safe upper boundary for Cr in sludge-treated soil using the maximum loading rates of Cr available in experimental studies of 3000 kg ha^{-1}, due to the absence of any detectable environmental effects of this element (Chang *et al.*, 1992; US EPA, 1992a). Consequently, the available evidence strongly suggests that the provisional soil limit value currently set for Cr in the UK (DoE, 1989a) of 400 mg Cr kg^{-1} (Table A2) provides a large margin of environmental protection particularly in relation to the limits set for other potentially more mobile elements in sludge-treated soil such as Zn (Smith *et al.*, 1992b).

Other Elements

It is unlikely that elements contained in sewage sludge other than Zn, Cu and Ni, will affect plant yields detrimentally. This is because their concentrations in sludge relative to Zn, Cu and Ni are small and/or because they are immobile in soil and so are not absorbed by plants. In other cases there is a need to control amounts applied below the phytotoxic threshold of crops to avoid concentrations accumulating in plant tissues potentially injurious to animals or humans (see Chapters four and six).

Pollard (1991) reported the concentration of Pb in ryegrass grown on a historically sludge-treated soil in the field was typically about 8 mg kg^{-1} for a soil containing approximately 1760 mg Pb kg^{-1}. Compared with published critical plant tissue concentrations (see Table 3.3) this value is small particularly in relation to the Pb concentrations in the soil. In addition, there is the possibility that soil contamination of grass could result in tissue concentrations of Pb being overestimated. For example, Stark *et al.* (1986) used Ti analysis of herbage as an indicator of the extent of soil contamination of grass samples and found that most of the Pb (90-95%) in grass grown on sludge-treated soil could be accounted for by soil contamination. Carlton-Smith and Stark (1987) later showed that very little Pb was taken up by ryegrass and concentrations in plant tissues were independent of the levels in soil which increased to a maximum value of 25 000 mg Pb kg^{-1} in the historically sludge-treated soils examined. Vigerust and Selmer-Olsen (1986) similarly confirmed that there is no crop uptake of Pb from sludge-treated soils (Table 3.1). Koeppe (1981) concluded that Pb has no toxic effect on plants due to low bioavailability and because any absorbed Pb is immobilized in the roots.

Carlton-Smith and Stark (1987) analysed herbage from ryegrass plots established on 12 historic sludge-treated sites for a selection of rarer elements which are present in sewage sludge (Tables 2.2 and 3.4). Total concentrations of rare trace elements in sludged soil were raised significantly compared with untreated control plots which contained only background levels (Tables 2.6 and 3.4). However, the concentrations of V, Ga, Y, Ag, Sn, Sb, W, Hg and Bi in ryegrass generally did not exceed the limit of analytical detection (Table 3.4). Plant tissues contained some detectable Ge, As, Se and Nb, but concentrations were small approximating to 1.0 mg kg^{-1} and were unlikely to result in phytotoxicity (Tables 3.2 and 3.3). The concentration of Se in the sludged soil was approximately 2.5 times the limit value for this element under grass (DoE, 1989a) whereas the sludge-treated soil studied contained less than the permitted maximum level of As (DoE, 1989a). In contrast, certain rare elements could accumulate significantly in crop tissues and relative to the soil levels. For example, concentrations of Rb and Sr in ryegrass were 40% of the soil levels and plant tissue content of Ba was about 20% of that in soil. The critical plant tissue concentration for Ba is 500 mg kg^{-1} (Table 3.3) whereas the maximum amount measured in ryegrass under these extreme conditions of Ba contamination in soil was 120 mg kg^{-1}. Pot trials using soils from historic sites have confirmed that rare elements in sludge-treated soil including W, Ce, Nb, La, Ag, Sb, Bi and also Co are unlikely to accumulate significantly in crop tissues or affect crop yields, particularly in relation to the concentrations of these elements normally found in sludge and to current statutory soil limits set for Zn, Cu, Ni and Cd (Stark and Whitelaw, 1986). Research using soils heavily contaminated with Hg applied in organomercurial fungicides has similarly demonstrated that large concentrations of Hg in soil are not toxic to plant growth. For example, Estes *et al.* (1973) showed that the yield of turf grass was unaffected by concentrations of Hg in soils as high as 455 mg Hg kg^{-1}.

Table 3.4. Maximum concentrations of rarer elements in ryegrass grown in historically sludge-treated soil.

Element	Symbol	Soil (mg kg^{-1})	Ryegrass (mg kg^{-1})
Vanadium	V	210	<=1.0
Gallium	Ga	23	<=0.4
Germanium	Ge	4	1.7
Arsenic	As	38	1.0
Selenium	Se	12	1.0
Rubidium	Rb	133	48
Strontium	Sr	115	48
Yttrium	Y	34	<=0.4
Zirconium	Zr	158	4.0
Niobium	Nb	13	0.6
Silver	Ag	41	<=2.0
Tin	Sn	210	<=2.0
Antimony	Sb	75	<=2.0
Barium	Ba	503	120
Tungsten	W	185	<=1.0
Mercury	Hg	15	<=0.4
Bismuth	Bi	110	<=1.0

<= concentration at or below limit of detection
Sources: Stark *et al.* (1986); Carlton-Smith and Stark (1987)

Davis *et al.* (1978) noted that the critical levels for animals of Cd, Hg, Mo, Pb and Se in herbage appear to be lower than those for crop plants so that animals could be poisoned by apparently healthy plants. Mercury and Pb do not accumulate to potentially zootoxic levels in plant tissues due to their low bioavailability in sludge-treated soil. However, Chaney (1983) listed Cd, Se and Mo as elements which escape the 'soil-plant barrier' because concentrations in edible plant parts potentially toxic to animals could accumulate before the phytotoxic threshold is reached. However, field experiments show that crops do not accumulate Se appreciably from sludge application to agricultural soils (Carlton-Smith and Stark, 1987; Logan *et al.*, 1987). In contrast, Mo can accumulate to potentially phytotoxic amounts in ryegrass if soil concentrations exceed 10 mg Mo kg^{-1} (Carlton-Smith and Stark, 1987). The UK limit for Mo in soil where sewage sludge is applied is 4 mg Mo kg^{-1} to protect grazing livestock from Mo-induced Cu deficiency (Davis, 1981; DoE, 1989a). In addition, large concentrations of F (>1000 mg kg^{-1}) can be tolerated in plant tissues with no apparent symptoms of phytotoxicity (WRc, 1979) such that the advisory limit (DoE, 1989a) is designed to avoid the transfer of potentially harmful amounts of this element to grazing livestock in herbage. However, Cd is the main PTE of concern in sewage sludge-treated agricultural land which circumvents the 'soil-plant barrier' potentially impacting the human food chain. The scientific basis to regulating Cd in sludge-amended farmland is discussed in Chapter four.

Chapter four:

Human Dietary Intake of PTEs from Agricultural Crops and by Soil Ingestion

Cadmium

Cadmium represents the human dietary poison of principal concern in relation to the utilization of sewage sludge on agricultural land. This is because Cd is readily bioavailable for plant uptake in sludge-treated soil (Table 3.1). Cadmium can accumulate in the edible portions of crop plants to levels which could potentially be injurious to humans, if consumed for long periods of time and in large quantities, whilst having no apparent detrimental effects on crops themselves. Consequently, the maximum permissible concentration of Cd in sludged agricultural soil (SI, 1989a) has been developed in the UK on the basis of human dietary impact potentially arising from increasing concentrations of Cd in crops grown on sludge-treated soil.

The experimental trials described by Carlton-Smith (1987) and discussed earlier in connection with phytotoxicity of Zn, Cu and Ni, were designed principally to assess dietary intake of Cd arising from the utilization of sludge in agriculture. The field experiments were established at three sites on soils broadly representative of UK agriculture and the trials were conducted over five growing seasons. The soil types included a sandy loam (Cassington A: pH 6.5, CEC 13 meq 100 g^{-1}), clay (Cassington B: pH 6.7, CEC 29 meq 100 g^{-1}) and a calcareous loam (Royston: pH 8.0, CEC 20 meq 100 g^{-1}). The six crops grown were selected to include major constituents of the human diet (wheat, potatoes and cabbage), sensitive indicator crops (red beet and lettuce) and ryegrass was also included. The concentrations of Cd in crop tissues approximated to a simple linear function of the total concentration of Cd in soil and amounts in the crops grown decreased in the order lettuce > red beet > wheat > ryegrass > potato > cabbage (Fig. 4.1). This pattern of crop uptake follows closely that expected from the 'league tables' of crop sensitivity to Cd in soil listed by Davis and Carlton-Smith (1980).

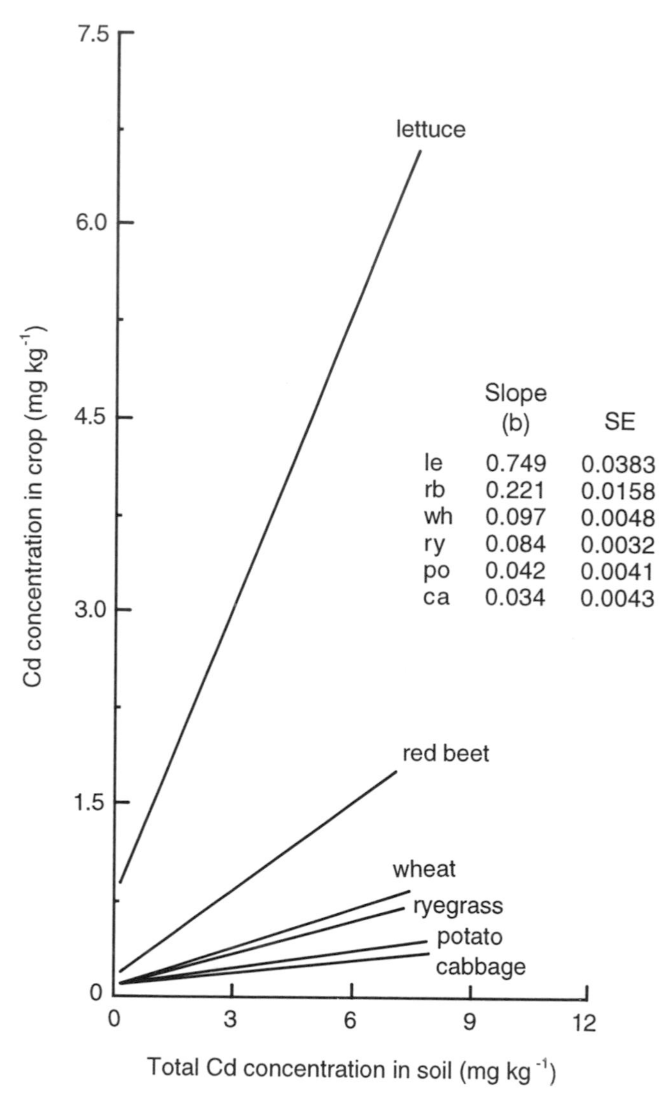

Fig. 4.1. Cadmium concentrations in various crops in relation to total content in a sandy loam soil, treated with liquid sludge (Carlton-Smith, 1987).

Carlton-Smith (1987) estimated that the background intake of Cd in plant foods by the average consumer in the UK was 16.2 μg Cd day^{-1}. This experimentally determined value compares closely with the background intakes of 20 μg Cd day^{-1} by UK citizens (DoE, 1980) and from normal US diets (Chaney, 1990b). The Cd exposure of an 'upper-range' (97.5 percentile) consumer in Great Britain has been recently estimated as 18.8 μg day^{-1} (MAFF/DoE, 1993b). The background intakes of Cd in the principal dietary plant food components are shown in Fig. 4.2. Cereals and potatoes represent the most important sources of dietary Cd for the average consumer because of the large quantities consumed compared with leafy salads, such as lettuce, which have little impact on Cd intake even though crop content can be significantly larger in comparison.

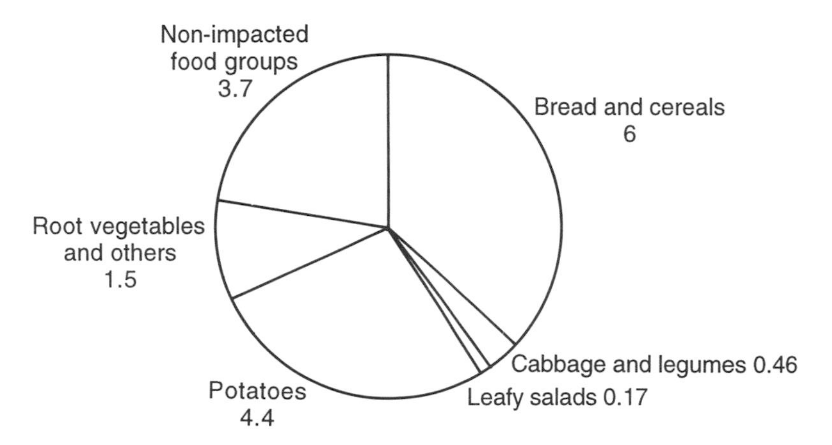

Fig. 4.2. Background cadmium intake (μg day^{-1}) in plant food components of the human diet (adapted from Carlton-Smith, 1987).

In the dietary model for sludge-treated soil, Carlton-Smith (1987) assumed that individuals obtain all plant food groups, other than bread and cereals, from soil at the CEC upper limit for Cd of 3.0 mg Cd kg^{-1} in sludge-amended agricultural land (CEC, 1986a). For bread and cereals the proportion which could be obtained potentially only from the contaminated locality was determined. This value was estimated to be 2% of the dietary intake of bread and cereal produce corresponding approximately with area of arable land which receives sewage sludge in the UK. The proportion of plant food in the diet originating from sludge-treated soil was only considered for this food group because the marketing and distribution of wheat make it unlikely that an individual would consume bread and cereals which originate solely from wheat grown on sludged land. In addition, the effect of milling wheat was accounted for since 95% of grain is milled in the UK which reduces the Cd concentration in white flour by 50% compared with that in grain (Zook *et al.*, 1970). The model consequently represented a 'worst-case' condition since calculations were only possible on this basis in the absence of further information on the distribution and consumer habits relating to other food groups. However, it is almost inconceivable that any individual would, in reality, obtain all their vegetable plant food from sludge-treated land.

Cadmium intake values for each food group, calculated for the three soil types at a soil concentration of 3 mg Cd kg^{-1}, are presented in Fig. 4.3. The Expert Committee on Food Additives of the WHO/FAO has set a recommended maximum average intake of Cd which equates approximately to 70 μg Cd day^{-1} for a 70 kg person (WHO/FAO, 1972). In addition, the US Environmental Protection Agency has independently estimated a safe daily intake limit of 70 μg of Cd (US EPA, 1979). However, none of the total dietary intakes estimated by Carlton-Smith (1987) at the upper permitted soil concentration for Cd (CEC, 1986a) exceeded the recommended safety threshold, even when the 95% confidence limit was included (Fig. 4.3). In addition, a further margin of safety exists between the recommended maximum value of 70 μg Cd, and the minimum intake which potentially may cause kidney damage in the most sensitive individuals set at 200 μg Cd day^{-1} for 50 years (US EPA, 1979).

The large dietary intake value obtained for the clay soil of 52 μg Cd day^{-1} was probably an overestimate and atypical of clay soils generally. There was strong evidence, for example, that metal availability in this soil was unusually high indicated by large background values of Cd in crops, due possibly to the low soil content of complexing iron and manganese sesquioxides. The sandy loam, however, was much more typical of farmland receiving sewage sludge in the UK.

Although the principal intake of Cd for non-smokers is food and is approximately 16 μg day^{-1}, the additional intake by smokers is around 20-40 μg for each packet of 20 cigarettes. Smokers are therefore in a high risk class, but since cigarette packets in the UK carry a Government health warning, adjustment of soil limits to avoid risks to heavy smokers is assumed to be unnecessary (Simms and Beckett, 1987). Other high risk classes could be consumers with dietary habits widely differing from the assumed average consumption. For example, certain individuals could conceivably consume twice the average amount of vegetables which would increase Cd intake for the sandy loam by 23 μg Cd day^{-1} providing a total intake of 57 μg Cd day^{-1}.

The contribution to dietary intake of Cd in potatoes and root crops calculated by Carlton-Smith (1987) was based on peeled, but uncooked plant material. Peeling reduces Cd levels although some consumers do not peel potatoes or root vegetables. Kampe (1984) reported a 50% reduction in Cd concentration would be achieved by peeling potatoes. More recently, however, Smith *et al.* (1992c) showed that potato peel itself contains twice as much Cd as peeled tubers, but in terms of dietary impact, peeling (according to normal

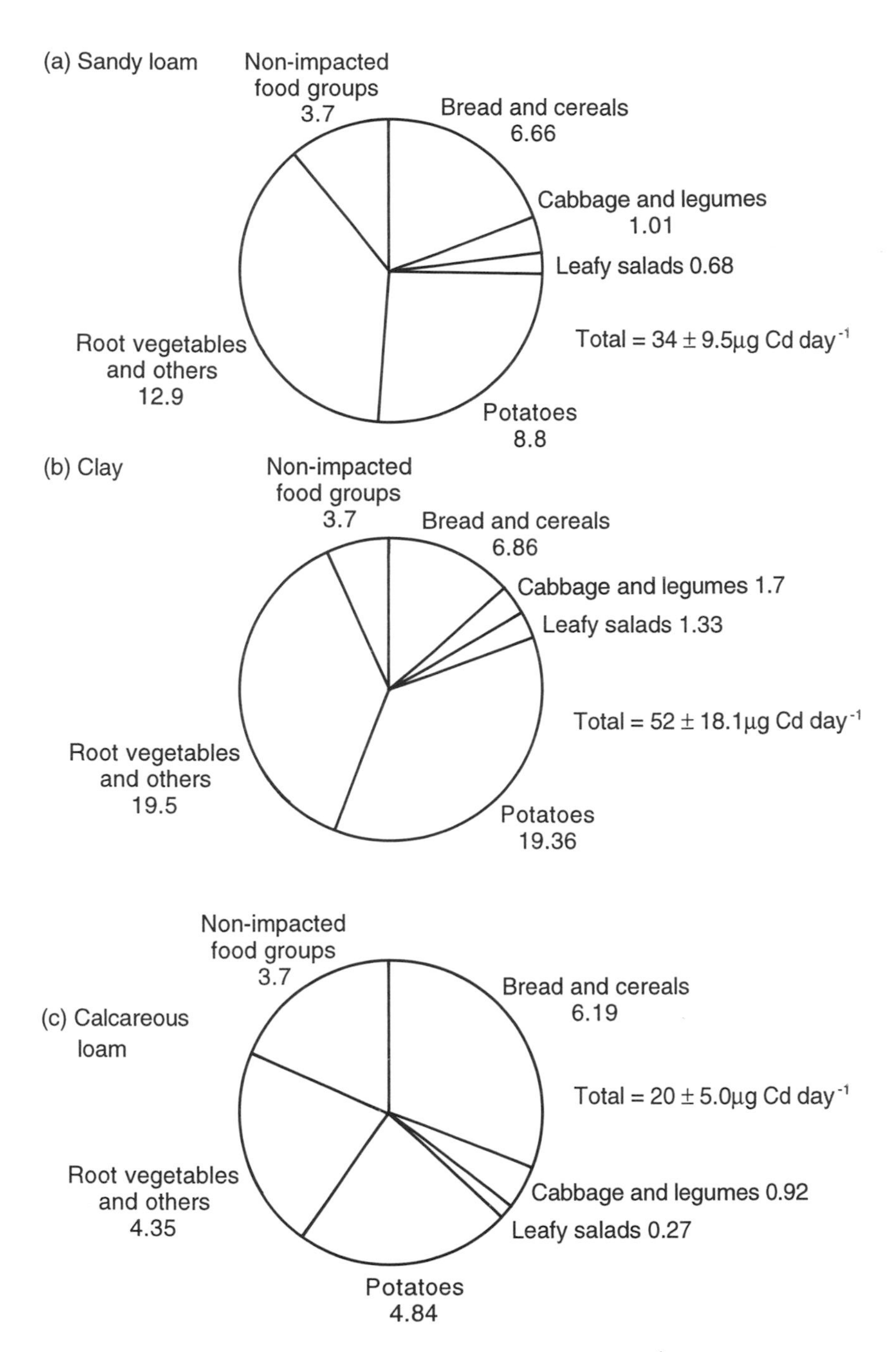

Fig. 4.3. Estimated dietary intakes of cadmium (µg day^{-1}) at a cadmium soil limit of 3 mg kg^{-1} (adapted from Carlton-Smith, 1987).

culinary practice) only reduces Cd intake from potatoes by about 5% because the quantity of peel consumed is relatively small compared with the amount of peeled tuber. Cooking can also reduce Cd concentration in food by up to 50% (Sherlock *et al.*, 1983). If the effect of cooking had been considered by Carlton-Smith (1987) then Cd levels in green and root vegetables and also in potatoes would have been reduced.

Carlton-Smith (1987) concluded that raising the concentration of Cd in soil by 1 mg kg^{-1} would increase Cd dietary intake by 6 μg day^{-1} for a sandy loam soil. The dietary model therefore demonstrated the large margin of environmental protection provided by the soil limit of 3.0 mg Cd kg^{-1}. Indeed, in an earlier analysis of these data, Davis *et al.* (1983) suggested a maximum concentration of Cd in sludge-treated soil (pH ≥7.0) of 6.0-12.0 mg kg^{-1} was compatible with the maximum safe limits set for dietary intake of Cd (Fig. 4.4). These field investigations clearly demonstrate the environmental protection provided by the upper CEC limit for Cd of 3.0 mg Cd kg^{-1} in sludge-treated soil (CEC, 1986a) implemented in current UK regulations where sludge is used in agriculture (SI, 1989a).

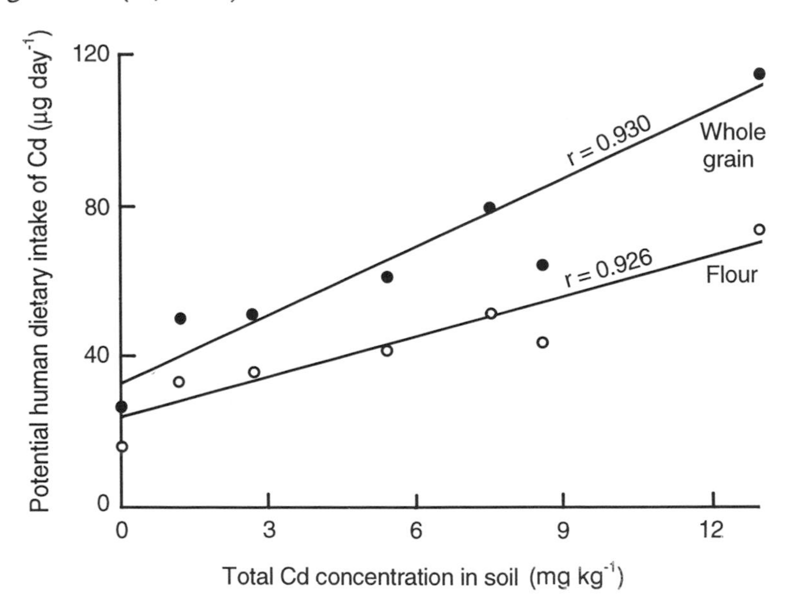

Fig. 4.4. The relationship between cadmium concentrations in soil and potential dietary intake of cadmium for an average consumer taking all his/her crops from sludge-treated soil (Davis *et al.*, 1983).

In contrast to the upper CEC soil limit for Cd of 3.0 mg kg^{-1} (CEC, 1986a) tested by Carlton-Smith (1987), the Peer Review Committee (Page and Logan, 1989) evaluating the US EPA Proposed Rule on sludge (US EPA, 1989a) recommended a cumulative application of Cd of at least 18.4 kg ha^{-1} (Table A7) which equates to a maximum soil concentration of approximately 10 mg Cd kg^{-1}. However, the cumulative application has been increased further to 39 kg Cd ha^{-1} in the US EPA Final Part 503 Rule (US EPA, 1993) which corresponds to a soil concentration of about 20 mg Cd kg^{-1}. This exceeds an earlier Cd loading rate limit of 5 kg ha^{-1} suggested by Ryan *et al.* (1982) and adopted previously in US regulations (US EPA, 1979) for soils receiving sludge. Webber and Monks (1983) similarly concluded that 5 kg Cd ha^{-1} added to agricultural soil represents little if any hazard to the human food chain. A loading rate of 5 kg Cd ha^{-1} is equivalent to an approximate soil concentration of 3.5 mg Cd kg^{-1} based on the precautionary assumptions that the sludge is cultivated into the soil to a depth of 0.2 m and the background concentration of Cd in the soil is 1.0 mg Cd kg^{-1} (the soil density being taken as 1.0).

Earlier suggested loading rates for Cd (Ryan *et al.*, 1982) were based on linear crop uptake response models similar to those determined by Carlton-Smith (1987). In developing the Final Rule, linear models were also used in the risk assessment analysis. However, the loading rate of Cd has been substantially increased for two principal reasons. Firstly, the potential small risk to the human food chain has been more accurately quantified with the improved database and understanding of crop uptake mechanisms for Cd. Secondly, re-analysis of food intakes has indicated that the apparent risk from sludge-borne Cd in the diet of US consumers was much smaller than the earlier dietary models had predicted (Chaney, 1990b). There is also evidence suggesting that Cd uptake by crops is curvilinear with increasing sludge application, attaining a plateau as a function of the metal adsorption capacity of sewage sludge (Chapter five), the sludge Cd concentration, and soil pH value as shown in Fig. 4.5 (Corey *et al.*, 1981; Logan and Chaney, 1983; Corey *et al.*, 1987; Logan and Chaney, 1987; US EPA, 1989a). Indeed, Logan and Chaney (1987) argued that regulations based on maximum application rates would not be required for sludges where risk assessment showed that the 'plateau' plant tissue Cd concentration for a particular sludge was low enough to preclude unacceptable food chain Cd contamination. However, the Risk Assessment Methodology (US EPA, 1989b) indicated that the plateau approach should be viewed with some caution at that time stating that: 'A thorough critical review and acceptance of this hypothesis by the scientific community and development of the data base will be necessary before it can become a basis for regulatory criteria.'

The potential occurrence of a 'plateau' response for metal uptake by plants emphasizes the level of environmental protection provided by the limit values derived from the risk assessment procedure. Indeed, Logan and Chaney (1983) have stated that: 'We believe that this new information should be used to demonstrate the existence of a larger safety factor when low Cd domestic sludges are utilized, rather than to support higher limits for Cd application to cropland.'

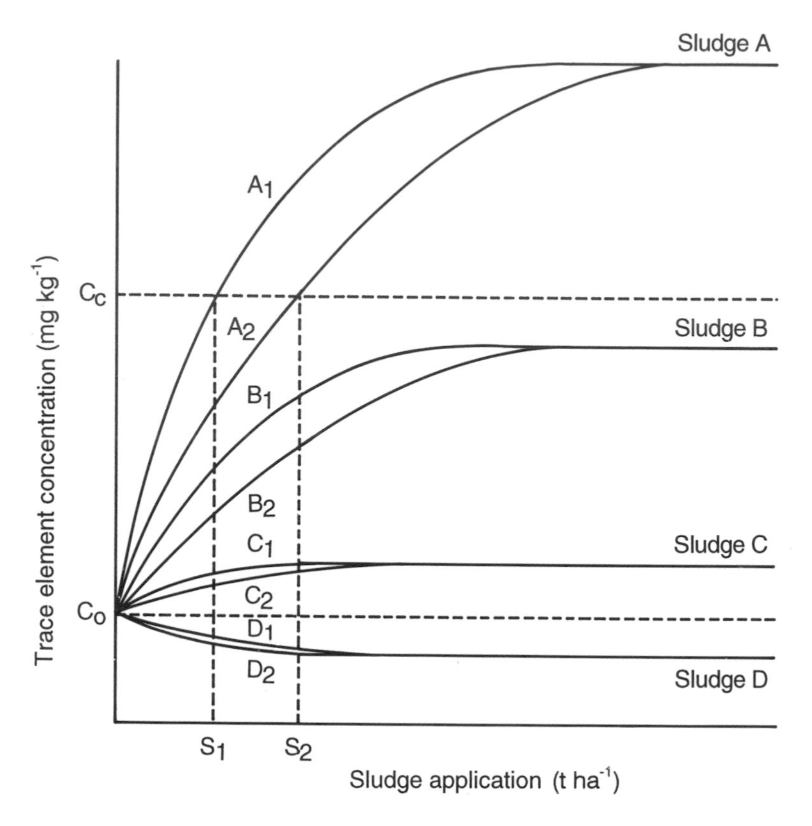

Fig. 4.5. Basis for differentiating sludges that do not require loading limits to prevent harmful trace accumulations in plants (sludges B, C and D) from one that does (sludge A). Two curves for each sludge represent that sludge applied to a soil of relatively low adsorption capacity (subscript 1) and a soil of higher adsorption capacity (subscript 2). C_o is the trace element concentration in plants grown on unamended soils. C_c is the critical concentration in the plant, and S_1 and S_2 are loading limits for sludge A applied to soils (1) and (2), respectively (Corey *et al.*, 1987).

Therefore, it may be argued that the loading rate for Cd derived by US EPA in the Final Part 503 (US EPA, 1993) from an analysis of sensitive environmental pathways represents a cautious estimate of human foodchain exposure to Cd because linear crop uptake models were used to evaluate the potential risk from Cd in sludge-treated soil.

Data from UK field trials (Carlton-Smith, 1987) show that the Cd uptake response of sensitive crops, such as lettuce, to sludge-borne Cd is approximately linear at rates of sludge addition to soil (up to 150 t ha^{-1} dry solids as liquid sludge) which produced apparently a curvilinear effect in relation to cumulative sludge applications in certain US studies (see Corey *et al.*, 1987). It is interesting that the range of soil Cd concentrations was achieved with four different sludges by Carlton-Smith (1987). Three of them were effectively identical except for their Cd content (Cassington sludges: S_1 69 mg Cd kg^{-1}, S_2 21 mg Cd kg^{-1} and S_3 33 mg Cd kg^{-1}), but the Cd contents of crops were related to the Cd concentrations in the soil and not the original Cd content of the sludge. Davis (1984a) concluded, therefore, that limits for Cd should be based principally on the Cd concentration in the soil rather than the Cd content of the sludge.

In contrast, Chang *et al.* (1987a) reported that the Cd uptake by Swiss chard approached an asymptote with cumulative additions of sludge up to 180 t ha^{-1} y^{-1} (dry solids) (Fig. 4.6). However, the uptake response was more or less linear within the range of soil Cd concentrations measured by Carlton-Smith (1987) and the maximum limit value for Cd allowed in soil under EU law (CEC, 1986a). Whether soil concentrations or rate of sludge addition explain the patterns in PTE accumulations by sensitive crop plants appears to depend upon the range of soil concentrations of interest and the extent of sludge applications in the field. Apparent differences in the uptake by crops may also be attributable to changes in the metal binding properties of sludges caused by different sludge treatment processes used in the UK and the US.

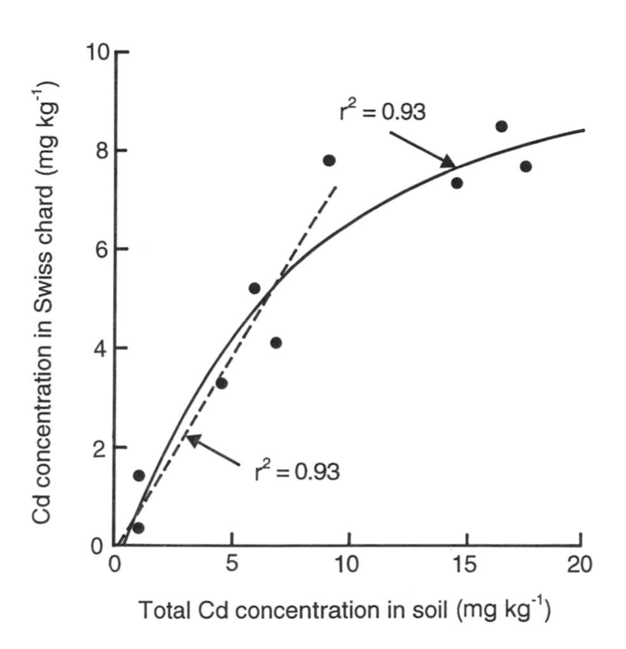

Fig. 4.6. Cadmium concentration in Swiss chard in relation to total content in a sandy loam soil, treated with cumulative additions of composted sludge at 180t ds ha^{-1} y^{-1} (adapted from Chang *et al.*, 1987).

In developing the 503 Rule, the US EPA (US EPA, 1992a) evaluated potential human consumption as the end-point of several sensitive environmental pathways of exposure to Cd from sewage sludge (Table A6). Pathway 1 assessed human exposure through ingesting plants grown in sewage sludge-amended agricultural land. This pathway is of immediate relevance to the discussion here. In contrast to EU and UK law, which only considers agricultural utilization of sewage sludge, the Final Part 503 also sets quality standards for sludge which facilitate the marketing of sludge to the domestic home gardener. Consequently, this was considered potentially as an important route of exposure due to the consumption of home produced vegetables grown with the aid of sludge-based products (Pathway 2). The possibility of child ingestion of undiluted sludge after application to the soil or from storage stock piles was also evaluated in the risk assessment (Pathway 3).

The Highly Exposed Individual (HEI) (see Appendix for explanation) in Pathway 1 is a consumer who ingests a mix of crops from sludge-treated agricultural land and from untreated land. It was estimated from the risk assessment that 2.5% of the plant food consumed by the HEI is taken from sewage sludge-amended farmland and that the duration of exposure is 70 years. To quantify potential dietary exposures resulting from the land application of sewage sludge, the amounts of various types of foods consumed over a lifetime were determined for the population as a whole. This involved estimating the quantities of unprocessed commodities consumed in complex prepared foods and calculating the average intake over a lifetime for each food category by weighting according to sex and age group. Relationships between plant tissue concentrations and cumulative pollutant loading rate were obtained from the scientific literature. Plant uptake slopes were determined for each of the following major plant food groups: grains and cereals, potatoes, leafy vegetables, legumes, root vegetables, garden fruits and peanuts. Finally, the Agency used the recommended maximum average intake of Cd of 70 μg Cd day^{-1} for an individual weighing 70 kg (i.e. 1.0 μg Cd day^{-1} kg^{-1} body weight) to calculate the allowable intake over and above the background level which was estimated to be 16.14 μg Cd day^{-1}. It is emphasized that the recommended tolerable daily intake of Cd appears to be without appreciable risk to human health during an entire lifetime on the basis of current knowledge. Therefore, US EPA estimated that an additional 53.86 μg day^{-1} of Cd could be ingested above the background level with minimal risk of adverse effects. If this value were exceeded, the Agency considered that adverse health effects might occur in the exposed individuals. Following this procedure, US EPA estimated that 610 kg Cd ha^{-1} could be applied to agricultural land in sewage sludge without affecting human health adversely.

Pathway 2 potentially offers greater exposure to Cd than Pathway 1 because home gardeners may consume greater quantities of certain important plant foods grown on sludge-amended soil compared with the agricultural utilization of sewage sludge. Indeed the HEI in this scenario is an individual who consumes 59% of their fruit and vegetable intake and 37% of the potatoes consumed for 50 years grown on a garden containing the maximum allowable cumulative Cd application. However, very few home gardeners in the US produce their own cereal crops, peanuts and certain dried commodities so the consumption of different food groups was adjusted to take various dietary trends into account. The US EPA used the same criteria followed in Pathway 1 to estimate the loading rate for Cd to protect the home gardener and determined that 120 kg Cd ha^{-1} could be safely added to garden soils in sewage sludge without adverse effects.

The regulations and the risk assessment approach to Cd adopted in the US for sludge utilization on agricultural land and in domestic gardens places in context the very large margin of protection of the human food chain provided by the current UK legislation for Cd in sludge-treated soil. Indeed, the US model estimated that garden soils could reach 60 mg Cd kg^{-1} before HEIs (home gardeners) would consume a lifetime average of 70 µg Cd day^{-1} (Ryan and Chaney, 1994), the WHO/FAO recommended maximum daily intake of Cd. Ryan and Chaney (1994) argued that, from a technical perspective, no individual would be harmed even if sludge-amended soils reached 60 mg Cd kg^{-1}. Even if all of the soil were to be eventually replaced by sludge and all of the organic matter contained in the sludge (50%) were to degrade, thereby doubling the Cd content in the growing media, this extent of Cd accumulation would be impossible to achieve in practice given the current quality of sludges recycled to agricultural land in the UK (Table 2.3). Whether this analysis is acceptable from a philosophical stand-point is another question. In particular, it is estimated that the Cd concentrations in the soils of Europe are increasing slowing on average at a rate of 4.5 µg kg^{-1} y^{-1} (van der Voet *et al.*, 1994). Allowing an appreciable, albeit only partial, filling of the margin between background intakes of Cd in the diet and the recommended maximum dietary level by sludge application could ultimately increase and accelerate the possible risk of human exposure to Cd in the long-term.

The risk assessment calculations performed in setting the Final Rule showed that the sludge→human ingestion pathway (Pathway 3, Table A6) was most limiting to the application of Cd to soil in sludge (US EPA, 1992a). Under this scenario a Highly Exposed Individual (HEI) is a normal child between the ages of one and six who ingests sewage sludge from storage piles or from the soil surface for a maximum of five years. The risk assessment stipulated that the HEI is not a pica child (one who exhibits excessive hand-to-mouth activity), because it is assumed that the parents of pica children will take precautions to prevent their children from eating sewage sludge. This pica child would be an example of a Most Exposed Individual (MEI) and is regarded as an unreasonable and unrealistic level of exposure for the purposes of risk assessment by the US model (see the Appendix). Since the risk to the human food chain arising from plant uptake of Cd is widely believed to be the principal cause for concern (Wagener, 1993), it is likely that there will be much continued debate in the future on this point and on the scientific basis to the numerical limits for PTEs in the Final Part 503 Rule (for example, see McBride, 1995). However, it could also be said that the risk assessment analysis performed for the Final Part 503 Rule is unique in that it has attempted to derive meaningful and realistic estimates of the potential risks from the application of pollutants to agricultural land in sludge by pooling the available scientific data on the different environmental pathways possibly affected by sewage sludge applications. It may become increasingly evident in the future that this process has usefully questioned many of the accepted views on the environmental issues concerning the land application of sewage sludge.

The large margin of safety built into the UK Cd limit in sludge-treated farmland is demonstrated further by a number of other dietary and health surveys of Cd intake in vegetables grown on contaminated soils arising from long-term sewage sludge applications (Sherlock, 1983) and also from geochemical origin through past mining activities (Sherlock *et al.*, 1983). For example, market gardeners and their families consuming plant foods grown on sludged soils containing up to 26 mg Cd kg^{-1} were considered by Sherlock (1983) as a 'critical group' in respect of exposure to Cd in vegetables. As expected, the average concentration of Cd in the diets was elevated (23 μg Cd kg^{-1}) compared with normal values (17 μg Cd kg^{-1}). However, the mean intake of Cd by these high risk individuals was the same as the calculated national average because the amount of plant food consumed by the test group was marginally smaller compared with the average national value. None of the participants consumed more than 60% of their plant food intake from contaminated soil, and even if all the plant component of their diet came from this source, Sherlock (1983) estimated that Cd intake would be around 44 μg Cd day^{-1} which is still well within the WHO/FAO recommended tolerable intake of 70 μg Cd day^{-1} (WHO/FAO, 1972).

In the second example, the dietary intake of Cd from plant produce grown on metal contaminated soil at Shipham, Somerset in the UK was studied by Sherlock *et al.* (1983). The soils around Shipham contain elevated concentrations of PTEs (Cd: range 2-360 mg kg^{-1}, median 91 mg kg^{-1}; Pb: range 108-6540 mg kg^{-1}, median 2340 mg kg^{-1}; Hg: range 0.015-12 mg kg^{-1}, median 1.27 mg kg^{-1}; Morgan and Simms, 1988) and chemical analyses of garden grown vegetables showed that Cd contents of crops were larger than normally expected values. For example, the Cd concentration in Shipham potatoes was about four times (0.13 mg Cd kg^{-1} fresh weight) that found in crops grown on non-contaminated soil. This is significant considering the important contribution potatoes make to the average diet as considered earlier. Potatoes comprised up to 50% of total home grown vegetables consumed at Shipham. Sherlock *et al.* (1983) estimated that the dietary intake of Cd at Shipham was 28 μg Cd day^{-1} compared with a national average estimate of 20 μg Cd day^{-1}. However, a number of the participants (6% of the population) had intakes exceeding 57 μg Cd day^{-1}. Medical examinations of the Shipham residents and of the families consuming vegetables from market garden soils considered earlier showed none of them to be suffering from health problems related to heavy metals (Sherlock, 1983; Sherlock *et al.*, 1983; Strehlow and Barltrop, 1988).

Finally, the bioavailability of Cd in the diet has an important influence on how much or how little Cd from plant food is actually absorbed by the gut. For example, Chaney *et al.* (1978) found that Cd did not accumulate in sensitive kidney or liver tissues of animals fed Cd-rich plant material grown on sludged soil compared with plant food from untreated soil. This is explained as being due to the interactions of Cd with other elements such as Zn, Cu, Ca and Fe also consumed in plant foods and which are antagonistic to the absorption of Cd by animals and man (Logan and Chaney, 1983; Fox, 1988; DH, 1991) (see Chapter six). The importance of such dietary factors in modulating Cd bioavailability to animals has been frequently reported (Logan and Chaney, 1983). Chaney (1990b) noted that individuals who actually consume model amounts of plant produce have a well balanced diet and will not be deficient in minerals which would otherwise potentially place them at risk from dietary Cd. Consequently, the individual exposed to a 'worst-case' condition of dietary Cd from plant produce grown on sludge-amended soil is no longer the most sensitive individual due to improved quality of diet.

Mercury and Lead

Mercury and Pb have been discussed in relation to toxicity to crop plants in Chapter three but are of concern in sludge-treated soil principally due to their potential zootoxic effects. However, these elements are held tightly in soil and there is generally no direct relationship between the amounts in soil and concentrations in crops (Davis, 1984a). For example, the summary of 117 sludge trials compiled by Vigerust and Selmer-Olsen (1986) showed no crop uptake of these elements overall for sludge-treated soil (Table 3.1). There is some evidence, however, that volatilization of Hg from soils and absorption by the aerial parts of plants could increase the concentration of Hg in plant tissues (Lindberg *et al.*, 1979). In contrast, studies using Hg salts applied to soil, representing a 'worst-case' of metal availability, demonstrated very little additional uptake of Hg even at large rates of addition to soil. For example, Bache *et al.* (1973) reported that edible portions of vegetables generally contained <0.1 mg Hg kg^{-1} when grown on soils containing up to 10 mg Hg kg^{-1} added as various mercury compounds. Kloke (1983) found the Hg concentration in grain of crops grown on soil containing 50 mg Hg kg^{-1} as $HgCl_2$ was similar to untreated controls and was generally $\leq$0.01 mg Hg kg^{-1}. In field studies with historically sludge-amended soils Smith *et al.* (1992b) detected no increase in the uptake of Hg by crop plants grown on soil containing up to 14 mg Hg kg^{-1}. The concentration of Hg in plant tissues was generally below the limit of analytical detection used of 0.5 mg Hg kg^{-1}. This confirms the earlier results of van Loon (1974) which similarly showed no uptake by crop plants of Hg from soil treated with Hg-rich sludge. Smith *et al.* (1992b) concluded, therefore, that availability of Hg in sludge-treated soil is very low and does not pose any risk to the human food chain through plant uptake.

Most field experiments with sewage sludge have failed to demonstrate any significant increase in the Pb content of food crops (Webber, 1981) as discussed already in relation to effects on plant growth. Further examples of the low bioavailability of Pb to food crops grown in sludged soil can be given. For instance, Chumbley and Unwin (1982) could find no correlation between soil and plant concentrations of Pb for commercially grown crops taken from fields with long histories of sludge spreading and containing up to 496 mg Pb kg^{-1} in soil. Similarly, the field trials with sewage sludge described by Carlton-Smith (1987) and the studies by Carlton-Smith and Stark (1987) with highly contaminated historically sludge-treated soils showed there were no relationships apparent between soil Pb concentrations and crop content. This consistent observation for sludge-treated soils is emphasized given that the Pb content of the historic site soils increased up to 25 000 mg Pb kg^{-1} (Carlton-Smith and Stark, 1987). In contrast, investigations with urban soils contaminated with Pb arising from aerial deposition, probably originating from domestic chimney ash and smoke, industrial emissions and vehicle exhaust, have shown some relationship may occur between soil and plant levels of Pb (Davies *et al.*, 1979). More recently, Davies (1992) also found that the Pb content of radish was significantly correlated with increasing total concentrations of Pb in soils contaminated by heavy metals from past mining activities. Davies *et al.* (1979) determined by linear regression analysis that a Pb-concentration for radish (root crops being potentially sensitive to Pb accumulation (Koeppe, 1981)) of 1.0 mg kg^{-1} fresh weight (the maximum legal limit for UK food (SI, 1985)) was achieved with a concentration of 345 mg Pb kg^{-1} in urban garden soils. However, Davis (1984a) noted that Pb applied to soil in sludge was probably less available for crop uptake than aerially deposited Pb in urban areas. For example, Sauerbeck and Rietz (1983) compared the step-wise extraction of Pb in a sludge-treated soil with that in two heavily Pb contaminated mineral soils and found the lowest proportion of Pb was released from the sludged soil. Adopting a similar sequential extraction approach, McGrath and Cegarra (1992) reported that the poor movement of Pb through the soil-plant pathway was reflected in the very low concentrations of Pb present in the soluble and exchangeable ($CaCl_2$) fractions (which were frequently below detection limits) of both the sludge-amended soil and control soil (which had received only mineral fertilizers) at Woburn. Furthermore, there was no change in the proportion of soluble and extractable Pb in sludge-amended soil with time, even though 20 years had elapsed since the last application of sludge to the field.

Most of the available data on Pb in sludge-treated soils indicate that there is minimal risk of contaminating the human food chain with Pb due to crop uptake. However, Aldridge and Alloway (1993) warned against possible complacency about Pb in sludge-amended soils after reviewing more than 400 references on Pb in soils and food crops. They emphasized that soils are dynamic systems responding to changes in environmental factors and that it may be inappropriate to assume that the current environmental and soil chemical conditions will persist indefinitely. It was suggested that long-term changes could conceivably result in increased bioavailability of Pb due to, for example, increased competition with Fe and Al for phosphate ions if soils were to become acidified in the future. Under these conditions it is possible that the formation of insoluble Pb phosphate minerals, which is an important mechanism controlling Pb solubility in soil, could be reduced. In spite of the arguably justified scepticism, however, Aldridge and Alloway (1993) concluded that the general consensus was that the transfer of Pb through the soil-plant-animal/human pathway does not constitute a serious problem for sludge-treated agricultural land.

In contrast to the minimal risk to the human diet from plant uptake of Pb, the direct ingestion of Pb contaminated soil by children is potentially a more serious problem. This exposure route for Pb has been considered in the US regulations for sludge (US EPA, 1992a; US EPA, 1993) because of concerns over possible encephalitis and resulting neurobehavioural impairment caused by higher than normal Pb intakes by children (Logan and Chaney, 1983). Probably the worst-case of soil ingestion by children may be as much as 5 g day^{-1} at certain times (Calabrese *et al.*, 1989) although the US EPA considered a soil ingestion rate of 200 mg day^{-1} (dry soil) was representative of children at highest risk (HEI) for calculating risk-derived standards for PTEs in sludge (US EPA, 1992a). In addition, sludge products are marketed in the US to domestic households often as fertilizers and mulches placing children exhibiting frequent hand-to-mouth activity potentially at risk of Pb poisoning from ingested sludge or sludge-treated soil. However, the bioavailability of Pb to children is largely dependent on the concentration of Pb in the sludge and the presence of sludge or soil in the intestine binds the Pb in the gut very strongly (Chaney, 1990b). The contents of Fe, P, Ca and organic matter in sludge reduce Pb absorption (Logan and Chaney, 1983). Chaney (1990b) therefore concluded that children are not at risk from Pb in median quality sludges. The risk assessment procedure (US EPA, 1992a) estimated a safe cumulative loading for Pb of 300 kg ha^{-1} and a pollutant concentration for Pb of 300 mg kg^{-1} to protect children from ingesting Pb in sludge (Table A7). The permitted cumulative loading corresponds with a maximum soil concentration of approximately 190 mg Pb kg^{-1} (Table A7). Consequently, it would appear that the US limit on Pb (US EPA, 1993) is more stringent than the current UK regulations (SI, 1989a) (Tables A2 and A7). However, it may be more appropriate to draw a comparison between the US EPA pollutant concentration for Pb in sludge and the UK maximum permissible soil concentration for Pb on the assumption that the risks from ingesting Pb in sludge or sludge-treated soil are broadly similar. On this basis the US and UK regulations are in agreement with respect to the limits on Pb. Furthermore, the direct ingestion route of exposure for sludge or for sludge-treated soil is less likely to be a problem with children in the UK under the current regulatory framework because sludge is applied principally to agricultural soils and the marketing of sludge products to domestic outlets is difficult under the current system (Hall, 1993). Nevertheless, the potential impact of Pb ingestion by animals grazing sludge-treated pasture is an important concern which is considered in Chapter six.

Zinc

Dietary intake of Zn from sludge-treated soils is not a concern from a human toxicological standpoint because the phytotoxic threshold of Zn in plant tissues is below that which could be potentially toxic to animals and man (Davis *et al.*, 1978). However, there is evidence that Zn deficiency may be widespread in Western diets (Bryce-Smith, 1986). Logan and Chaney (1983) also noted that many humans consume low dietary Zn and that increased crop Zn could be beneficial. In particular, the dietary deficiency of Zn has been linked to the conditions of anorexia nervosa and bulimia which can be corrected with oral Zn supplementation (Bryce-Smith and Simpson, 1984; Schauss and Costin, 1989). Comparison of the nutritional status of the diets of British adults (MAFF/DH, 1990) with recommended dietary reference values for nutrients (DH, 1991) shows on average that UK diets are adequately supplied with Zn. However, a proportion of the population consume less than the recommended Lower Reference Nutrient Intake for Zn which is considered an inadequate level of Zn intake for most individuals (DH, 1991). Females in particular have lower dietary intakes of Zn compared with males of similar age group which may be linked to the development of behavioural eating disorders in high risk individuals. Therefore, increasing the availability of Zn in soil and uptake of Zn by food crops through the application of sewage sludge to agricultural land may have a positive and beneficial effect on human health although realistically the overall contribution of Zn to the diet from sludge is likely to be small.

Chapter five:

Factors Influencing the Bioavailability of PTEs to Crop Plants

Soil pH Value

Soil pH has been comprehensively identified as the single most important soil factor controlling the availability of PTEs in sludge-treated soils (Davis and Coker, 1980; Webber, 1980; Williams, 1980; Logan and Chaney, 1983; Sommers *et al.*, 1987; Alloway and Jackson, 1991; Berrow and Burridge, 1991; US EPA, 1992a). With the exception of Mo, Se and As, trace element mobility decreases with increasing soil pH due to precipitation as insoluble hydroxides, carbonates and as organic complexes (Kiekens, 1984). Adsorption onto clay minerals and organic matter is also raised by increasing soil pH value (Kiekens, 1984). Therefore, increasing soil pH by liming, for example, reduces the absorption of PTEs by crop plants (Davis and Coker, 1980). Indeed, modulating soil pH conditions by applying lime is the only practicable method available which can be used to control metal activity in soil. In contrast, the availability of Mo, Se and As for plant uptake increases with increasing soil pH value (Davis, 1981; Pierzynski and Jacobs, 1986; Edwards *et al.*, 1995; Neal, 1995). This is explained because Mo, Se and As exist in soil as anions whereas the other principal elements of concern are present in cationic form.

Many examples can be quoted showing the lower uptake of PTEs by crops grown on calcareous soils compared with acid soils (Mahler *et al.*, 1980; Mahler *et al.*, 1982; Carlton-Smith, 1987; Alloway *et al.*, 1990) and the effect of lime addition to soil on decreasing trace element content of crops (Bolton, 1975; John and van Laerhoven, 1976; Bingham *et al.*, 1979; Narwal *et al.*, 1983; Pepper *et al.*, 1983; King, 1988a; Eriksson, 1989; Jones and Johnston, 1989; Jackson and Alloway, 1991). These effects are consistent with the reduction in PTE extractability (Hall *et al.*, 1988; Sanders and Adams, 1987) and soil solution concentration (Cavallaro and McBride, 1978; Cavallaro and McBride, 1980; Sanders *et al.*, 1986; Hall *et al.*, 1988) observed as soil pH is raised due to increased sorption of trace elements by organic and mineral fractions in soil and through precipitation processes (Farrah and Pickering, 1978; Hatton and Pickering, 1980; Harter, 1983; Christensen, 1984a; Cline and O'Connor, 1984; Kuo and McNeal, 1984; Tipping *et al.*, 1986; Sanders and El Kherbawy, 1987; Stahl and James, 1991). However, metal absorption by plants is also directly influenced by pH value. For example, Hatch *et al.* (1988) concluded that the uptake of Cd by plants grown in solution culture was markedly suppressed by acidification due to increased competition with hydrogen ions. Consequently plants grown in soil take up less Cd with decreasing pH than would be expected from the apparent increase in solubility.

Certain elements are much more sensitive to changing soil pH conditions which is reflected in their overall availability for uptake by crop plants (see Table 3.1). Thus Zn, Ni and Cd tend to be influenced strongly by soil pH whereas Cu, Pb and Cr are little affected by changing soil pH conditions (Bolton, 1975; Dijkshoorn *et al.*, 1981). The low availability of Cr in sludge-treated soil, for example, is explained because the solubility of Cr(III) ions in soil decreases above pH 4.0 and above pH 5.5 complete precipitation occurs (McGrath, 1995). However, Kiekens and Cottenie (1985) reported that Cr availability increased significantly only as soil pH value decreased below pH 3.5. Lead is similarly only available for plant uptake under very acid soil conditions (Davies, 1995).

Early guidelines for the use of sludge on agricultural land (Chumbley, 1971) recommended that soil pH value be maintained close to pH 6.5. This value was also adopted in US regulations for sludge (US EPA, 1979). However, there is little specific basis for selecting this pH (Logan and Chaney, 1983) and Davis (1984a) concluded that soil metal limits should be graded to take account of increased availability under low soil pH conditions. Under the current CEC regulations (CEC, 1986a) the metal limit values set out (Table A1) apply to soils with a pH value in the range 6.0-7.0. Where sludge is used on soil of pH <6.0, however, Member States are required to consider the increased availability of heavy metals to crops and reduce the limit values accordingly. The maximum permissible amounts of Zn, Cu and Ni may be raised in soil with pH >7.0 provided concentrations do not exceed the fixed values by more than 50%.

To determine appropriate pH-related soil limits for PTEs, Carlton-Smith *et al.* (1988) and Smith *et al.* (1992c) conducted field experiments using historically sludge-treated soils (Coker and Davis, 1979) of intrinsically low soil pH which were limed to provide a range of soil pH treatment levels (pH 5.0-7.0). The concentrations of Zn, Ni and Cd in the ryegrass test crops grown at each site decreased as simple linear functions of increasing soil pH across the range of pH values measured, whereas uptake of Cu was virtually independent of soil pH (Fig. 5.1). Pot trials with ryegrass (Sanders *et al.*, 1986; Carlton-Smith *et al.*, 1988) grown on soils amended with mono-metallic sludges showed patterns of uptake similar to those obtained in the field studies on historic sites for these elements. Sanders *et al.* (1986) only detected a significant increase in the Cu content of ryegrass grown on sludged soil (containing 160 or 192 mg Cu kg^{-1}) compared with unsludged soil at pH values <5.5.

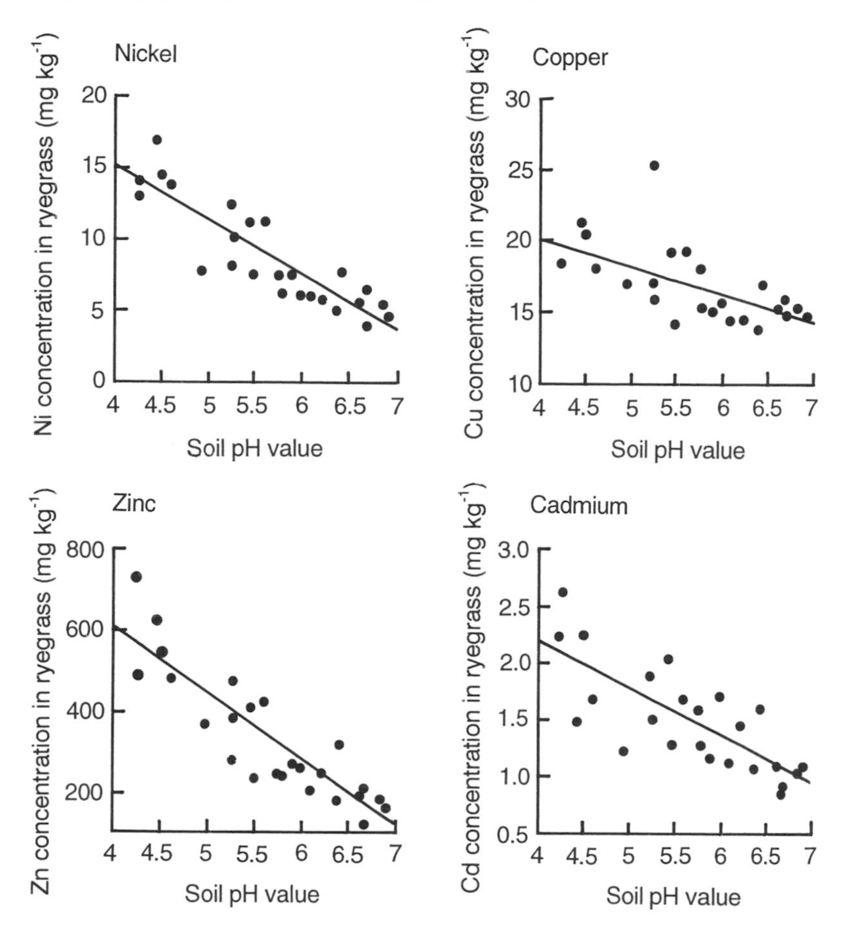

Fig. 5.1. Metal concentrations in ryegrass in relation to soil pH value (Smith *et al.*, 1992c).

Effects of soil pH value on metal solubility have also been examined using chemical extraction procedures and by studying the release of metals into solution from sludges of high metal content prepared according to the procedure of Davis and Carlton-Smith (1981). For example, Adams and Sanders (1984a) found that the concentrations of heavy metals released from metal-rich sludges into the aqueous supernatant liquid of a sludge-water suspension increased markedly as the pH decreased below certain threshold values. The threshold pH values were approximately 5.8 for high-Zn sludge (26 250 mg Zn kg^{-1} dry solids), 6.3 for high-Ni sludge (4219 mg Ni kg^{-1} dry solids) and 4.5 for high-Cu sludge (12 174 mg Cu kg^{-1} dry solids). However, the excessive concentrations of these elements in the experimental sludges is apparent compared with the quality of sludges currently recycled to agricultural land in practice (see Table 2.3). The concentrations of Zn, Ni and Cu in $CaCl_2$ extracts of soils contaminated with the same metal-rich sludges also increased rapidly as pH decreased below certain threshold ranges of soil pH value (Sanders and Adams, 1987). The pH ranges established were 6.2-7.0, 6.2-7.2 and 4.7-5.7 for Zn, Ni and Cu, respectively. Above these pH values the metal content of the sludge supernatent or soil extract was small and relatively constant. Interestingly, Adams and Sanders (1984a) reported that the threshold values for metal release into solution from the control sludge (containing 1540 mg Zn kg^{-1}, 96 mg Ni kg^{-1} and 448 mg Cu kg^{-1}, dry solids) were consistently lower compared with the metal-rich sludges at pH 5.4 for Zn and pH 5.5 for Ni. Furthermore, the proportion of total Zn in the solution phase of extracts from high-metal sludge-soil mixtures was very much higher (2.5-4.5 times at pH 6.0) than the proportion extracted from control sludge-soil mixtures, which was itself higher (1.8-4.0 at pH 6.0) than the proportion extracted from soil alone (Sanders and Adams, 1987). As soil pH value decreased there was a relative increase in the proportion of total Zn released into solution from the high-metal sludge compared with the control sludge. This behaviour implied that the binding strength of Zn originally on the high-metal sludge was weaker than the strength of binding on the low-metal control sludge. Chaney and Ryan (1993) argued that these observations were consistent with the idea that the specific binding sites for metals on the sludge were saturated by the high metal content of the sludge causing the release of loosely bound metals more readily compared with sludges of normal quality as the pH declined. However, Carlton-Smith *et al.* (1988) also found that the concentrations of $CaCl_2$ extractable metals in a historically sludge-treated soil generally increased markedly as soil pH declined although the relationship for Cd was linear over the same range of pH values studied. In contrast, the concentrations of metals in the tissues of crop plants and in displaced solutions from sludge-treated field soils appear to decrease in linear relation with increasing soil pH value (Carlton-Smith *et al.*, 1988; Hall *et al.*, 1988; Smith 1994a,b). This suggests that neither the behaviour of metals in sludges of controlled metal content, or $CaCl_2$ extraction precisely reflect the effects of soil pH value on potential metal availability to crops in sludge-treated agricultural land. Several workers (Mullins *et al.*, 1986; Hall *et al.*, 1988) have shown that metal concentrations in displaced soil solutions provide a much closer relation to plant availability compared with soil extraction. Soil extractants provide some information on the degree of metal bioavailability, but the results of such procedures should be interpreted cautiously (Beckett *et al.*, 1983; Jackson and Alloway, 1991).

Carlton-Smith *et al.* (1988) and Smith *et al.* (1992c) estimated the proportional changes in tissue concentrations of Zn, Cu, Ni and Cd in ryegrass at soil pH values of 5.0, 5.5 and 7.0 relative to pH 6.0 from linear regression models as a basis for calculating pH adjusted metal limits. The trials were separated both temporally (1986-87 and 1989-90) and spatially (Harrogate, North Yorkshire and Swinton, Greater Manchester in the UK), but the proportional changes in crop metal contents were remarkably consistent for each element and are listed in Table 5.1. The plant tissue concentration ratios showed that the order of decreasing sensitivity to changing soil pH conditions was Ni=Zn>Cd>>Cu>Cr=Pb. The relative insensitivity of Cu to soil pH has been reported by several workers (Sanders *et al.*, 1986; Sims, 1986) and has been attributed to the strong adsorption of Cu to soil surfaces and organic matter complexation in solution (Sanders and Adams, 1987). It also appears that plants control their internal Cu concentration much more closely than their Zn and Ni concentrations (Sanders *et al.*, 1987). In contrast, soluble Zn, Ni and Cd may occur almost exclusively as free metal ions below pH 6.0 and are increasingly complexed above this pH (Kiekens, 1984; Sanders and Adams, 1987).

Table 5.1. Proportional changes in metal concentrations in ryegrass relative to pH 6.0 for trials conducted at Harrogate 1986-87 (HA) and Swinton 1989-90 (SW).

Element	Soil pH						
	5.0		5.5		6.0	7.0	
	HA	SW	HA	SW		HA	SW
Zn	0.76	0.63	0.86	0.77	1.0	1.48	2.45
Cu	0.95	0.89	0.97	0.94	1.0	1.06	1.14
Ni	0.71	0.65	0.83	0.79	1.0	1.70	2.15
Cd	0.70	0.76	0.83	0.86	1.0	1.70	1.46

Sources: Carlton-Smith *et al.* (1988); Smith *et al.* (1992c)

Appropriate pH-related maximum soil concentrations were calculated by Carlton-Smith *et al.* (1988) and Smith *et al.* (1992b) by multiplying the proportional tissue concentration ratios and the established and tested soil limits set for pH 6.0-7.0 (SI, 1989a). Thus, maximum permissible soil limits were estimated which accounted for increased availability at lower pH ranges (pH <6.0) and reduced availability at higher pH (pH >7.0). Average soil maxima for Zn, Cu, Ni and Cd at pH 5.0-5.5, 5.5-6.0 and >7.0 obtained from both trials are presented in Table 5.2 (also see Smith, 1994a and b).

Smith *et al.* (1992c) also assessed the effect of soil pH on Cd uptake by potatoes and oats to determine potential impacts on human dietary intake of Cd from sludged soil. These crops were selected for study because they represent

Table 5.2. Estimated maximum permissible concentrations of Zn, Cu, Ni and Cd in soil (mg kg^{-1} dry soil) for ryegrass grown at different soil pH values.

Element	Soil pH			
	5.0-5.5	5.5-6.0	6.0-7.0	>7.0
Zn	210 (200)	246 (250)	(300)	591 (450)
Cu	126 (80)	130 (100)	(135)	147 (200)
Ni	51 (50)	61 (60)	(75)	145 (110)
Cd	2.3 (3)	2.6 (3)	(3)	4.7 (3)

Values in brackets denote UK statutory limits (SI, 1989a)

principal components of the plant part of the human diet such that small changes in Cd content can have large effects on the total consumption of Cd. In addition, potatoes and oats are often grown commercially on soils with low pH value whereas previous studies of metal uptake by crops from sludged soils only considered soil at pH 6.5 (Carlton-Smith, 1987) which is the optimum pH for most arable crops (MAFF, 1981). These crops could therefore be potentially at risk from excessive Cd accumulation from sludge-treated soil. However, no significant relationship between Cd concentration of oat grain and soil pH value was detected reflecting the known low accumulation of Cd by seeds compared with other plant parts (Davis, 1984b). In contrast, the concentration of Cd in potatoes increased with decreasing soil pH value. Potato peel accumulated more Cd than peeled potato tubers or the ryegrass test crop. Proportional changes in tissue concentrations reported by Smith *et al.* (1992c) indicated that the dietary intake of Cd from unpeeled potatoes could be raised by approximately 30% to 12 µg day^{-1} for crops grown on soil containing the statutory limit concentration for Cd of 3 mg kg^{-1} (SI, 1989a) at pH 5.0 compared with soil at pH 6.0. Proportional changes in crop tissue concentrations indicated the permissible amounts of Cd at pH 5.0<5.5 and 5.5<6.0 should be adjusted to 2.0 and 2.5 mg Cd kg^{-1} respectively to account for the increased uptake of Cd under low soil pH conditions and thus further protect the human food chain (Smith, 1994b).

The Steering Group on Chemical Aspects of Food Surveillance (MAFF/DoE, 1993b), reviewing the rules for sewage sludge application to agricultural land in the UK, argued that the differences in dietary intake for soils with Cd concentrations between 2 and 3 mg kg^{-1} were small. They showed that, where only 1% of plant food consumed originates from sludged land, very little change occurs in the dietary intake of Cd up to a soil concentration of 3 mg kg^{-1}. The intake for the average consumer was only approximately 6.9% higher than the upper background range of dietary intake of 18.8 μg Cd day^{-1}, where 10% of food is grown on sludge-amended land at the maximum permitted concentration of Cd in soil of 3 mg kg^{-1}. Consequently, there would appear to be little justification in supporting a recommendation that soil Cd limits should depend on pH (MAFF/DoE, 1993b). Smith (1994b) had suggested a reduction in the soil limit for Cd according to soil pH value may be appropriate, based on the assumption that all of the plant food consumed was from sludge-treated agricultural land, acknowledging that this was an unrealistic estimate of reasonably anticipated exposure. For the purposes of dietary modelling, assuming that 10% of food consumed originates from sludge-treated agricultural land (MAFF/DoE, 1993b) is a more realistic estimate of the exposure which may be anticipated in practice. Since this is a more appropriate estimate of food consumption from sludge-amended agricultural land, the author concurs that there is little justification in proposing that the soil limit for Cd should be adjusted according to pH value. Nevertheless, even 10% is probably a highly conservative estimate of the actual plant food intake from sludged soil. For example, US EPA estimated that a HEI under Pathway 1 (Sludge → Plant → Human) of the risk assessment for developing the Final Part 503 regulations (US EPA, 1993) was an individual consuming 2.5% of their plant food intake from sewage sludge-treated farmland (US EPA, 1992a).

Soil pH value was also considered by US EPA (US EPA, 1992a) in the exposure pathway analysis on Cd (and the other pH sensitive metals which may cause phytotoxicity) although pH control was considered unnecessary in the final Part 503 Rule (US EPA, 1993). This is because good agricultural practice requires the soil pH to be greater than 5.5 to avoid natural Al and Mn phytotoxicity to crops. Consequently, it is argued that sludge will be rarely applied to agricultural land with pH below 5.5 under US conditions. Nevertheless, the data set on plant uptake included information from studies in which the pH was as low as 4.5. Overall, 40% of the total data set was from studies in which the pH was less than 6.0. Thus, the acid soil system was considered to be well represented in the data set used to derive plant uptake relationships. The remaining data came from studies in which the pH ranged from 6.0 to 8.0. The Agency consider that the Part 503 numerical limits protect human health and the environment under most US soil conditions without requiring pH control for all agricultural land practices. The result of not regulating minimum soil pH is that under some 'unreasonably worst-case' conditions, the limit values are not as protective as in the 'reasonably worst-case' conditions modelled in the risk assessments for the final rule.

Despite accommodating the important effects of soil pH on Cd availability to crop plants, the risk assessment (US EPA, 1992a) identified soil ingestion by children as the most critical pathway of exposure to Cd from sludge. This pathway required greater protection than dietary intake of Cd from plant foods even when a large proportion of consumed crops are grown in sludged soil as was the assumption for the home gardener considered in Pathway 2 (Table A6).

The numerical limit for the maximum loading rate of Cd in the Final Part 503 Rule (US EPA, 1993) approximates to a soil concentration of 20 mg Cd kg^{-1} (Table A7). If the argument is followed that a comparison between the US EPA pollutant concentration and the equivalent UK soil limit is appropriate when direct ingestion is the limiting pathway, it becomes apparent that the UK statutory limit for Cd in sludge-treated agricultural land (Table A2) is a highly conservative value in relation to protecting the human diet.

In contrast to the US approach, the maximum concentrations of Zn, Cu and Ni permitted in sludged soil by the current UK regulations (SI, 1989a) are adjusted downwards according to pH bands of 0.5 pH units for soil with pH <6.0 to account for increased availability of metals in acid soils (Table A2). The spreading of sludge on soil with pH <5.0 is not allowed because bioavailability can increase markedly below this pH threshold. However, larger additions of PTEs can be safely applied to soils with pH value >7.0 which is reflected in the higher maximum soil concentrations given in the legislation. There is very close agreement between the estimated permissible soil concentrations for Zn and Ni at pH <6.0 derived by Carlton-Smith *et al.* (1988) and Smith *et al.* (1992c) listed in Table 5.2 and the UK soil limit values given in Table A2 substantiating the environmental protection provided by the regulations (SI, 1989a) for these elements in acid soils. However, these data indicate that larger concentrations of Zn and Ni in soil could be safely permitted under alkaline conditions. In addition, the statutory limits for Cu are highly precautionary since the estimated permissible concentrations were considerably larger than the regulatory values for soil at pH <6.0.

The intrinsic safety of the soil limit concentrations for Zn, Cu and Ni in protecting crop yields is emphasized because they are based on values set below phytotoxic thresholds for sludged soil with pH in the range pH 6.0-7.0. These metal concentrations have been substantiated under 'worst-case' conditions of metal availability by experimental studies using sensitive indicator crops, coarse-textured soils of low metal adsorption capacity and pot culture techniques (Davis and Carlton-Smith, 1984; Sanders *et al.*, 1987) and have also been tested under field conditions (Carlton-Smith, 1987). However, the pH-dependent soil processes which reduce metal availability may be reversible to some extent (Christensen, 1984a,b; Eriksson, 1989) and future decreases in soil pH value after sludge application has ceased could potentially increase metal accumulations by plants. Therefore, land use and management should be undertaken responsibly once soil concentrations have been increased to avoid potential problems. Of particular concern is the deterioration of soil pH status of pasture soils in traditional grassland areas of England and Wales. For example, Skinner *et al.* (1992) reported a net increase of 9% in the number of fields showing a decrease in soil pH over the past five years. However, grassland production is generally centred in the UK on soils of low pH, but under operational practice the area of land in the lowest statutory pH range 5.0<5.5 receiving sludge is relatively small corresponding to <7.0% of the total area used for sludge application (CES, 1993).

The Independent Scientific Committee reviewing soil fertility aspects of PTEs in sludge-treated agricultural land in the UK recommended that the scientific basis for different soil limits for Ni and Cu according to soil pH value should be re-examined (MAFF/DoE, 1993a). The Committee recommended that soil limits for these potentially phytotoxic elements should be determined without pH qualification to bring the regulatory basis for control in line with the highly precautionary single numerical value which they considered appropriate for Zn in all soils of pH 5.0-7.0 (see Chapter eight and the Appendix).

There is no evidence of any major shift in pH on fields which remained in arable or ley-arable rotation over the past five years (Skinner *et al.*, 1992). However, fields currently in arable cropping, including those which were not in arable production five years earlier, have improved in overall pH status. This is encouraging since soils in the statutory pH range 6.0-7.0 represent almost 50% of the total land area receiving sludge (CES, 1993). Nevertheless, the pH of these soils could deteriorate in the future as more agricultural land is managed under set-aside programmes, and also with possible changes in land use (Blake *et al.*, 1994). However, it is unlikely that even significant changes in soil pH would be detrimental to crop yields or the human diet given the precautionary scientific basis adopted in setting the soil metal limit concentrations.

Cation Exchange Capacity

The cation exchange capacity (CEC) is a measure of the negative charge density of a soil as a function of the soil's ability to adsorb cations (Mott, 1988). Soil cation exchange is also dependent on other soil factors including pH, organic matter content and soil texture (Brady, 1990). As soil pH is increased, hydrogen held by the organic and inorganic colloids becomes ionized and is replaceable. In addition, the adsorbed Al hydroxy ions are removed releasing additional exchange sites on soil minerals. The net result is that CEC increases with increasing soil pH value. Furthermore, the CEC increases with organic matter and clay content of soil because these soil fractions provide a large negative charge density. Thus sands and sandy loams have small CECs because they are low in clay content and probably contain little organic matter. In comparison, finer textured soils contain significantly more clay, and also organic matter such that their cation-adsorption capacities are usually higher.

On the basis of these properties, many authors recognize CEC as one of the soil properties controlling the retention and toxicity of metals in sludge-treated soil (Logan and Chaney, 1983). Consequently early US regulations for spreading sludge on agricultural land established differing limits for toxic metals depending on soil CEC. For example, in 1976 the North Central Regional Research Committee on Land Application of Sewage Sludge in the US (Logan and Chaney, 1983 quoting US EPA, 1980) arbitrarily divided soils into three groups by CEC (<5, 5-15, >15 meq 100 g^{-1} soil) with cumulative metal additions as listed in Table 5.3.

Table 5.3. Cumulative metal additions to soil (kg ha^{-1}) according to CEC.

	Cation exchange capacity (meq 100 g^{-1})		
	<5	5-15	>15
Zn	280	560	1120
Cu	140	280	560
Ni	56	112	224
Cd	5.6 (5)[(1)]	11.2 (10)	22.4 (20)
Pb	560	1120	2240

[(1)] Figures in brackets denote regulatory amounts of Cd set by US EPA (US EPA, 1979) for soil at pH≥6.5. The permitted loading was set to 5 kg Cd ha^{-1} for soil at pH<6.5 irrespective of CEC
Source: Logan and Chaney (1983)

However, research has shown the link between soil CEC and metal bioavailability is a relatively tenuous one. For example, Latterell *et al.* (1976) reported that the uptake of Cd, Zn, Cu, Cr and Pb by soybean shoots remained constant as the soil CEC increased. Similarly, King (1981) found no consistent effect of CEC on the metal content of fescue grass grown on soil amended with swine manure and sewage sludge and concluded that a better criterion than CEC was needed for determining acceptable metal loading rates. Applying stepwise multiple regression analysis to the accumulation of Cd by vegetables grown on contaminated soils, Alloway *et al.* (1990) showed that CEC was only important for two soils out of nine probably because of their larger organic matter content since in each case these had been treated with sewage sludge. However, King (1988a) had earlier used a similar statistical approach to determine the effect of soil properties on Cd availability to plants using tobacco, which is a sensitive indicator crop to Cd in soil, and found that CEC did not appear as a significant factor in any of the regression models. King (1988b) also determined that the retention of metals by soils was independent of CEC, whereas the presence of Fe oxides and clay content in soil were more important.

In contrast, other studies apparently show soil sorption and plant uptake are related in some way to CEC. For example, King (1988c) found that CEC could be correlated with the concentration of Cd in soil extracts, although the problems of using extractants to estimate plant availability have already been discussed in relation to soil pH value. Reddy and Dunn (1986) showed that clay soils, with high CEC, had a larger capacity to adsorb Ni and Zn from equilibrium solutions of metal salts compared with a low CEC sandy soil. In that study, there was apparently no consistent relationship between soil sorption capacity and soil pH and organic matter. Larger concentrations of Cd and Zn reported by Chang *et al.* (1982a) in barley cultivars grown on coarse-textured soils compared with fine-textured soils could also be attributed to differences in soil CEC.

The inconsistency in metal uptake by crops in relation to CEC is explained largely because the effects of CEC are compounded by the specific sorption characteristics of soil organic matter, and Fe, Mn and Al oxides (Logan and Chaney, 1983). Organic matter binds metals by chelation as well as by cation attraction. Iron, Mn and Al oxides can specifically adsorb metals yet do not have sufficient net negative charge to bind metals by cation exchange.

Clay content of soils is important in determining CEC, but clay minerals vary in their exchange properties. Bittell and Miller (1974) reported average selectivity coefficients for the exchange of Pb^{2+} and Ca^{2+} were 0.60, 0.44 and 0.34 for montmorillonite, illite and kaolinite respectively although Pb^{2+} adsorption was favoured over Ca^{2+} with most preference being for kaolinite. In contrast Ca^{2+} and Cd^{2+} more or less compete on an equal basis (selectivity coefficients $\approx$ 1.0) for exchange sites on the clay minerals such that clays may not bind as much Cd as Pb. In addition, raising pH increased CEC, but an increase in metal retention may also be due to decreased solubility of the metals adsorbed to solid phase minerals. For example, Mullins and Sommers (1986) measured the diffusion coefficients for Zn and Cd in sludge-treated soil and established that potentially diffusible amounts of these elements may be controlled by processes other than simple exchange including solid phase dissolution and/or release from chelation sites on organic matter. Furthermore, studies with metal salts suggest that sandy soils with low CEC may be less suitable for receiving metal contaminated sludges compared with finer textured soils with high CEC. However, Korcak and Fanning (1985) showed this relationship did not hold on the basis of crop tissue concentrations where sewage sludge was applied to such soils. They concluded that low CEC soils were no less suitable for receiving sludge than soils of high CEC.

Sommers *et al.* (1987) stated that the relationship between CEC and metal uptake from sewage sludge-amended soils had not been conclusively demonstrated under field conditions. Indeed, no phytotoxic effects have been reported at metal levels which greatly exceed the limits set according to CEC (Table 5.3) (Chang *et al.*, 1983; Hinesly *et al.*, 1984; Vlamis *et al.*, 1985). Sommers *et al.* (1987) therefore concluded that CEC should be abandoned in regulating the agricultural use of sewage sludge. Consequently, in contrast to soil pH, it is unlikely that CEC is important in considering the environmental impact of sludge recycling to agricultural land.

Organic Matter, Metal Speciation and Time

The binding of PTEs in soil by native and sludge organic matter, and the changes in metal speciation which may occur with time, particularly once sludge application has ceased, are intimately linked and are critical factors in assessing the potential long-term environmental effects of spreading sewage sludge on farmland. For example, Beckett and Davis (1979) put forward the concept of a 'time bomb' effect whereby sludge-applied metals may become more available over time (at constant pH) once the organic matter supplied in sludge has decomposed. More recently, Alloway and Jackson (1991) concluded that changes in the bioavailability of heavy metals during the residual period after sludge application had stopped was incompletely understood and posed important questions for the longer term. They also stressed the importance that changes in climatic conditions and land management practices could have.

Very large additions of sludge to soil are necessary to obtain any detectable increase in soil organic matter content (Guidi and Hall, 1984). This is because the amount of sludge organic matter added to soil is typically very small in operational practice (2-5 t ha^{-1}) compared to that already present in the soil profile. For example, a mineral soil with 3% organic matter contains in the plough-layer approximately 100 t ha^{-1} of organic matter. A typical application of 5 t ha^{-1} (dry solids) of sewage sludge adds to the soil about 2.5 t of organic matter and would raise the soil organic matter content to 3.07%. Thus, single large applications of sludge of 50-100 t ha^{-1} (dry solids) are recommended to achieve significant improvements in the physical properties of impoverished soils (Guidi and Hall, 1984).

Nevertheless, relatively small quantities of organic matter normally applied to soil in sewage sludge under operational practice can significantly increase the adsorption capacity of soil for metals (Karapanagiotis *et al*., 1991). Increasing the complexation capacity of soil and the formation of stable organo-metallic complexes by applying sludge organic matter reduces the mobility of PTEs in soil (Ram and Verloo, 1985) and thus lowers their availability to plants. However, Karapanagiotis *et al.* (1991) showed that a six-fold increase in the rate of sludge application from an initial rate equivalent to 3.3 t ha^{-1} did not produce a proportional increase in complexation capacity, measured by a pyrophosphate extraction method, following a five-month incubation period with soil. Senesi *et al.* (1989) similarly observed a threshold saturation effect when the metal-loading of the soil by sludge application was high since increasing the total metal content of sludged soil, was not paralleled in the corresponding organic fractions. On balance, these data would therefore suggest that other soil and/or sludge properties may ultimately become more important than the organic matter content in controlling the lability of heavy metals in sludge-treated soil following repeated applications of sludge in the long-term.

Karapanagiotis *et al.* (1991) estimated that the decreasing stability of organo-metal complexes in sludge-amended soil at pH 5.6 was

Cu>Cd>Zn=Pb>Ni indicating the stronger binding of Cu with organic matter compared with Cd or Zn. Nickel was the most weakly bound metal after sludge addition whereas Zn appeared to ascend in the series of relative metal affinity for organic ligands as sludge addition increased. Over half of the Cd was in the organic phase, compared to less than 20% of the Pb. The proportion of extractable Pb relative to the total amount was unaffected by the rate of sludge addition, whereas Cu extractability decreased with increasing sludge application and for Ni and Zn the extractable proportion increased. However, Karapanagiotis *et al.* (1991) did not attempt to separate the complexing ability of the different soil organic fractions. These organic fractions have been separated into so-called humic and fulvic acids depending on their solubility characteristics and both groups interact with heavy metals in different ways. However, this categorization represents a gross over simplification of the form and range of organic compounds actually responsible for binding metals in sludge-treated soils (Keefer *et al.*, 1984; Dudley *et al.*, 1987) although it has been quoted widely in the literature and is used here to provide a convenient basis for simplifying and describing the behaviour of organo-metal complexation in soil. The various classes of organic compounds typically found in soils which can react with metals are listed in Table 5.4 on the basis of their affinity for water and acidity-basicity reaction.

Table 5.4. Classification of hydrophilic and hydrophobic compounds present in soils.

Compound	Description
Hydrophilic acids	Uronic acids, simple organic acids, polyfunctional acids, or polyhydroxy phenols
Hydrophilic neutrals	Carbohydrates, polysaccharides, polyfunctional alcohols, or phosphate salts
Hydrophilic bases	Most amino acids, amino sugars, low molecular weight amines, and pyridine
Hydrophobic acids	Aromatic phenols, certain organic acids, anionic detergents, and aromatic acids
Hydrophobic neutrals	Hydrocarbons, fats, waxes, oils, resins, amides, phosphate esters, chlorinated hydrocarbons, high molecular weight alcohols, amides, esters, ketones and aldehydes.
Hydrophobic bases	Complex polynuclear amines, nucleic acids, quinones, porphyrins, aromatic amines, and ethers

Source: Keefer *et al.* (1984)

Kiekens (1984) provided a useful account of the behaviour of fulvic and humic acid components in relation to metal availability in sludged soils. Fulvic acids mainly form chelates with metal ions over a wide pH range which are soluble in both acid and alkaline conditions, thus increasing the solubility and mobility of heavy metals in soil. A large proportion of the heavy metals present in soil solution appear to be associated with a yellowish compound with fulvic acid properties. For example, Verloo (1979) reported that 93% of the soluble Cu was bound in this way and the proportions of bound Pb and Zn were 59% and 16%, respectively. According to Stevenson and Ardakani (1972) the stability constants (K) of metallo-fulvic complexes decrease in the order:

	Cu >	Ni >	Pb >	Zn
log K at pH 3.5	5.8	3.5	3.1	1.7

However, the stability of the complexes formed increases as pH is raised according to the following order:

	Cu >	Pb >	Ni >	Zn
log K at pH 5.0	8.7	6.1	4.1	2.3

Other workers (Schnitzer and Skinner, 1966; 1967) have measured a similar affinity series by fulvic acid for metals at pH 5.0. From a review of literature, Lake *et al.* (1984) reported that neither Cd nor Ni was associated strongly with fulvic acid whereas complexation between Cu and Zn with fulvic acid was apparent. Dudley *et al.* (1987) incubated anaerobically digested sludge for a period of 30 weeks and found that Cu in $CaCl_2$ extracts was associated almost exclusively with soluble organic compounds following chromatographic separation. Nickel became evenly partitioned between organic and inorganic complexes after four weeks' incubation whereas Zn was complexed mainly as inorganic forms. The increase in complex stability which occurs as pH is raised may be explained through the dissociation of functional -COOH and -OH groups in the fulvic acid molecule and a reduction in H^+ competition for adsorption sites. Furthermore, it is apparent that fulvic acids complex Cu^{2+} and Pb^{2+} ions preferentially over the other metals present in sludged soil.

In contrast to fulvic acids, the solubilities of humic acids increase as pH is raised resulting in metal complex formation and a concomitant decrease in metal ion activity in the soil solution (Kiekens, 1984). An affinity series similar to that for fulvic acid was reported by Stevensen (1977) for humic acids which followed the order: Cu>Pb>>Cd>Zn at pH 4.0-6.0. However, at pH 4.0-7.0 the series was found to be Zn>Cu>Pb by Verloo and Cottenie (1972) probably due to differences in experimental conditions employed between the studies.

The forms of organic matter also differ in that humic acids behave as a colloidal system being flocculated principally by the cations Ca^{2+} and Mg^{2+} present in soil, and also by Fe^{3+} and Al^{3+} as pH declines, forming insoluble complexes. Kiekens (1984) therefore considered that humic acids provided an 'organic storage for many metals' in soil. Petruzzelli *et al.* (1978) demonstrated that humic compounds of high molecular weight played an essential role in retaining heavy metals irreversibly in soil. This view is also supported by Stevenson and Chen (1991) who showed that humic acids form stronger complexes with Cu^{2+} than the more mobile fulvic acids and thus concluded that humic acids act as a sink for Cu^{2+} thereby reducing Cu availability to plants. Indeed, Emmerich *et al.* (1982a) predicted that Cu in the solution phase of sludged soil was almost exclusively in organically complexed forms. Elliott *et al.* (1986) agree that increased soil organic matter content should restrict mobility and bioavailability of Cu and also of Cd, at least under acidic conditions (pH 5.0) where soluble metal complex formation is limited. Furthermore, the humic acid fraction of sludge-amended soils has a well-defined selectivity in binding trace metal ions introduced by sludge application (Senesi *et al.*, 1989). As the metal loading of the soil increases, the metal-humic acid adsorption-desorption equilibria shift preferentially binding Cu, Ni, Zn and Cr which form stronger complexes with readily available sites on humic acid, compared with more labile metals such as Mn, U, Ti and Mo which are desorbed and replaced.

Bjerre and Schierup (1985) showed that plant uptake of Cd from soil was inversely correlated to soil organic matter content in response to the increased capacity of the soil to absorb Cd. Solution culture studies (Cabrera *et al.*, 1988) also provide evidence that Cd-organo complex formation in soil can reduce Cd uptake by plants. However, using the GEOCHEM model, which combines known chemical equilibria and properties of soils to determine metal speciation in soil solutions, Mahler *et al.* (1980) predicted that free ionic Cd could account for 64-72% of soluble Cd in sludged soil while organic complexes of Cd were <10%. Keefer *et al.* (1984) did not detect Cd in any organic fraction extracted from a sludge-amended soil three years after sludge treatment, which agrees with other GEOCHEM predictions (Lake *et al.*, 1984). Using chromatographic techniques, Alloway and Tills (1984) similarly concluded that Cd exists principally in a cationic form in soil solution and that organic species are relatively unimportant. Fractionation studies (Keefer *et al.*, 1984) of metal-organic components extracted from sludge-treated soil showed that organic Zn, Ni and Pb were partitioned principally in hydrophobic base fractions (complex amines) whereas Cu was most often associated with the hydrophilic bases (amino acids). Keefer *et al.* (1984) concluded there were at least two binding mechanisms for Cu (weak and strong) and four each for Zn, Ni and Pb in sludge-amended soil. Alloway and Tills (1984) similarly reported that there may be more than one site or mechanism for Cd sorption in soils since they found two linear parts to the adsorption isotherm describing sorption at low and high Cd concentrations. The distribution of Cd between the solid and aqueous phases of the soils tested was controlled by soil pH, organic matter and oxide content.

The behavioural patterns of metal binding to organic matter in soils are clearly complex and vary according to the amount of sludge applied and the metal under consideration. It is clear, however, that most elements are bound to a greater or lesser extent to organic matter contained in soil and applied in sewage sludge. For example, King and Dunlop (1982) showed that large organic matter contents in soils could substitute for high pH value in limiting the uptake of PTEs applied to soil in sewage sludge. However, mineralization of sludge organic matter and nitrification of released and applied ammoniacal nitrogen can lower soil pH thus increasing the bioavailability of PTEs despite the high buffering capacity of sludge organic matter (Logan and Chaney, 1983; Dudley *et al.*, 1986).

On application of sewage sludge to soil there is an immediate addition of heavy metals to the soluble and exchangeable pools resulting in the greater extractability of PTEs (Chang *et al.*, 1984a; Mashhady, 1984; Mullins and Sommers, 1986; Carlton-Smith, 1987; Sanders and Adams, 1987; Rappaport *et al.*, 1988), increased concentrations in the soil solution (Adams and Sanders, 1984b; Sanders *et al.*, 1986) and elevated concentrations in crop tissues (Kelling *et al.*, 1977a; Schauer *et al.*, 1980; Carlton-Smith, 1987) compared with untreated soils (also see Chapters three and four). A further portion is then released into potentially bioavailable forms by the decomposition of inorganic compounds contained in the sludge that are unstable in soil and of organic matter that is moderately biodegradable (Schauer *et al.*, 1980; Adams and Sanders, 1984b). Both of these reactions may take place relatively rapidly and within 1-2 years following sludge application in the field at ambient temperatures. For example, Dudley *et al.* (1987) separated and quantified water-soluble organic compounds by gel filtration chromatography and infrared spectroscopy from sludge-amended soils incubated at 25 °C for 1, 2 and 4 days and 1, 2, 4, 10 and 30 weeks. The soluble organic material was initially composed of a large molecular-size fraction principally containing carboxylic acids and polysaccharides, and a smaller molecular-size fraction composed mainly of amide-containing moieties. The short-term trend was an immediate decrease in the molecular size of polysaccharide, polypeptide and organic phosphate compounds with a corresponding loss in the amount of material eluting in regions associated with smaller molecular size fractions. Over longer periods of time the water-soluble material became more homogeneous with respect both to chemical composition and molecular size. The molecular size distribution changed little between 10 and 30 weeks, with the 30-week extract being composed primarily of polysaccharides and of aliphatic and carboxylic acids. Observed changes in the metal complexation with soluble organic compounds with time were apparently consistent with the changes in the nature of soluble organic ligands. Low soluble C levels correlated with decreases in the fraction size of complexed species from 100-92% to 71-86% for Cu and from 100-47% to 39-30% for Ni during the 30 week incubation period. Zinc showed variable results: for one soil organic complexes accounted for 100% of the total soluble metal initially measured which decreased to 39%, whereas no consistent trend was observed in Zn complexation in another soil which varied from 2 to 100%. These properties were qualitatively associated with concomitant increases in the fraction size of bioavailable inorganic species in the soil solution. However, the observed effect on bioavailability could be largely explained because pH value declined concomitantly with time and the decreasing level of soluble organic matter.

Adams and Sanders (1984b,c) similarly measured an increase with time in the proportion of Zn and Ni which occurred as free metal ions in solutions displaced from a sandy loam soil following the addition of sewage sludge. This was partly attributed to the decreasing levels of soluble organic matter observed during incubation for periods of 1, 3, 7 and 11 months. However, it was concluded that such increases in the proportion of soluble metals present as free ions were unlikely to seriously affect phytotoxicity of Ni, Cu and Zn. Total concentrations of Ni and Zn in soil solutions were initially high due to complexation with soluble organic matter, but the total levels of Ni and Zn in solution decreased with time as the organic matter content declined. Soil pH also had a major influence here and declined towards the end of the study thus increasing metal solubility again. In contrast, total Cu and Cu^{2+} concentrations in displaced soil solutions were small and did not vary much. As expected, concentrations of all three metals in solutions displaced from clay soil were smaller than those from the sandy loam and remained constant or declined with time. This was explained by the greater adsorption capacity and higher pH value of the clay compared with the sandy loam soil.

Despite the addition to soil of applied and released soluble and exchangeable heavy metals in sewage sludge, Lewin and Beckett (1980) considered the potential increase in these pools was held down by the massive, but temporary, increase in microbial biomass which occurs in response to the application of sludge organic matter (see Chapter eight). The capacity for micro-organisms to immobilize heavy metals is well recognized and has potential biotechnological applications for the detoxification and/or recovery of valuable metals (Gadd and White, 1989). In soils and with respect to biogeochemical cycling of metals in the environment, micro-organisms are being increasingly seen as important components of models designed to predict metal mobility (Mullen *et al.*, 1992) since they can significantly immobilize and mineralize soluble metals in the environment (Beveridge, 1989). This is especially true since immobilized metal products are difficult to remobilize (Flemming *et al.*, 1990). Heavy metals are retained on and in the biomass by processes of adsorption and precipitation, and are released slowly as the biomass gradually diminishes on further microbial cycling. The rise in metal bioavailability in sludge-amended soil is also buffered, and contained, by the transfer of heavy metals into less labile organic complexes, adsorption complexes and insoluble salts. However, some fractions of the heavy metals still remaining in the sludge are resistant and are only released very slowly, if at all. Lewin and Beckett (1980) considered that the complex interactions between these opposing processes probably account for the highly inconsistent reports of the short-term changes in metal availability which occur following the addition of sludge to soil.

This raises the question, however, as to what happens to the availability of heavy metals in the long-term and particularly after sludge application has ceased and organic matter levels have declined through decomposition. For example, Hohla *et al.* (1978) reported that 60% of the organic carbon applied in digested sewage sludge over a period of six years to soil held in lysimeters could not be accounted for and had presumably decomposed. Other studies (Miller, 1974) suggest much slower rates of decomposition than this for digested sludges depending on environmental and edaphic conditions. For example, Johnston *et al.* (1989) reported that the half-life of organic carbon applied in sewage sludge to soil at the Woburn experimental field site was about 20 years.

Lewin and Beckett (1980) imply that bioavailability could decrease with time because heavy metals released from their more mobile forms apparently accumulate as non-biodegradable organic complexes or become occluded in insoluble precipitates of more common elements. In a more recent review of the long-term effects of applying sludge on agricultural land, Chang *et al.* (1987b) concluded that there was no evidence for the bioavailability of metals in sludge-treated soils increasing with time after terminating sludge applications when chemical conditions of the soil remain constant. They reported that concentrations of Cd and Zn in plants grown in soils no longer receiving sludges either were not significantly different from pretreatment levels or decreased with time. Research with metal salts applied to organic soils or soils with pH>6.5 (Levesque and Mathur, 1986; Payne *et al.*, 1988) also demonstrated the conversion of added inorganic sources of Cu and Zn to plant-unavailable forms over time. Metal speciation studies (Emmerich *et al.*, 1982b) suggested that the forms of Cd, Ni and Zn applied to soil in sewage sludge similarly shift towards the residual stable form with time. The more stable the solid phase form of a heavy metal the less likely it will be to dissociate into the solution phase and to speciate into a potentially available form. Furthermore, decreases in concentrations of metals in soil extracts (Kelling *et al.*, 1977a; Korcak and Fanning, 1985) and in plant tissues (Kelling *et al.*, 1977b; Bidwell and Dowdy, 1987) with time reported by some workers also indicate reversion to less available forms is possible. Interestingly, Bidwell and Dowdy (1987) showed that uptake of metals by corn diminished with consecutive croppings for constant levels of Cd and Zn extracted by diethylenetriaminepentaacetic acid (DTPA) and 1 M HNO_3. Hence, the reliability of these extractants in the prediction of Cd and Zn uptake in subsequent croppings during the residual period is questionable.

During the third and the fourth year after terminating sludge application to an experimental field site, Chang *et al.* (1982b) found that the Cd and Zn contents of wheat grain and straw grown on the treated soil (which contained 1.22 mg Cd kg^{-1} and 177 mg Zn kg^{-1}) were slightly higher than the control. However, concentrations of these elements in plant tissues were within normally expected ranges for wheat grown on uncontaminated land. The sludge application did not result in any detectable increase in the concentration of water-extractable Cd and Zn in the soil.

In contrast to the cases demonstrating the reversion of metals to less available forms, other reports show crop uptake of PTEs remains relatively constant with time once sludge application has been terminated (McGrath, 1987). At Cassington and Royston, for example, Carlton-Smith (1987) could not detect any significant change in the plant availability of sludge-derived metals over the five-year duration of the study after sludge application. Similarly, Dowdy *et al.* (1978) measured increased concentrations of Zn and Cu in snap beans in response to increasing rates of sludge application; concentrations in the crop reached an apparent maximum value and remained at that level once sludging had ceased.

Other work reported by Berrow and Burridge (1984) from the Luddington and Lee Valley experiments indicated that PTEs in sludge-amended soil can persist in available forms for long periods of time. Ten years after the application of large dressings of highly contaminated sludge at these sites the metal contents of grass and clover were increased and reflected the high levels of extractable Zn and Ni in the soils compared with untreated controls. However, even under such extreme conditions of sludge treatment, the results suggested that Cu availability had decreased with time and that concentrations of Zn, Cu, Ni, Cr and Cd in acetic acid extracts of soils had also declined between 1972 and 1981. The reduction in Cu availability was explained because the proportion bound to low molecular weight species in soil solution decreased whereas that bound to high molecular weight species in unavailable form increased in soils containing 200 and 400 mg Cu kg^{-1} (Berrow *et al.*, 1990). These changes occurred over a period of 17 years following sludge application, in spite of the significant reduction in soil organic matter content which would be anticipated over this time period.

The evidence for metal reversion is contradictory, however, and Sauerbeck and Styperek (1986) suggested that overall, no generalizations could be drawn about the potential reduction in metal availability as a function of time, particularly where sensitive crops such as sugar beet are grown. Chaney (1988) similarly commented that there was little demonstration of an explanation for apparent reversion of heavy metals in sludge-treated soils. This contrasts with an earlier report by Chaney (1973) stating that metals revert with time to chemical forms less available to plants. Furthermore, it is likely that the strength of the reversion processes may vary for different metals. Cadmium in particular does not appear to revert to unavailable forms permanently (Lloyd *et al.*, 1981; Eriksson, 1989). However, Emmerich *et al.* (1982b) postulated that metals would continue to slowly shift towards residual forms with time since they found that this was the dominant form in all cases in sludged soil, except for Cd.

The possibility that metals may revert with time to biologically inert forms or that they remain at more or less constant availability contrasts directly with the notion of the so-called potential 'time-bomb' effect of toxic metal release due to the decomposition of sludge organic matter. However, few studies have assessed these changes for long enough (10-20 years or more) for confident predictions about long-term behaviour to be made, apart from one notable exception. McGrath and Cegarra (1992) determined the chemical fractions of Cd, Zn, Ni, Cu, Pb and Cr in archived soil samples from the Woburn Market Garden Experiment using a sequential extraction procedure to study how these fractions changed in samples taken over a 40-year period at more or less constant soil pH. The four extracts employed were $CaCl_2$ (water-soluble and exchangeable fractions of metals), NaOH (organically bound metals), EDTA (mainly carbonate forms) and aqua regia (residual fraction). These showed that large increases in the proportions of Pb, Cu, Zn, Ni and Cd in at least one of the first three fractions occurred during the first ten years of sewage sludge additions. Chromium always remained in the residual form. For 30 years after this, including a period of more than 20 years after application of sludges to the field had ceased, there was very little change in the proportion of each metal extracted by each reagent. As would be expected most metal was found in the residual and EDTA fractions in both sludge-treated and control soils. However, significant differences between the fractionation of the metals was apparent. Lead had the largest fraction of any metal extracted principally by EDTA, Cu by NaOH and Cd by $CaCl_2$. These patterns of extractability, however, apparently remain independent of the organic matter status of the soil since the organic matter content had significantly declined during the period after sludging had stopped (Fig. 5.2).

Clearly, there is no evidence which substantiates the 'time-bomb' theory whereby sludge-applied metals become more available over time as organic matter degrades, and although implicated in metal speciation, the organic matter content of soils is apparently unimportant in relation to the bioavailability of heavy metals in the long-term and when sludge application has ceased. Research has shown that the processes which may potentially revert heavy metals to non-available forms occur in soils through the reduction in formation of soluble ligands due to microbial action and the slow reaction of metals with minerals in solid-solution reactions (Lewin and Beckett, 1980; Alloway and Jackson, 1991). Nevertheless reversion has not been demonstrated universally in sludge-treated soil. However, what is known is that availability does not increase with time provided soil pH value does not decline. Consequently, sludge-treated agricultural land which contains elevated concentration of PTEs must be managed for the foreseeable future to minimize metal uptake by crops (Logan and Chaney, 1983). Alloway and Jackson (1991) also point out that with the long half-lives of metals in soils, it cannot be assumed that conditions will

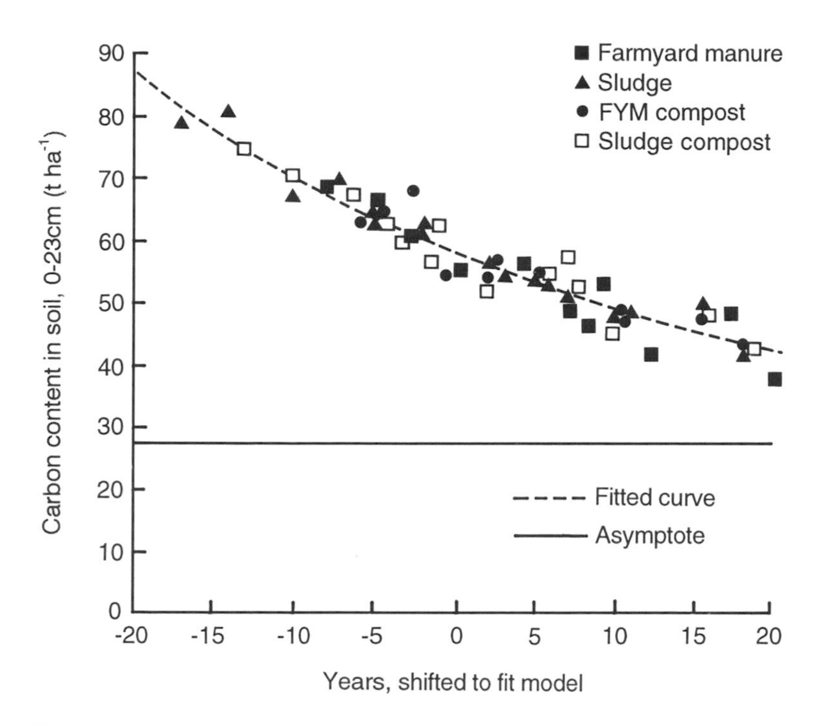

Fig. 5.2. Decline in carbon content in a sandy loam soil, Market Garden experiment, Woburn. All treatments shifted horizontally to fit model (Johnston *et al.*, 1989).

remain constant for all the time that the soil will remain significantly contaminated. In particular, they stress that the climatic changes predicted over the next 30-50 years and increasing acidification of soil from air pollution could result in changes in metal bioavailability in soils. Decreasing soil pH value would have a predictable effect on metal availability (see earlier). However, potentially warmer environmental conditions in the future could also increase the uptake of metals by crops due to higher rates of plant metabolic activity (Haghiri, 1974) although, on balance, inconsistent effects of temperature on metal absorption by crops have been reported (Giordano *et al.*, 1979; Sheaffer *et al.*, 1979).

Sludge Properties

Scientists in the US are emphasizing increasingly the important role of sludge properties in controlling metal availability to crop plants in sludge-treated soil (Corey *et al.*, 1987). Sludge chemistry is thought to explain the observation that plant uptake of metals appears to approach a plateau response with increasing rates of sludge application (Fig. 4.5) rather than following the usual linear pattern which occurs with increasing level of metal salt addition to soil. Chaney and Ryan (1993) considered that the specific metal adsorption capacity added to the soil by sludge itself plays a significant role in controlling the phytoavailability of metals of concern regarding phytotoxicity or food chain contamination. In principle, Corey *et al.* (1987) reasoned that equilibrium trace element concentrations of a sludge depend on: (1) the presence of trace element precipitates in pure or co-precipitated form with Fe, Al, Mn or Ca; (2) the strength of bonding to organic and mineral adsorption sites; (3) the proportion of potential adsorbing sites filled; and (4) the presence of dissolved ligands capable of complexing the trace metals. If the sludge matrix is constant then plant availability of any particular metal increases with increasing concentration of that element in the sludge. It was further hypothesized that if a sludge supports a higher equilibrium solution concentration (since the biologically significant fraction is the free-ion form of the metal in solution) of an element than does a soil, then combining both together will give an intermediate equilibrium level. As increasing amounts of sludge are applied to soil then the soil's adsorptive capacity becomes progressively saturated until the equilibrium concentration approaches that of the sludge alone. Further sludge applications above a particular level depending on soil properties, such as pH, should therefore result in little if any change in equilibrium concentration as shown in Fig. 4.5. Below this critical sludge rate, soil adsorption characteristics affect the equilibrium concentration at a given level of sludge treatment. Above that critical sludge rate, the equilibrium concentration should be controlled by the sludge. Corey *et al.* (1987) go on to suggest that if the equilibrium heavy metal concentration of the sludge is less than that which will result in excessive concentrations in plant tissues or damage to the plant then there is no basis for limiting application rates of that sludge in relation to metal content. If, on the other hand, the equilibrium metal concentration of the sludge at a particular pH value is large enough to cause a detrimental impact on plant metal content, maximum loading rates based on both soil and sludge characteristics will be required.

The inorganic component of sludge is considered the principal site of the specific metal adsorption capacity of applied sewage sludge (Chaney and Ryan, 1993). In particular, it is the Fe, Al and Mn oxides in soil and sludge which exhibit the specific metal adsorption properties. Chaney and Ryan (1993) noted that, even though sludge organic matter may decompose over time, the ability of crops to accumulate soil metals decreases with time (provided soil pH does not decline). This implies that the non-organic matter adsorption sites are adequate to protect against metals added in sludges. Furthermore, the evidence reviewed by Chaney and Ryan (1993) indicated that the specific metal adsorption capacity added to the soil with sludge will persist as long as the heavy metals of concern persist in the soil. However, McGrath and Cegarra (1992) did not observe a shift in sizes of extractable metal fractions from the organic phase to the inorganic phases in the residual period after sludge applications had ceased at Woburn even though the organic matter content of the soil had declined due to decomposition. An alternative explanation might therefore be that the specific metal adsorption capacity and strength of metal binding of the organic phase increases as the organic matter ages and is transformed into highly stable forms by decomposition processes.

In addition to the US reports, there are other data suggesting that sludge characteristics may be important in modulating bioavailability of PTEs in sludge-treated soils. For example, Carlton-Smith (1987) determined that the availability of metals for crop uptake was generally lower in soil treated with bed-dried sludge compared with liquid sludge. In particular, Cd availability from bed-dried sludge was less than half of that from liquid sludges. Carlton-Smith (1987) explained that this was probably because physically discrete particles of bed-dried sludge remained in the soil during the trial reducing the likelihood of contact with crop roots. However, chemical extraction of sludge-treated soil with EDTA indicated that metals were also in a less extractable form in this sludge than for the liquid sludge types suggesting that there were also differences in the chemical properties of the sludges influencing bioavailability. Physical effects would be expected to decrease with time, particularly on cultivated land.

If heavy metals are held in different forms in sewage sludges, their release into available forms, or immobilization into unavailable forms could occur at different rates. For example, Jing and Logan (1992) applied 16 anaerobically digested sludges of widely varying Cd content (range: 7.87-229.3 mg Cd kg^{-1}) at a uniform rate of 1.23 mg and 2.47 mg Cd kg^{-1} to soil in a greenhouse pot experiment and measured Cd uptake by *Sorghum* sp. (a sensitive accumulator crop) grown over a 28 day period. Plant uptake of Cd was positively correlated with total Cd concentration in the sludge, but better correlations were obtained with sludge Cd bioavailability measured by chemical extraction. This agrees to some extent with Riffaldi *et al.* (1983) who determined that the strength of bonding of Cd by sludge decreases as the total Cd concentration increases. Jing and Logan (1992) concluded, therefore, that plant uptake was controlled at least in part by sludge Cd chemistry, but that factors other than Cd content alone determined the potential for sludge Cd to be transferred to the food chain. In particular they considered that the improved correlation of Cd/P contents of the sludges with uptake by *Sorghum* compared with that obtained with sludge Cd content alone supported the suggestions of Corey *et al.* (1987) that sludge P may reduce the solubility of sludge Cd by coprecipitating Cd as various phosphates. John and van Laerhoven (1976) similarly showed that the availability of heavy metals to crop plants also varied between sludges in a short-term growth chamber study (12 weeks) with successional cropping of lettuce followed by red beet. According to Jing and Logan (1992) future regulatory approaches should therefore incorporate the effects of differential sludge composition on plant uptake. This contrasts with the conclusion reached by Davis (1984a) from a five year study, that crop uptake of Cd is more generally dependent upon the concentration of Cd in the soil rather than the concentration of Cd in the sludge as applied to the soil (see Chapter four). The key question, then, is whether differences in metal availability measured in the short-term persist over longer periods and once sewage sludge has equilibrated with the soil.

The metal sorption properties of Al and Fe oxides in soil are known to modulate the availability of heavy metals applied in sewage sludge (Alloway *et al.*, 1990). Kuo (1986) also showed that hydrous Fe oxide concurrently retained phosphate and Zn, Cd or Ca. The selectivity of the oxide for metals decreased in the order Zn>Cd>Ca and was apparently not affected by the absolute amounts of phosphate adsorbed. Ageing the oxide for two years did not adversely affect the sorption of Zn or Cd. It follows, therefore, that the application of Fe and Al to soil in sewage sludge may lower metal availability in treated soils. For example, Bell *et al.* (1991) found that the retention of Cd in an Fe oxide pool was enhanced, relative to the soluble and exchangeable forms determined by a sequential extraction procedure, nine years after the application of sewage sludge to a sandy loam soil. They suggested that sludge Fe additions to soils can be substantial. For example, an application of 100 t ha^{-1} of sludge could increase soil Fe content by 9% for sludge containing 25 000 mg Fe kg^{-1}. The median concentration of Fe in UK sludges (Table 2.2) is probably less than half this value (Sleeman, 1984). Furthermore, the added Fe could be more reactive than indigenous soil Fe.

In the US and Canada, tertiary treatment of sludge is practised increasingly to improve final effluent quality through removal of P and solids, and enhance the dewatering characteristics of sludge (Soon *et al.*, 1980; Gestring and Jarrell, 1982). The chemicals commonly used to reach these objectives are $FeCl_3$, $Al_2(SO_4)_3$ (alum), CaO or $Ca(OH)_2$. Such tertiary treatment of sludge is not currently widely applied in the UK although lime may be added to raw sludge to improve dewatering characteristics and reduce odour.

Iron salts added during sewage treatment (25-100 g Fe kg^{-1} sludge, Bell *et al.*, 1991) form reactive, paracrystalline oxides and hydroxides resulting from the precipitation of Fe with hydroxyl ions. With no added Fe, however, anaerobic digestion of sewage will result in the reduction and solubilization of indigenous Fe forming hydrous Fe oxides on mixing with soil.

In contrast to the data of Bell *et al.* (1991), however, Riffaldi *et al.* (1983) reported that Cd was retained by digested sludge in exchangeable and complexed forms involving carboxyl and phenolic hydroxyl groups and that sludge saturation by Fe and Al ions resulted in significantly lower sorption of Cd than untreated or Ca-sludge. Riffaldi *et al.* (1983) therefore concluded that sludges treated with Fe and Al salts could actually adsorb less Cd when applied to soil resulting in greater availability to soil colloids and to plants compared with unamended sludge. Other research (Soon *et al.*, 1980; Gestring and Jarrell, 1982) has shown that differences in the availability of metals in soils amended with Fe-, Al- or Ca-treated sludges were explained principally by the effects of the sludges on soil pH value suggesting that the presence and form of the coagulating ion may not be particularly important in directly modulating crop uptake of metals. Soon *et al.* (1978a) demonstrated that the Ca-sludge increased pH, the Al-sludge had no effect whereas the Fe-sludge decreased soil pH value. In contrast, Gestring and Jarrell (1982) found that all the chemically treated sludges generally reduced soil pH but that the Al-treated sludge lowered the soil pH to the greatest extent and thus increased metal availability.

Fractionation studies of heavy metals in sewage sludges show widely contrasting results. For example, Oake *et al.* (1984) fractionated heavy metals in nine representative sludges (primary, secondary, digested) by a sequential extraction procedure using 1M KNO_3 (exchangeable), 0.5 M KF (sorbed), 0.1 M $Na_4P_2O_7$ (organic), 0.1 M Na_2-EDTA (carbonate) and 6 M HNO_3 (sulphide) and found no major differences according to sludge type. Metals in the exchangeable, sorbed and organically-bound fractions were likely to be comparatively mobile in soil following sludge application. In the fractionation scheme adopted by Oake *et al.* (1984), Ni showed the highest concentrations in the exchangeable and sorbed fractions whilst for Zn, Pb, Cd and Cr the organically-bound fraction was dominant. Holtzclaw *et al.* (1978) had shown earlier that Cd, Ni and Zn were likely to be the most mobile elements in sludged soil based on an analysis of the distribution of heavy metals among the humic and fulvic acids, and pH precipitable fractions extracted with NaOH from soil amended with four sewage sludges. Here also no significant difference in the proportions of extractable Cd, Cu, Ni and Zn was detected between the sludges examined.

Other studies indicate that the forms of metals can differ markedly between sewage sludges but not necessarily in the way anticipated from the earlier discussion. For example, Steinhilber and Boswell (1983) found that the majority (80%) of Zn in one air-dried sludge, containing 1130 mg Zn kg^{-1} (fresh weight), was EDTA extractable. In another more contaminated air-dried sludge, containing 21 000 mg Zn kg^{-1} (fresh weight), the extractable Zn was divided among the EDTA (adsorbed and carbonate fraction), sodium hypochlorite + ammonium acetate (organic fraction) and acid ammonium oxalate (metal oxide fraction) extracts (21, 30 and 21%, respectively). In contrast to Riffaldi *et al.* (1983) and Bell *et al.* (1991), however, Steinhilber (1981) reported that Zn was eight times more plant available in a low Zn sludge than that in a high Zn sludge (Zn uptake/sludge Zn) despite the low Zn sludge also containing over four times the amount of Fe (31 700 mg kg^{-1} fresh weight).

Beckett *et al.* (1983) considered that heavy metals in activated sludges would probably be complexed in different forms compared with those present in anaerobically digested sludges. Furthermore, Berrow and Webber (1972) found that for most elements the ratio of extractable and total concentrations varied widely between sludges from different works. Stover *et al.* (1976) also suggested that heavy metals may be held in different forms even in sludges produced by the same treatment but at different works.

In spite of potentially large variations in the complexation of metals in sewage sludges, Mitchell *et al.* (1978b) showed that various biological and chemical properties of soil treated with aerobic and digested sludges converged with time. Similarly, Beckett *et al.* (1977) mixed activated, liquid digested and dewatered digested sludges with an acid loamy soil. After one year of repeated cropping the quantities of Cu, Ni or Zn extracted from the soils or absorbed by a test crop of young barley or present in the soil solution, depended only on the concentrations of Cu, Ni or Zn in the soil and not at all on the nature of the applied sludge. In addition, Berrow and Burridge (1984) observed that after eight years following the application of sewage sludge at the Luddington experimental field site, the treated soils contained very similar concentrations of Zn and Ni extractable by acetic acid and by EDTA. However, the acetic acid extractable concentrations of these elements in the original sludges were about three times greater than the concentrations extracted with EDTA (Berrow and Burridge, 1980). Thus the originally acid-soluble forms of Zn and Ni in the sludges had changed considerably in the soil over a period of eight years. The ratio between the amounts of Zn and Ni extractable by acetic acid and by EDTA also tended to reach a single value at the Lee Valley site, irrespective of the amounts of Zn applied. However, the value of the ratio for Zn differed between the two sites (1.2 at Luddington and 0.8 at Lee Valley), indicating differences in soil properties. The attainment of a common ratio between the amounts of Zn extracted by acetic acid and by EDTA during the residual period, following treatment with five different sludges, suggested that it was the soil rather than the source of the sludge which ultimately controlled the forms of metals that eventually persisted in the sludge-treated soil under aerobic conditions. There is also evidence (Beckett and Brindley, 1983) that heavy metals tend towards the same forms of combination during periods of prolonged incubation under anaerobic soil conditions irrespective of the forms initially applied in sludge.

Clearly there is some contradiction apparent in the literature concerning the importance of sludge properties in modulating the bioavailability of heavy metals in treated soils. Alternatively, monitoring the accumulation of PTEs in soils directly provides an absolute measure of the extent of soil contamination from sludge (as well as other sources of PTEs) and appropriate soil limits can be developed on this basis for normal sludges spread on agricultural land in operational practice. Controls limiting the PTE content of operationally produced sludges are generally less important following this approach except under certain circumstances such as when sludge is surface-applied to grassland and there is a potential risk of direct ingestion of sludge by grazing livestock. This approach to regulating the agricultural use of sludge has been adopted by the CEC Directive for the protection of soil where sludge is applied (CEC, 1986a) and has been implemented in the UK through the *Sludge (Use in Agriculture) Regulations 1989* (SI, 1989a). Much has already been done to reduce metal contamination of sewage effluents from industrial sources and levels of Cd in particular have declined markedly (Figs 2.1 and 2.2 and Table 2.3) such that most sludges being spread on agricultural land contain background concentrations of Cd and are unlikely to place the human food chain at risk. It has also been suggested that sludge treatment could also include metal removal stages for agricultural use (Smith and Hadley, 1990). For example, 70-100% of Cd, Zn and other metals in sludge may be removed by acid extraction and dewatering techniques with complexing agents or with micro-organisms of the genus *Thiobacillus* (Logan and Feltz, 1985; Rulkens *et al.*, 1989; Couillard and Zhu, 1992). So far metal removal has not been practised operationally for economic reasons, principally because reagent costs are high and an acidified effluent may require additional treatment and disposal. Given the improved quality of sewage effluents in recent years it is arguable whether the additional treatment costs of metal removal are justified. Ironically, the recycling of sewage sludge to agricultural land will be constrained in the future by the beneficial attributes of sludges, particularly their nutrient content (N and P), and not by their PTE content (see Chapter nine).

Chapter six:

Ingestion of PTEs by Grazing Livestock: Implications for Animal Health and the Human Food Chain

The Importance of Grassland as an Outlet for Sewage Sludge

Approximately 20% of UK sludge currently spread on agricultural land is applied to long-term pasture (Carlton-Smith and Stark, 1985). However, the importance of grassland as an outlet lies in its year-round availability. Long-term pasture is more likely to be available for the spreading of sewage sludge during the winter period as it is often relatively well drained and the presence of fibrous grass roots in the surface layers of the soil give some protection against compaction from sludge application vehicles. Thus, pasture may be the only available outlet for sludge in the winter months when arable land cannot be used due to problems of soil compaction or the presence of a young established crop (Fig. 6.1). There is also a period in the summer when almost all arable land is supporting a standing crop and is not available for sludge application. At this time, grassland used for rotational grazing becomes important either following a silage cut or following grazing. However, the grassland outlet has been regarded as environmentally sensitive due to potential PTE problems.

Metal Limits and Soil Sampling Depth

Surface applications of sludge to grassland lead to the accumulation of PTEs in the surface layers of the soil profile (Davis *et al.*, 1988) and higher PTE soil limits are permitted in the UK (Table A3) provided that the depth of sampling is to 7.5 cm (DoE, 1989a) compared with the statutory sampling depth of 25 cm (SI, 1989a) (Table A2). Smith *et al.* (1992d) showed that ryegrass grown in a soil column experiment was relatively insensitive to large concentrations of PTEs in sludged soil (up to 620 mg Zn kg^{-1} and 270 mg Cu kg^{-1}) when metals were restricted to the upper 7.5 cm of the soil profile. In contrast, uptake of Zn, Cu, Ni and Cd by grass increased with increasing soil concentration when soil was contaminated to a depth of 25 cm. As expected, soil Pb content had no effect on crop uptake of this metal reflecting the low availability to plants in sludge-treated soil. The differences in metal uptake response relative to location in the soil profile were probably explained by the extent of root development of ryegrass in the soil columns. For example, Garwood and Sinclair (1979) determined that the effective rooting depth of *Lolium perenne* was approximately 80 cm although the distribution of root mass through depth showed considerable variation, with a large density of roots in the upper 10 cm of the soil profile. Consequently, plants grown in the 7.5 cm metal-incorporated soil extended roots below this zone into low metal soil beneath and therefore absorbed less PTE compared with rooting into contaminated soil. Smith *et al.* (1992d) demonstrated that soil sampling to 7.5 cm in grassland avoids excessive concentrations of PTEs accumulating at the soil surface compared with deeper sampling regimes which are comparatively insensitive to the extent of metal contamination in the surface layers.

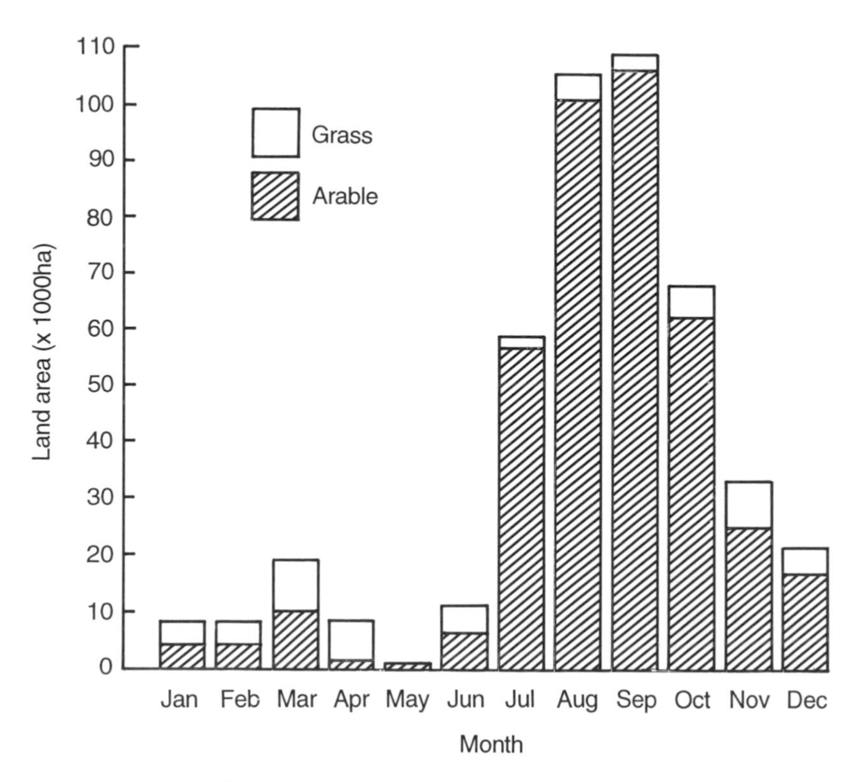

Fig. 6.1. Seasonal land availability in Lincolnshire in the UK (Smith and Powlesland, 1990).

Nevertheless, due to the distribution patterns of PTEs in the profiles of grassland soils (i.e. principally uncultivated 'permanent' pasture), Davis *et al.* (1988) estimated that the concentrations of PTEs in the top 2.5 cm may be twice the concentration measured by sampling to the recommended depth of 7.5 cm (DoE, 1989a) (Table A3). Metal accumulation at the soil surface is important because of direct ingestion of soil by grazing animals which represents potentially a significant source of metals in the diet of ruminants. However, in the practical grazing situation the concentrations of PTEs in the surface layers of soil may be only approximately 8% higher than the values observed when sampling soil to 7.5 cm (Smith *et al.*, 1992d). This is because, in practice, poaching by livestock, soil movement by earthworms and other fauna, and the occurrence of some sward cultivations cause soil mixing aiding the dispersal of PTEs in the surface soil. The adherence of sludge solids to treated herbage is another important mechanism that may potentially result in the transfer of sludge-borne PTEs to grazing livestock.

Sources of PTEs to Grazing Livestock

Soil Ingestion

The concentrations of trace elements in plants grown on sludge-treated soil are influenced by many factors including total soil content, soil type, soil pH value, fertilizer practice, environmental conditions, plant species, variety and also the vegetative state of the plant (see previous chapters). However, grazing animals can also ingest significant quantities of sludged soil and sludge particles, adhering to the surface of the sward, which potentially may represent a larger intake of PTEs from surface-treated pastures than that contained in the herbage. The relative importance of plant material and soil as sources of PTEs in the ruminant diet depend on the quantities of each material ingested and on the concentrations of elements present.

Grazing livestock can be returned to sludge-treated pasture following a minimum no-grazing period of three weeks to avoid potential infection from pathogenic agents in the sludge (DoE, 1989a and see Chapter eleven). However, the grazing animal is still potentially at risk of taking in significant quantities of heavy metals and organic compounds through direct ingestion of soil and sludge solids from the surface of the sward and from sludge adhered to herbage after this minimum exclusion period.

The extent of soil ingestion by grazing animals was reviewed by Stark (1988). In general soil intake occurs in one of two ways: (1) the soil is present on the surface of plant leaves which are eaten, or (2) it is ingested directly from the soil surface, from earthworm casts or attached to plant roots (Healy, 1973). Various studies have found that intake of soil is dependent on the amount of herbage which is available for grazing. For example, Field and Purves (1964) found that soil ingestion by adult sheep increased from less than 0.5% of total dry matter intake in late May to 14% in mid-December. From a review of the literature on soil ingestion by sheep, Logan and Chaney (1983) determined that soil was normally only 1-2% of sheep's diet, but could reach 24% in 'worst-cases'. Fewer studies of soil ingestion have been carried out for cattle, but experimental evidence indicates that soil appears to comprise a lower proportion of total dry matter intake than that observed for sheep, possibly because cattle generally graze longer pasture than sheep. Thornton (1974) observed that cattle ingest 1-11% of their dry matter intake as soil during the winter, decreasing to a maximum value of 7% in May. However, Fries (1982) considered that soil ingestion by dairy cattle rarely exceeds 3% of dry intake under good management conditions. Stark (1988) estimated that the ingestion of soil by cattle and sheep could typically be of the order of 5-10% and 10-15% (dry matter) of the total intakes, respectively on an annual basis. Soil ingestion may also vary markedly between farms due particularly to differences in soil structure and stocking rate (Healy, 1967). A heavy stocking rate tends to result in an open grass habit (Healy and Ludwig, 1965) and large soil intakes at high stocking rates can be due to low pasture availability (Arnold *et al.*, 1966).

In practice the extent of soil ingestion by grazing animals is highly variable. This is because many different factors can influence soil ingestion rate including soil properties, stocking rate and pasture availability, earthworms, plant type and sward characteristic, management practices, seasonal and climatic variations and individual animal behaviour (Healy, 1973; Kirby and Stuth, 1980; Thornton and Abrahams, 1983; Fleming, 1986). Nevertheless, the mineral element content of soil clearly has an important impact on animal nutrition both through its indirect effects on herbage composition and also directly on animal intake of mineral elements in soil. Furthermore, soil can act as a potent antagonist modulating availability of micro-elements in feedstuffs (Suttle *et al.*, 1975).

Ingested soil contains a variety of mineral elements, often at relatively high concentrations compared with those in herbage growing in the soil. Effects on the animal of ingested trace elements depend on the concentration present in the diet and, perhaps more importantly, also on their biological availability. Furthermore, complex synergistic and antagonistic interactions between two or more elements can beneficially enhance absorption and rectify a potentially deficient intake, or reduce absorption of a potentially toxic intake by modulating availability in the animal. However, the opposite may also occur producing a mineral imbalance leading to an induced deficiency or toxicity of a particular element apparently not at deficient or toxic levels in the diet.

Adherence of Sludge Solids to Herbage

When liquid sludge is sprayed onto foliage, treated plants become covered, but even where smothered, the visual appearance of grassland usually recovers within a few days. There have been several quantitative studies investigating the persistence of the sludge on herbage after spreading. Chaney and Lloyd (1979) assessed the adherence of a liquid digested sewage sludge to tall fescue grass and found that neither rainfall nor detergent washing removed much of the adhering sludge once it had dried on to the surface of the leaves. Only growth dilution or winter die-back followed by spring regrowth appeared to reduce the proportion of sludge in the herbage substantially. This is in agreement with the work reported by Jones *et al.* (1979) who found that sludge could only be effectively washed off the herbage immediately after application, before it had dried on the leaves. In addition, Chaney and Lloyd (1979) also reported that the sludge content declined more rapidly in swards which had been mown prior to application, apparently because less forage was contaminated and rapid growth diluted any contamination present.

More recent work by Klessa and Desira-Buttigieg (1992) related herbage metal concentrations from adhered sludge to dietary thresholds which may pose a risk to the health of grazing animals and potentially increase the entry of PTEs into the human food chain. Their results indicated that intakes of Cr, Mn and Ni never exceeded dietary thresholds. This was also true for Cd, Pb and Zn when a sludge containing 2% dry solids was applied. With this sludge, most of the adhering metals declined to non-zootoxic concentrations within the minimum three week no-grazing period given in the UK Code of Practice (DoE, 1989a). The extent of potential herbage contamination with metals was substantially reduced by decreasing the rate of sludge application, and particularly by spreading sludge on cut swards. Application of sludge to cut or short swards provides an inherent level of protection through growth dilution since adequate grass regrowth is also necessary before grazing animals can be returned. As expected, Klessa and Desira-Buttigieg (1992) found that metal retention was greater for a thickened sludge (12% dry solids) applied to uncut pasture and that Cd, Cu, Pb and Zn required 3-7 weeks to reach concentrations below the suggested limits for zootoxicity to grazing livestock. In particular, 5-6 weeks were needed for herbage concentrations of Cd and Pb, which present the greatest risk of entry into the food chain, to decrease to dietary concentrations considered 'safe' for grazing livestock with this sludge. For Cu, a longer period of 6-7 weeks was necessary to protect sheep, the grazing ruminant most sensitive to Cu toxicity, against potentially harmful intakes of this element resulting from the application of high dry solids sludge to uncut pasture.

To place the results of Klessa and Desira-Buttigieg (1992) into context, however, few thickened liquid sludges are applied to agricultural soils on a national basis. A more typical dry solids content range would be 2-5% for liquid digested sludges. On this basis 'safe' concentrations of PTEs in herbage may be achieved at or near the permitted no-grazing interval, especially if sludge is applied to cut swards. Unfortunately, Klessa and Desira-Buttigieg (1992) did not examine the effects of applying a high dry solids sludge to a cut sward on the extent of herbage contamination with PTEs. Therefore, it is not possible to quantitatively assess the efficacy of this straightforward management practice in reducing the potential exposure of grazing animals to undesirable intakes of PTEs from high dry solids sludges surface applied to grassland soils based on their data. Furthermore, the zootoxicity thresholds used in the calculations of Klessa and Desira-Buttigieg (1992) were based on empirical toxicological values which assume a fixed intake level and do not consider bioavailability and antagonistic relations between elements in ingested sludge, soil or herbage which are known to modulate absorption and toxicity (Logan and Chaney, 1983). In practice, animals are not exposed to a constant level of PTE intake from sludge-treated pasture during their life-time so a short exposure to high concentrations is unlikely to present much risk overall to the animal or to the human food chain.

A detailed assessment of the extent of pasture contamination from the application of liquid digested sewage sludge to the surface of grassland soils has been reported by Sweet *et al.* (1994). They described a series of field and glasshouse experimental studies where the persistence of sludge solids and PTE contamination on grass swards was examined in relation to management and environmental factors. Sward conditions and sludge dry solids content were shown to be the principal factors controlling the initial level of sludge adherence (Table 6.1). The extent of sward contamination was raised markedly with increasing sludge dry solids content at comparable rates of application (100 m^3 ha^{-1} fresh sludge). Indeed, there was a disproportionate increase in the contamination level resulting from the application to grass of sludges with high dry solids content compared with those of low dry solids content. Thus, doubling the dry solids content of sludge increased the level of sward contamination by a factor of 3. This further emphasizes the need for caution when estimating the potential impacts on grassland of sewage sludge applications based on experimental studies with atypical sludges with high dry solids content. However, the condition of the sward at the time of sludge application was also critical to the extent of initial herbage contamination with sludge. Short swards were less prone to contamination than were long swards confirming the earlier results of Chaney and Lloyd (1979) and Klessa and Desira-Buttigieg (1992). Indeed, the results were consistent with the observation that applying sludge to recently mowed or grazed fields, and waiting to allow the crop to grow and dilute the adhering sludge, can keep the sludge to 2-5% of dry forage (Logan and Chaney, 1983; Chaney *et al.*, 1987). The presence of broad-leaf species or grass mixtures with hairy-leaf species increased the extent of contamination because these offer a larger leaf area or provide surfaces with a greater capacity to adhere sludge compared to grasses with smooth leaves. Sweet *et al.* (1994) went on to develop a mechanistic model of the processes controlling the rate of disappearance of PTEs from sludge-treated pasture. The model was developed on the basis of the experimental data using Cr as an indicator of direct crop contamination since this element is barely absorbed by crops from sludge-treated soil.

Table 6.1. Initial levels of sward contamination from the application of sludge to pasture.

Sludge dry solids content (%)	Sward conditions	Herbage contamination with sludge (%)
11.0	Long sward after hay cut	15
8.1	Long sward after late silage cut	13
10.6	Short sward (sparse): first year ley	3
8.1	Long grass sward	15
4.0	Long grass sward	5
2.0	Long grass sward	1
8.1	Long grass sward with clover	24
4.0	Long grass sward with clover	8
2.0	Long grass sward with clover	3
8.1	Short grass sward	11
4.0	Short grass sward	2
2.0	Short grass sward	1.5
8.1	Short grass sward with clover	15
4.0	Short grass sward with clover	5
2.0	Short grass sward with clover	1

Source: Sweet *et al.* (1994)

The decay in contamination levels after surface treatment with sludge depended upon the length of the sward at the time of application and the growth rate of the crop. The seasonal differences in grass growth show a dynamic pattern increasing in the spring attaining maximum growth rate in May and subsequently declining generally through the summer and autumn period with a secondary small peak in growth occurring in July (Brockman, 1988). In general, the growth rate of grass is likely to be relatively slow from late October until early March in the UK.

Consequently, during periods of rapid dry matter production by the crop, the contamination disappeared at a faster rate due to growth dilution compared with early spring and autumn or winter applications when growth rates were low. However, even under conditions of minimal dry matter production, Sweet *et al.* (1994) demonstrated that there was a basal rate of decline in contamination due to the normal turnover of leaf tissues in grass under low temperature conditions in the field. Ambient weather conditions apparently had no direct effect on the extent of pasture contamination with sludge. In particular, no link could be established between the level of sward contamination and the incidence of heavy rainfall after sludge application. This provided further evidence that, once dried onto the leaves of grass, sludge is strongly adhered (Chaney and Lloyd, 1979), emphasizing that the process of growth dilution and sward turnover are critical for minimizing the potential transfer of pollutants to grazing animals from surface-applied sludge on grassland.

Examples of the typical patterns in grass contamination with sludge are shown in Fig. 6.2 in relation to time (days) after sludge application (Sweet, unpublished). The model simulations utilize experimentally defined values for the initial level of sward contamination with sludge and then estimate the decline in grass contamination due to crop growth, on the basis of daily incremental increases in dry matter production for typical growth rates at different times of the year. The model predictions showed very good agreement with the experimental observations (Sweet *et al.*, 1994), suggesting that plant growth was the principal factor controlling the disappearance of sludge contaminants. The model showed that herbage contamination disappears more rapidly during periods of rapid crop growth for sludge applied in May, for example (Fig. 6.2a), compared with March application (Fig. 6.2b) when growth rates are relatively slow during the early part of the growing season. Indeed, the extent of sward contamination is likely to decline to a low level (≈0.5%) and within the required minimum no grazing period (DoE, 1989a) for high dry matter sludge (10% dry solids) applied to a short sward (e.g. 3-5 cm) during the phase of rapid crop growth. As expected, the simulations showed that liquid sludges of typical dry solids content (5%) have a smaller effect on sward contamination overall, compared with higher dry matter sludge. In both cases, however, sward contamination may persist for a long period of time when sludge is applied to long grass (e.g. >10 cm) during the early spring period when growth rates are low, which may coincide with utilization of the pasture later in the season. A similar pattern of growth dilution is observed in the autumn period as growth rates decline. In contrast with early spring applications, however, the persistence of sludge solids on the herbage may be less critical in the autumn since the pasture is less likely to be used or grazed for longer periods, extending beyond the minimum no grazing requirement, until the following spring. It is emphasized that the model only considered the effects of growth dilution on sward contamination and probably underestimated the actual rate of disappearance of the sludge during the critical periods when dry matter production is low. During periods of minimum crop growth, Sweet *et al.* (1994) showed that sward contamination decreased at a rate of between 0.4-1.2% per day due to the turnover of leaf tissue.

In overall assessment, the potential transfer of sludge-borne contaminants to grazing animals is minimized from soil surface treatment when sludge is applied to short swards irrespective of the dry solids content of the sludge or application time. It is emphasized that sludge application to short grass provides an inherent level of protection in practice because the land cannot be subsequently grazed until the crop has had the opportunity to regrow. Careful

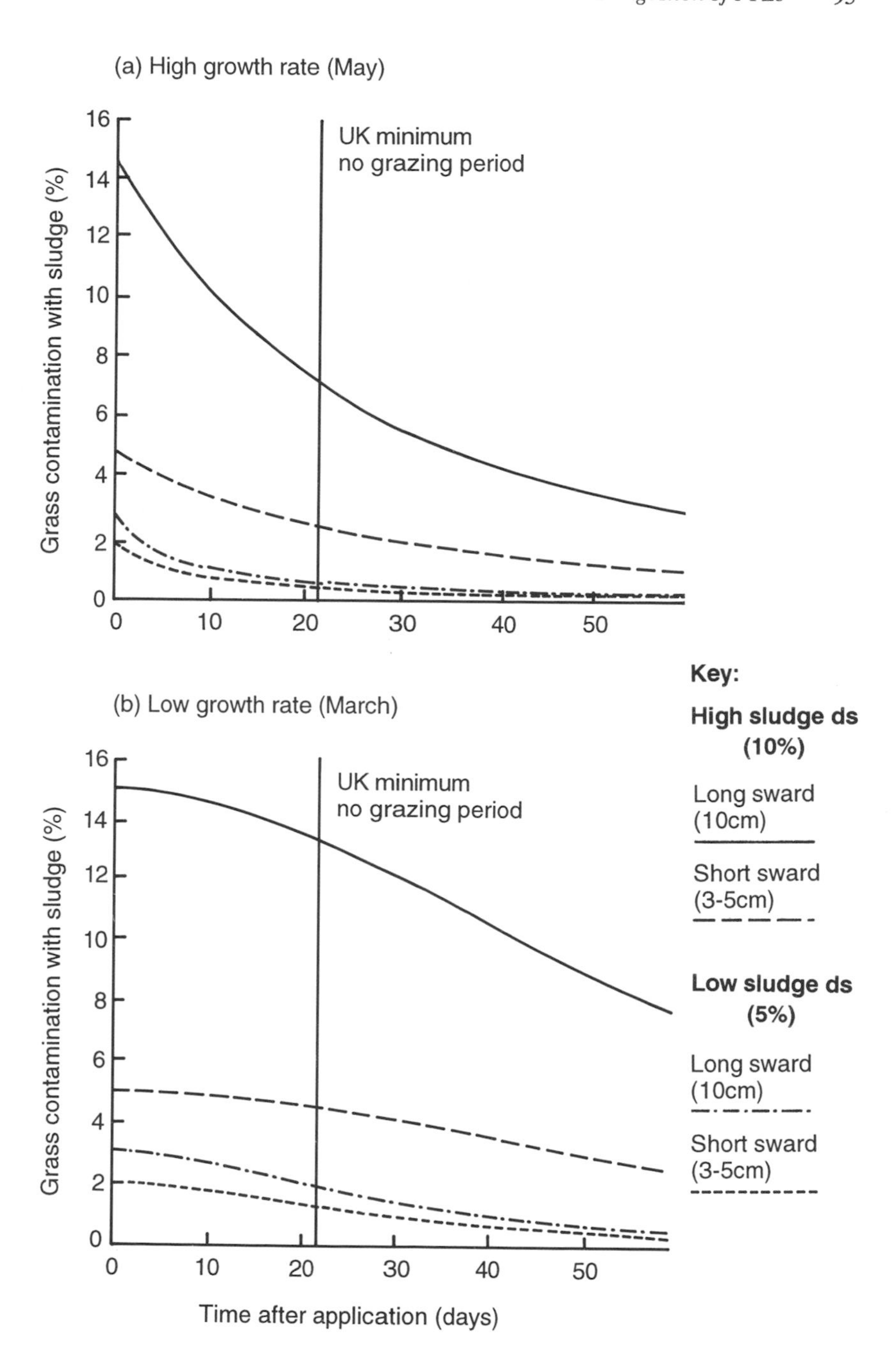

Fig. 6.2. Extent of grass sward contamination with surface applied sewage sludge in relation to growing conditions, sludge dry solids (ds) content and sward height.

consideration of grass growing conditions and subsequent grazing regimes is necessary when sludge containing more than 8% dry solids is surface applied to grassland soils (Sweet *et al.*, 1994). The no-grazing interval of three weeks is only the *minimum* period required in the UK to reduce the potential risks to grazing animals from sludge-borne pathogenic organisms (Chapter eleven). A larger exclusion period may be necessary depending upon local conditions and especially when crop growth is slow during the early spring and when sludges high in dry matter are applied to long grass. However, the dry solids content of liquid digested sludges are typically in the range 2-5% and generally constitute a relatively low risk of pasture contamination. Finally, sewage sludge is an effective grassland fertilizer (Chapter nine). The increased dry matter production resulting from sludge application will in itself contribute to the rapid disappearance of sludge solids from the grass sward.

In contrast to thickened liquid sludges that adhere strongly to herbage, finely dispersed sludge composts and cake sludges do not adhere to plant leaves. Consequently, the ingestion of PTEs by grazing livestock from composted and dewatered sludges surface applied to grassland is lower than with liquid sludge applied with a conventional splash plate (Decker *et al.*, 1980). The UK Code of Practice (DoE, 1989a) permits the surface application of dewatered digested sludges or composts to pasture provided that they are finely divided so as to minimize direct ingestion by livestock.

Pasture contamination following application of sewage sludge can be minimized using appropriate application technology. As discussed, the rate of disappearance of sludge contaminants from the sward following conventional surface spreading is largely dependent on dilution by growth which is influenced by weather conditions. However, soil injection of sludge can overcome the potential problem of herbage and soil surface contamination with PTEs by ensuring any metals present are well distributed within the soil profile and thus diluted. Conventional injection (15-20 cm depth) is highly effective in improving pasture hygiene, but can cause sward damage through substantial root disruption reducing the yield benefit from the plant nutrients supplied in sludge (Hall *et al.*, 1986). Shallow slot injection techniques (5-10 cm depth) are being introduced (Sneath and Phillips, 1992) which enhance pasture hygiene whilst avoiding extensive sward damage associated with deep injection methods. Results from recent studies (Sweet *et al.*, 1994) have shown that shallow injection can reduce initial sward contamination to less than 50% of that found when sludge is spread on the surface and that subsequent reductions in herbage PTEs occur more rapidly.

Alternatives to surface application have been more widely developed in continental Europe and are reviewed by Sneath and Phillips (1992). Machinery which allows the application of sludge into a growing crop by means of flexible tubes dragged along the soil surface, and drag shoe applicators are widely used for spreading sludge and animal slurries on grassland in a number of European countries (Hall, 1995). These machines place the sludge on the surface of the soil, beneath the sward canopy thus minimizing contamination of the herbage. However, many of the machines have been developed in The Netherlands and Northern Germany where soil types are generally light textured and have been specifically designed for the relatively low application rates (20-40 t ha^{-1}) dictated by legislation in those countries. Many would need adapting to UK conditions of heavier soil types and larger application rates.

In future it is likely that environmental concerns over the problems of sward contamination and, to a lesser extent, of odour associated with surface application of sludge will increase. However, sludge application technology is already available that will ensure the minimal environmental impact of PTEs applied to grassland in sewage sludge.

Effects on Animals of PTEs Ingested in Sewage Sludge

Information specifically relating to the mineral nutrition of livestock has been lacking for many inorganic elements in sludge particularly with respect to upper limits for dietary tolerances to and interactions between different components of the diet (Stark, 1988). Published data on the trace element requirements and tolerances of grazing animals are sparse and heterogeneous and were reviewed by Stark (1988). In the UK, the Agricultural Research Council (ARC, 1980) and an Interdepartmental Working Party (MAFF, 1983) provide recommendations for dietary allowances and tolerances of a limited number of trace elements. A more comprehensive review of dietary tolerances for a range of elements is provided by the US National Research Council (NRC, 1980) (Table 6.2). However, most information relates to cattle and sheep, and few data are available for other grazing livestock.

A recent review for the UK Government on food safety aspects of the rules for sewage sludge application to agricultural land (MAFF/DoE, 1993b) acknowledged that the exposure of animals to PTEs in sludge-amended soils is unavoidable. Evaluation of the potential risk to animals of sludge-applied trace elements is very complex. However, despite the difficulties, there is a need for assurance that the current regulations on the agricultural use of sewage sludge do not compromise animal health or the human food chain at the levels of soil intake by grazing animals which may be reasonably anticipated in the field. Animal species and breeds differ significantly in their tolerance, and tolerance is also influenced by age, with young animals generally being more sensitive than old. Dietary intake of PTEs by animals raised in a sludge-amended environment is a composite of intakes in plant material, sludge and soil, all of which are highly variable. The principal risk, however, is from direct ingestion of sludge and sludge-treated soil, but sludges and amended soils differ in their absolute and relative concentrations of elements, interactive effects and bioavailability for the animal. Because these interactions are often the basis for any observed toxicity effects, they become very important in assessing the risk

Table 6.2. Maximum tolerable dietary levels of trace elements for livestock.

Element	Maximum tolerable dietary levels ($mg\ kg^{-1}$)[1]		
	Cattle	Sheep	Horses
Aluminium	1000	1000	(200)
Antimony	-	-	-
Arsenic			
Inorganic	50	50	(50)
Organic	100	100	(100)
Barium	(20)	(20)	(20)
Bismuth	(400)	(400)	(400)
Boron	150	(150)	(150)
Bromine	200	(200)	(200)
Cadmium	0.5	0.5	(0.5)
Chromium			
Chloride	(1000)	(1000)	(1000)
Oxide	(3000)	(3000)	(3000)
Cobalt	10	10	(10)
Copper	100	25	800
Fluorine	Young 40 Mature dairy 40 Mature beef 50 Finishing 100	Breeding 60 Finishing 150	(40)
Iodine	50[2]	50	5
Iron	1000	500	(500)
Lead	30	30	30
Manganese	1000	1000	(400)
Mercury	2	2	(2)
Molybdenum	10	10	(5)
Nickel	50	(50)	(50)
Selenium	(2)	(2)	(2)
Silicon	(2000)	2000	-
Silver	-	-	-
Strontium	2000	(2000)	(2000)
Sulphur[3]	(4000)	(4000)	-
Tin	-	-	-
Titanium	-	-	-
Tungsten	(20)	(20)	(20)
Uranium	-	-	-
Vanadium	50	50	(10)
Zinc	500	300	(500)

[1] Values in brackets are extrapolated from toxicity data for other species. Dashes indicate insufficient data to recommend a tolerable level. Values are assumed to be on a dry solids basis

[2] Concentration may be undesirably high in milk at this intake

[3] Not trace element; included due to relevance to copper

Source: NRC (1980)

implied by any sludge-borne PTEs (Mills, 1970; Hegsted, 1971; Mills, 1974; Lee, 1975; Suttle, 1975; Bremner and Mills, 1979; ARC, 1980; Logan and Chaney, 1983; Mills, 1986; Stark, 1988). Despite the general recognition that dietary interactions occur, in a detailed summary of investigations on trace element deficiency and excess in ruminants, the ARC (1980) commented that most experiments had studied only the effects of different levels of a single element. Except for Cu, Mo and S, even an approximate quantitative estimate of the influence of antagonists on availability could not be provided. In addition, several other dietary components such as protein, fats and organic acids may also influence trace element absorption and metabolism (Hegsted, 1971; Butler and Jones, 1973; Underwood, 1977; Bremner and Mills, 1979; Underwood, 1981; Georgievskii, 1982a).

Logan and Chaney (1983) and more recently Stark (1988) have considered the various possible interactions which occur between elements contained in sewage sludge and sludge-amended soil, which are many and complex. The principal interactions of importance in relation to animal ingestion of sludge, listed in Table 6.3, are the antagonistic effects of an element, or group of elements, decreasing the absorption of another element thus reducing its toxicity or in extreme cases inducing a deficiency. For example, absorption of sludge-borne Cu by grazing animals is strongly influenced by, and the absorption of Cd and Pb in sludge appears to be influenced by, antagonist agents in sludge, soil and grass within the diet. Stark (1988) also highlighted the wide differences apparent in published values for requirements and tolerance levels of several elements. Since the absorption of certain trace elements is influenced by both inorganic and organic constituents of food, and marginal or excess supplies of one trace element may influence the tissue retention and metabolism of another, the definition of a single dietary tolerance level for many trace elements may not be appropriate especially for assessing the risks to grazing animals of sludge-derived PTEs. When intakes of several trace elements are relatively high, as may occur with sewage sludge, recommended dietary requirements and tolerance levels such as those published by ARC (1980) and NRC (1980) (Table 6.2) are considered to be of doubtful relevance (Logan and Chaney, 1983). Until recently (Stark and Wilkinson, 1994; Stark *et al.*, 1995), the almost complete lack of studies with grazing animals during which concentrations of interacting elements in the diet have been recorded means that requirements and tolerances have not been properly considered in relation to the presence of possible interacting elements.

The adverse effects on livestock of high trace element intakes can be placed into three categories corresponding to acute toxicity, chronic toxicity and imbalances. Stark (1988) considered that the main effects of sewage sludge were likely to be manifest as problems of chronic toxicity or trace element imbalances rather than of acute toxicity. Table 6.4 (Stark, 1988) illustrates deficiency and toxicity symptoms caused by a range of different elements. With some elements there are clearly identifiable dietary intakes below or above which adverse effects are observed in an acute response. In this case, requirements or toxicity levels may often be determined by feeding various quantities and observing the level at which the symptoms of toxicity or deficiency appear or disappear. However, with other elements, or in other dietary situations, chronic toxicity or deficiency may occur and initial subtle effects such as reduced growth rate may not be immediately obvious. Diagnosis of mineral imbalances may be particularly difficult because non-specific adverse effects may occur at much lower dietary levels than those associated with chronic toxicity. In addition, many symptoms of chronic toxicity and mineral imbalances are common to several elements.

Table 6.3. Antagonists to the absorption of food chain (Cd and Pb) and animal (Cu) PTEs by livestock grazing sludge-treated pastures.

Element	Principal antagonists	Comments
Cu	S, Fe, Mo, Zn, Cd	Sludges over 4% Fe have caused Cu deficiency Cu deficiency induced by excessive dietary Zn (300-1000 mg kg^{-1}) Dietary Cu of 50 mg kg^{-1} from sludge has caused no toxicity problems
Cd	Fe, Zn, Ca	Sludge Cd content is now decreasing relative to Zn (Table 2.3) and low Cd : Zn sludges present little risk of accumulation
Pb	Fe, P, Ca, organic matter	Antagonism reduces accumulation

Sources: Logan and Chaney (1983); Suttle *et al.* (1984)

A knowledge of the extent of absorption of trace elements by animals is required to predict the relative toxicity of different sources of the elements. Data for most trace elements in livestock feeds are sparse and even less is known about the form or the availability of the inorganic components of soil and sewage sludge to grazing animals. Large differences observed in the concentrations of many inorganic elements in animal tissues suggest that elements present in herbage, soil or sludge are likely to be absorbed to different extents and that numerous interactions may occur. In many cases, therefore, a combination of production trials together with clinical, pathological and biochemical studies are needed to provide conclusive evidence of deficiency or toxicity. Not surprisingly, Mills *et al.* (1980) concluded from a review of

Table 6.4. Main symptoms of inorganic element deficiencies and excesses.

Element	Deficiency symptoms	Toxicity symptoms
Major minerals		
Calcium	Abnormalities of bones and teeth	-
Phosphorus	Impaired reproduction, bone and teeth abnormalities, depressed appetite, reduced growth rate and milk yield, depraved appetite	-
Magnesium	Depressed growth, anorexia, hyper-irritability, tetany, vaso-dilation, convulsions, muscular uncoordination	-
Sodium	Depressed appetite, unthriftiness	-
Potassium Chlorine	Observed only in laboratory studies, not with livestock on farms	-
Sulphur	Depressed growth and milk production	Secondary copper deficiency
Trace elements		
Arsenic	Reduced growth rate and milk yield, reproductive disorders, sudden death	Acute: restlessness, rapid breathing, muscular and visual disorders, inflammation of digestive tract, death. Chronic: reduced growth rate, weakness, haemorrhages, muscular disorders, tissue inflammation
Cadmium	-	Kidney malfunction, gastric disorders, reproductive disorders, osteomalacia, reduced growth rate and feed conversion efficiency
Chromium	Impaired growth, decreased life expectancy, eye disorders	Depressed growth, liver and kidney damage, scouring, nervous degeneration
Cobalt	Ruminants: anaemia, muscular atrophy, listlessness, loss of appetite, depressed growth, reduced viability of young	Blood disorders, anaemia, loss of appetite, impaired growth

Table 6.4 continued

Element	Deficiency symptoms	Toxicity symptoms
Copper	Anaemia, impaired reproduction, depressed appetite and growth rate, ataxia, bone disorders, cardiovascular disorders, depigmentation, defective keratinization, scouring (cattle), swayback (sheep)	Retarded growth, weight loss, anorexia. Other symptoms vary with species. Terminal stage is haemolytic crisis
Fluorine	-	Chronic: depressed food intake, impaired reproduction, impaired teeth and bone structure, lameness. Acute: digestive and muscular disorders, abdominal pain, collapse.
Iodine	Impaired reproduction, stunted growth, enlarged thyroid gland, reduced growth rate and milk yield	Impaired reproduction, reduced growth rate, nasal discharge, coughing, reduced milk yield.
Iron	Anaemia, depressed growth, lethargy, lowered resistance to infections	Reduced feed intake and growth rate,weight loss, scouring, reduced milk yield. Acute: diphasic shock, vascular congestion, anorexia, diarrhoea
Lead	-	Vary according to species. e.g. stiff gait, fractures, osteoporosis, kidney disorders, impaired vision, reproductive disorders, neurological disorders. Acute: blindness, excessive salivation, hyper-irritability, convulsions, death
Manganese	Impaired reproduction, depressed growth, skeletal disorders, ataxia	Anaemia, depressed growth rate, leg stiffness
Molybdenum	Not reported for grazing livestock	Secondary copper deficiency

Table 6.4 continued

Element	Deficiency symptoms	Toxicity symptoms
Nickel	Not reported for grazing livestock. Laboratory animals show non-specific symptoms including retarded growth and anaemia	Kidney damage, hyperglycaemia, respiratory disorders, reduced growth rate, increased mortality
Selenium	Impaired reproduction, muscular dystrophy, ill-thrift	Chronic: anaemia, dullness, rough coat, hair and hoof loss, stiffness, lameness
Silicon	Stunted growth and bone formation	Depressed digestibility, growth rate and reproductive performance, kidney stones
Tin	-	Ataxia, muscle weakness, anorexia
Vanadium	Impaired growth and reproduction, disturbed lipid metabolism, reduced milk yield and milk fat content	Chronic: depressed growth rate. Acute: diarrhoea, dehydration, haemorrhage, emaciation, prostration
Zinc	Impaired reproduction, severe inappetance, depressed growth, skin abnormalities	Anaemia, depressed intake, reduced liveweight gain and feed conversion efficiency

(1) Source: Stark (1988)
Quoting: Underwood (1977, 1981); ARC (1980); NRC (1980); Reid and Horrath (1980); Goyer (1981); Georgievskii (1982b); Clark and Stewart (1983); Grace (1983); Roche (1985-87) and Mertz (1986)

information on the influence of dietary composition on heavy metal toxicity, that the effectiveness of assessing the influence of inorganic pollutants upon the health of livestock depended greatly upon experimental design. In particular, there is a need to consider the effects of variations in dietary composition on heavy metal retention and toxicity. Mills *et al.* (1980) also noted that the design of effective experiments was often limited by insufficient knowledge of the form in which many heavy metals occur in the environment.

The review on food safety (MAFF/DoE, 1993b) concluded that current evidence was insufficient to warrant a recommendation that limits for Pb, Cd and Cu in sludge-amended soils be reduced. However, research was underway on PTE concentrations in tissues of animals grazing on sludge-amended pastures and the Steering Group on Chemical Aspects of Food Surveillance (MAFF/DoE, 1993b) recommended that the situation should be kept under review in the light of these studies.

Two research programmes (funded separately by the UK Ministry of Agriculture, Fisheries and Food and the Department of the Environment) have now been completed, during which diets comprising dried grass and sludge-amended soils or sludge/soils mixtures were fed to sheep to simulate, under controlled conditions, livestock grazing sludge-amended pasture (Stark and Wilkinson, 1994; Stark *et al.*, 1995). Sheep were used as the experimental animal since, while they are more susceptible than cattle to Cu toxicity, in general the digestion and metabolism of the two species is similar.

In the long-term studies reported by Stark *et al.* (1995), sheep were fed fixed quantities of dried grass either alone (control animals) or mixed with one of three sludge-amended soils from selected historically treated fields. Soil inclusion rates in the dietary dry matter fed to finishing lambs for up to six months were 2.5, 5 and 10%. Ewe lambs for breeding were fed 0 or 5% of the three soils from weaning to the production of their second lamb crop three years later. Aged ewes, expected already to have a high body burden of Cd and Pb were fed 10% of one of the soils from mid-pregnancy for 10 months.

The soils contained a range of concentrations of Cu, Cd and Pb, maximum values of which approached, or were in excess of, those representative of the surface couple of cm of sludged grassland at the maximum soil limit values (sampled to 25 cm) for these elements (Tables A2 and A3). The highest concentrations studied were twice the statutory limits for Cd and three times for Pb. To reduce the risk of Cu toxicity compromising the study of Cd and Pb residues, the concentrations of Cu in two of the soils were relatively low. However, the third soil contained very high concentrations of Cu at seven times the 25 cm statutory limit (SI, 1989a) and four times the 7.5 cm monitoring limit (DoE, 1989a).

In two separate feeding trials, Stark and Wilkinson (1994) examined the effects firstly of fresh additions of sewage sludge to soils (unsludged with contrasting physical properties) on PTE accumulation by finishing lambs fed *ad libitum,* and secondly of aged sludge-amended soils obtained from the Cassington (sandy loam - pH 6.6) and Royston (calcareous loam - pH 8.0) experimental sites described by Carlton-Smith (1987). The sludge used in the first experiment was highly contaminated with Cd and Pb (20.4 mg kg^{-1} and 1658 mg kg^{-1} dry solids, respectively), exceeding the UK 90th-percentile concentrations in sludge in both cases (Table 2.3) (CES, 1993). Indeed, the Pb concentration in the sludge exceeded the recommended limit of 1200 mg Pb kg^{-1} for surface application to grassland by nearly 40% (DoE, 1989a). The concentration of Cu in the sludge (605 mg kg^{-1}) exceeded the 50th-percentile level of sludge used for agricultural recycling in the UK. The sludge was selected to contain relatively low concentrations of Cu to avoid Cu toxicity compromising the study of Cd and Pb residues. The sludge was mixed with uncontaminated soil to provide a range of PTE concentrations from background levels (soil only) to approximately twice the statutory limits for Cd and Pb. The maximum concentration of Cu in the sludge/soil mixtures exceeded the statutory limit (SI, 1989a), but was less than the 7.5 cm monitoring value (DoE, 1989a). In the second experiment, four soils were fed from Cassington and three soils from Royston; one soil from each site was from a plot which had not received sewage sludge in the past. The other soils were selected on the basis of their Cd content and were from plots which had been treated with different amounts of sewage sludge. Sludge-treated soils from the experimental field sites were contaminated principally with Cd in relation to the other PTEs and the soil limit values. The Cd concentration in the soils increased to 2-3 times the permitted maximum for this element. Copper increased approximately to the statutory limit (SI, 1989a) in Cassington soil. Royston soil also contained a similar amount of Cu although here the Cu concentration was significantly below the level permitted in soil with pH >7.0. The Pb concentration reached less than half the soil limit for this element, but was probably realistic of normal practice when sludge is used in agriculture given current sludge quality and soil limit values.

Despite the extreme exposure of sheep to PTEs in the diet from historically sludge-treated soil, there were no adverse effects reported by Stark *et al.* (1995) or Stark and Wilkinson (1994) on animal liveweight gain or on reproduction. With the recently mixed sludge and soil the overall animal performance was also relatively good although the dry matter intake and growth rate were both reduced by sludge additions (Stark and Wilkinson, 1994). There have been very few reports of significant detrimental effects of sludge on animal growth and performance (Johnson *et al.*, 1981; Hogue *et al.*, 1984). An interesting observation by Hogue *et al.* (1984) was that male sheep grazing a sludge-treated pasture soil had significantly larger testes as a proportion of total body weight compared to an unsludged control! Of crucial importance to the human diet, however, is that no effects of sludge amendment were observed on the concentrations of Cd, Pb and Cu (as well as Zn and Fe) in the muscle (meat) of ewes or lambs with the soil intakes reported by Stark *et al.* (1995) or Stark and Wilkinson (1994). In both research programmes the concentrations of Cd and Pb in muscle tissue were below the analytical limits of detection ($\leq$0.04 mg Cd kg^{-1} and $\leq$0.4 mg Pb kg^{-1}, freshweight) and below levels of concern for carcass meat entering the human food chain. The results provide reassuring confirmation of earlier reports on the absence of significant accumulations of PTEs in muscle tissues of sheep grazing sludge-treated pastures or consuming sludge solids (Hogue *et al.*, 1984; Sanson *et al.*, 1984) and demonstrate the minimal potential for transfer of PTEs to the human diet by carcass meat from sewage sludge surface applied to grassland soils.

In contrast to muscle tissue, however, elevated concentrations of Cd, Pb and Cu were measured by Stark *et al.* (1995) in the livers of some of the lambs and ewes at slaughter. Cadmium and Pb also accumulated in the kidneys. Stark and Wilkinson (1994) also observed the accumulation of these PTEs in the offal of sheep. Concentrations of Pb in the blood generally reflected dietary Pb intakes and tissue levels in both studies.

Stark *et al.* (1995) detected no marked interactive effects of other inorganic elements identified except for Mo, S and Fe on Cu, although the accumulation of Pb appeared to be influenced by some dietary factor(s) as tissue Pb concentrations did not always follow the pattern of Pb content in the soil. In other words, soil with the highest Pb did not always give the highest tissue Pb. Stark and Wilkinson (1994) also observed dietary influences on PTE accumulation. In particular, marked reductions were measured in tissue concentrations of Cu, and to a lesser extent of Cd and Pb, with increasing age of sludge/soil mixtures (Fig. 6.3). Subtle effects of soil type were apparent on PTE accumulation although these were less important than PTE concentrations in the sludge-amended soil and the period of time after soil treatment with sludge. Nevertheless, at the higher accumulated intakes of Cd, tissue concentrations were lower for the Royston soils than for the other soils (historically treated and fresh mixtures) examined by Stark and Wilkinson (1994). The distinctive features of the Royston soils were a high Ca concentration (2.2%) and a relatively high pH value (8.0).

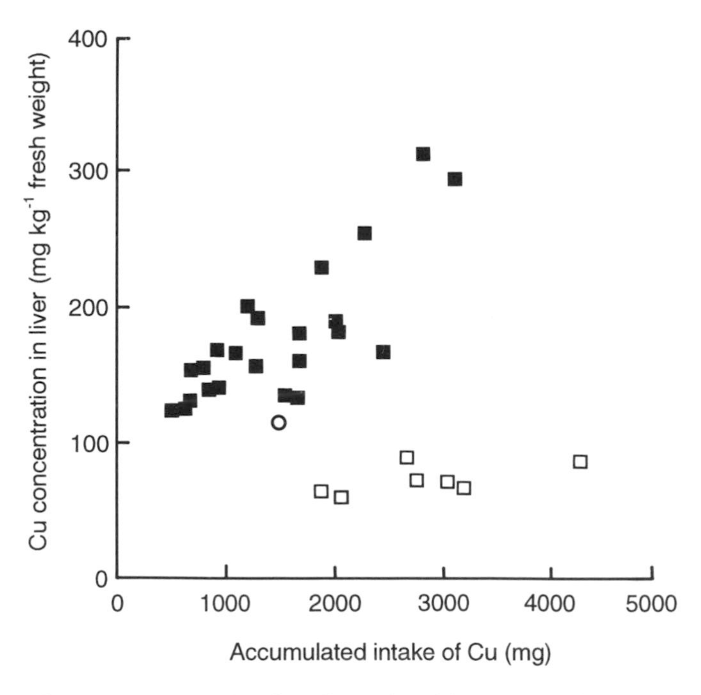

Fig. 6.3. Copper concentration (median values) in the liver of sheep fed grass with 10% inclusion of soil amended recently with contaminated sewage sludge [■] or from aged sites (10 years) treated with sludge in the past [□]. The open circle [○] denotes control animals receiving a diet without soil addition (Stark and Wilkinson, 1994).

In practice, under operational conditions, soil PTE concentrations increase very slowly taking many decades of sludge treatment to attain the soil limit values (Table 2.7). During this time an increasing fraction of sludge-applied PTEs will tend to the forms which appear to be less available to grazing animals. Consequently, under the conditions of this experiment (Stark and Wilkinson, 1994) it could be argued that the PTE levels achieved by soil amendment with large quantities of fresh sludge represented an extreme 'worst-case' of exposure to PTEs in sludge-treated soil, particularly when the very large concentrations of Cd and Pb in the sludge are also considered. Nevertheless, the concentrations of PTEs at the highest rate of inclusion of sludge in the diet could possibly reflect those dietary levels which might occur if sludge adhered to ingested forage. Using Pb as an example, the concentration of this element in the diet (approximately 50 mg kg^{-1} dry matter) could equate to a level of adherence to pasture at the 90th-percentile concentration of Pb in sludge (585 mg kg^{-1} dry solids) of 9% sludge in the dry forage. Whilst this level of exposure to Pb could be envisaged for short periods of time with a small proportion of sludges of relatively high Pb content, it would not be maintained for the same duration as in the controlled experimental feeding trials. At the 50th-percentile concentration of Pb in sludge (217 mg kg^{-1} dry solids), the high Pb treatment would be equivalent to an adherence level of 23% sludge in the dry forage which is unlikely to occur beyond the three week no-grazing period for sludge-treated pasture in practice (Sweet *et al.*, 1994).

Copper toxicity in livestock is of particular concern, especially for sheep grazing sludged pasture because traditional toxicological studies with added metal salts suggest that intakes of Cu in sludge-contaminated herbage and sludge-treated soil could reach levels potentially hazardous to sheep at current soil limits (NRC, 1980; Stark, 1988; Smith *et al.* 1992d) (Table 6.2). However, the bioavailability of metal salts is potentially much higher than for elements complexed in sludges or soils such that maximum tolerances for sludge or soil can be underestimated due to the presence of antagonistic effects from other elements (Logan and Chaney, 1983; Chaney *et al.*, 1987). Indeed, Chaney *et al.*, (1987) considered that the most consistent potential problem from sludge ingestion was *reduced* Cu concentration in the liver (Table 6.3). In other studies, Bremner (1981) found no cases of Cu toxicosis or any major increases in liver Cu content of sheep grazing pasture severely contaminated with Cu (50-150 mg kg^{-1} dry matter) from treatment with pig slurry for up to three years. McGrath (1981) and Poole (1981) suggested that Cu accumulation by pasture species, and intakes of Cu in soil, were unlikely to present a hazard to grazing animals. The recent feeding trials where large amounts of sewage sludge were freshly added to soil in the diet of sheep could suggest that the current soil limit for Cu may be unsafe to grazing livestock (Stark and Wilkinson, 1994). By contrast, however, Stark and Wilkinson (1994) showed that liver Cu was independent of intake for animals ingesting soil treated ten years previously with sludge at similar or higher accumulated intakes of Cu, compared with the freshly amended soil, at the maximum permissible concentration for this element in sludge-treated agricultural land (Fig. 6.3). Indeed, the control diet which did not contain soil, and provided the lowest intake of Cu, actually gave a higher liver Cu concentration compared with the relatively Cu-rich soils from historically sludge-treated field experiments (Stark and Wilkinson, 1994). This supports the suggestion that animal feeding trials with high inputs of contaminated sludges may significantly overestimate the accumulation and potential toxicity of PTEs in sludge-treated soil in practice and should be interpreted with caution.

The studies reported by Stark *et al.* (1995) also showed that soil *per se* depresses liver Cu concentrations with no toxic effect even when dietary concentrations (43.5 mg kg^{-1} dry matter) exceeded the NRC (1980) maximum tolerable level for this element (25 mg Cu kg^{-1} dry matter). Indeed, the results of Stark and Wilkinson (1994) and Stark *et al.* (1995) suggested that Cu toxicity in sheep is unlikely to be a problem at twice the statutory limit concentration for this element which may be encountered in surface soil of sludge-treated pasture. The excessive accumulation of Cu from Cu-rich diets can be suppressed in ruminants due to the intakes of inhibitory elements, including Fe, Zn, S and Mo in soil and grass, which are antagonistic towards Cu absorption by the animal (Logan and Chaney, 1983; Suttle *et al.*, 1984; Stark *et al.*, 1995). However, factors other than Fe antagonism apparently explained the differences in Cu accumulation observed by Stark and Wilkinson (1994) for sheep ingesting historically sludged soils compared with soils recently amended with sludge, because liver Cu concentrations from the recently sludged soil were much greater than with the aged soil yet some of the soils had similar Fe concentrations.

On balance, there appears to be little reason for concern over Cu toxicity to grazing animals from sludge-amended grassland soil (Bremner, 1981; Poole, 1981; Logan and Chaney, 1983; Stark *et al.* 1995). In contrast, Cu deficiency is potentially a widespread phenomenon in England and Wales (McGrath and Loveland, 1992) and may be a particularly acute problem in Scotland (Reaves and Berrow, 1984; Berrow and Reaves, 1985). Under these circumstances, Poole (1981) argued that it was reasonable to consider the beneficial effects of Cu in Cu-rich manures. In Northern Ireland, for example, Todd (1978) correlated the reduction of Cu usage in agriculture with the increased incidence of bovine Cu deficiency. The concentrations of Cu reported by Stark *et al.* (1995) in the liver of sheep fed a diet of grass and soil low in Cu were only about half of those for the control animals (no soil). However, the addition of Cu to the soil from sludge application reduced the Cu-depleting effect of soil *per se* and improved Cu status in the soil-fed animals. Thus, rather than being seen as an environmental problem, sewage sludge application could provide an important and essential source of dietary Cu for grazing ruminants.

Cadmium has been a much studied element in relation to the ingestion of sludge PTEs by grazing livestock because of its potential mobility in the food chain. However, none of the published reports considered in this review revealed levels of Cd in the tissues of grazing animals which would be generally constituted as a health risk. Extensive research indicates that sludge application to grassland presents little or no risk of entry of Cd into the human food chain *via* animals (Bertrand *et al.*, 1980 and 1981; Johnson *et al.*, 1981; Baxter *et al.*, 1982; Smith *et al.*, 1985). Nevertheless, almost all the studies reported the accumulation of Cd and other PTEs in liver and kidney in proportion to the amounts fed (Telford *et al.*, 1982 and 1984; Sanson *et al.*, 1984; Bray *et al.*, 1985; Brams *et al.*, 1989). The recent feeding trials with sheep (Stark and Wilkinson, 1994; Stark *et al.*, 1995) also indicated that concentrations of Cd remained at an 'acceptable' level (i.e. <0.5 mg Cd kg^{-1} fresh weight in offal - maximum concentration considered tolerable by MAFF Food Science, personal communication) for liver and kidney in finishing lambs fed diets with up to 10% inclusion of sludge-treated soil (dry matter) and containing two to three times the soil limit for this element, simulating a 'worst-case' of ingesting the top 1 to 2 cm of soil. Although the increases in offal Cd content with sludged soil were generally small and always below the tolerable level in offal, Stark *et al.* (1995) warned of the possible additive effects on accumulation in body tissues, particularly in kidney, of other sources of Cd in the diets of grazing animals and the duration of exposure to elevated concentrations of Cd. For example, the Cd concentration of grass itself may be increased to 0.3 mg Cd kg^{-1} (dry matter) at the soil limit for Cd depending critically on soil pH value (Smith, 1994b) which would markedly increase Cd intake by the animal.

Lead also represents potentially a cumulative poison. In contrast to Cd, however, Stark *et al.* (1995) estimated that, after six months of exposure, Pb concentrations in the kidneys of an average lamb may be close to the UK statutory limit for Pb in food of 1 mg kg^{-1} (fresh weight) (SI, 1985) from ingesting soil containing Pb at the statutory limit for this element at a rate of ingestion of 10% of the total diet (dry matter). Even with a more realistic 'worst-case' soil ingestion rate of 5% in the dry matter intake, median Pb concentrations in kidneys of both ewes and lambs approached or exceeded the food limit for Pb at soil Pb concentrations representative of the top 1 to 2 cm of soil sludged to statutory limits (i.e. up to twice the soil limit value in a 'worst-case'). On this basis, some kidneys would be considered unacceptable for human consumption. Lead concentrations also increased in liver compared to the control treatments, but because the food limit for Pb in liver is higher than in kidney at 2 mg kg^{-1} (fresh weight) (SI, 1985), this tissue would be considered acceptable. The pattern of Pb accumulation in offal showed a curvilinear response which plateaued with increasing Pb intake. Stark *et al.* (1995) therefore suggested that additive effects from other sources of Pb in the diet and the duration of exposure will be less important than for Cd. Furthermore, unlike Cd, Pb is not taken up by crops from sludge-treated soil so this will not be an important source of Pb in the diet of grazing animals. Whilst the period of exposure may be generally less critical for Pb, short-term exposure to high Pb levels is likely to increase tissue Pb to a greater extent than long-term exposure to low dietary Pb given an identical cumulative intake (Stark and Wilkinson, 1994).

The liver Pb concentrations measured by Stark and Wilkinson (1994) were generally in a similar range to those reported by Stark *et al.* (1995) and were below the statutory limit value for liver in the human diet. However, the statutory limit value for Pb in food (1 mg kg^{-1} fresh weight) was approached in the kidneys of animals after only two months of feeding recently sludge-treated soil at a rate of 10% inclusion in the diet (dry matter) and containing approximately twice the soil limit for Pb. The food quality standard for Pb was exceeded in kidneys of animals fed the sludge/soil mixtures for four months at twice the soil Pb limit at the 10% dietary inclusion rate (dry matter).

In practice, neither Pb nor Cd generally limit sludge recycling to agricultural land in the long-term at the current statutory soil limit values for PTEs since the soil limits for Cu and/or Zn are always reached first (Table 2.7). The Pb content in sludge has declined by 50% in the last decade (Table 2.3), and the long-term trend is for the decrease to continue. The Cd concentration in sludge has decreased by 60% over the same period.

Consequently, when the soil limits for Zn and Cu have been reached, the concentrations of Pb and Cd will typically be only 30% and 50% of their soil limits, respectively. Thus it is unlikely that the 'worst-case' scenarios tested by Stark *et al.* (1995) and Stark and Wilkinson (1994) would even be approached, let alone exceeded in context of the agricultural recycling of sewage sludge. Exceptions to this which might be envisaged could include sites where sludge has been disposed of historically, or on a dedicated basis and where soil concentrations of PTEs are significantly elevated. Nevertheless, on a strict interpretation of the Lead in Food Regulations (SI, 1985), there is some risk of kidney concentrations exceeding the statutory limit value of 1 mg Pb kg^{-1} (fresh weight) in food (SI, 1985) at the maximum permissible soil concentration of 300 mg Pb kg^{-1} in sludge-treated agricultural land (SI, 1989a). Stark *et al.* (1995) considered that such an interpretation of the data would be excessively cautious for the kidney comprises a very minor part of the human diet and the consumption of kidney is probably declining. Therefore, even taking extreme individuals consuming significant amounts from animals only reared on land of maximum soil limit values, the increased intake of Pb will be very small. Proportionately much larger amounts of liver are consumed compared to kidney yet liver is permitted a higher Pb value. Stark *et al.* (1995) recommended that the basis for different limit values of Pb in offal is re-examined. In view of the very small intakes of offal in the human diet generally (0.5% fresh weight; Davis *et al.*, 1983), there is apparently little overall risk to the human food chain from Pb (or Cd) applied in sludge to grassland soils under normal agricultural management.

Chapter seven:

Mobility of PTEs in Soil and Impacts on Water Quality

General Comments

The movement of PTEs in sludge-amended soils is of concern due to potential impacts on the environment through contamination of ground water supplies by leaching and of surface waters by both surface run-off and sub-surface movement of sludge-borne metals. The distribution of metals in permanent grassland soils (Chapter six) following surface applications of sewage sludge is also important due to the potential ingestion of PTEs by grazing animals and resulting effects on animal health and also on the human food chain (Stark, 1988). Factors influencing surface run-off are more pertinent to nutrient losses from sludge-treated soils and are discussed in detail in Chapter nine.

Factors Influencing Downwards Movement

Heavy metals may potentially move through the soil profile as a function of various physical and chemical mechanisms. The chemical properties influencing the mobility of heavy metals in sludge-amended soils are essentially those already described in Chapter five which modulate bioavailability. Thus, decreasing soil pH conditions increase the solubilization of metals and laboratory studies indicate potentially greater movement of metals down soil profiles as soil pH declines (Emmerich *et al.*, 1982c; Gerritse *et al.*, 1982; Welch and Lund, 1987). However, long-term studies of metal mobility in field experiments suggest that metal movement within the soil profile is not affected by increased soil acidity even at pH values <pH 5.0 (Williams *et al.*, 1987).

Welch and Lund (1987) found that soil texture also modulated Ni movement in sludged soil which correlated positively with sand content and negatively with silt and clay contents. However, Davis *et al.* (1988) showed equal penetration of Cd, Cr, Cu, Ni and Zn from surface sludge-treated grassland in both a sandy loam and a calcareous loam soil. This suggests mobility was largely independent of soil type which agrees with work by Brown *et al.* (1983) for Cd, Cu, Ni and Zn. However, when the overall amount of sludge metal found in the sampled profile was compared with theoretical additions, Davis *et al.* (1988) detected some apparent losses of the elements analysed from the 30 cm profile on the sandy loam soil, but not on the calcareous loam. There was no evidence from this study that rainfall was associated with greater movement of metals down the profile of the sandy loam soil since the yearly rainfall was higher on the calcareous loam. Welch and Lund (1987) similarly showed in soil columns that the amount of Ni leached was independent of the volume of water applied, but was related to the moisture status that existed in the sludge-amended soil layer. Considerably more Ni was apparently leached to deeper depths with substantially smaller volumes of water in sludge-amended soil maintained in the unsaturated condition. However, Davis *et al.* (1988) considered that the principal mechanism of metal movement in sludge-treated grassland was probably by infiltration of sludge particles in soil water and movement by earthworms.

Lund *et al.* (1976) reported that the distribution of metals with depth was closely related to the changes in chemical oxygen demand of soil samples with depth beneath sewage sludge disposal lagoons. This suggests that under certain extreme conditions metals may also move as soluble metal-organic complexes, but this will probably vary with the complexation characteristics of different metal species. For example, Welch and Lund (1987) found no correlation between organic carbon content and Ni concentration in the soil solution since inorganic reactions principally control the solubility of Ni. In contrast, movement of Cu in soluble organic form could be anticipated due to the significant complexation of Cu with organic matter in sludge-treated soils (Chapter five). Mobility of metals by this mechanism would potentially be more pronounced in neutral or alkaline soils due to increased complexation in soil solution. Furthermore, the presence of soluble metal chelators with heavy metals in sewage sludge may increase metal mobility in soil particularly for ligands that form stable, non-adsorbing complexes with metals (Elliott and Denneny, 1982).

Quantitative Assessment of PTE Movement in Soil

Significant downwards movement of metals through the profile of sludge-treated soil has been reported only where large applications of sludge have occurred substantially above suggested appropriate agronomic rates (WRc, 1985). Dowdy *et al.* (1991) applied 765 t ha^{-1} (dry solids) of sludge over a period of 14 years and found that Cd concentrations in the subsoil of sludge-amended areas were significantly higher than levels present in the untreated control. Similarly, Zn concentrations in the 32-51 cm region of the soil profile were also elevated by sludge application, but interestingly no increases in Cu levels were detected. In contrast, Darmody *et al.* (1983) found that Cu was particularly mobile in a silt loam soil amended with 150 and 300 t ha^{-1} (dry solids) of composted sewage sludge. The concentrations of both Cu and Zn were significantly increased to a soil depth of 75 cm at the higher rate of addition and most of the increase with depth occurred within the first year, but continued after the third year. Dowdy *et al.* (1991) considered that desiccation cracking below the cultivation zone of the highly structured silt loam soil studied, and the resulting movement of water down the profile in macro pores, explained the transport of metals to lower depths. In other work, Robertson *et al.* (1982) applied large rates of liquid sludge up to 335 t ha^{-1} to fine sandy loam soils and found that sludge applications increased soil organic matter content to a depth of 90 cm. Metal ion accumulation or movement also occurred in the top 90 cm of the profile, but the mobility patterns for the metals differed. Copper remained in the upper 30 cm whereas Zn had moved down to 90 cm.

Soil profile sampling of old sewage farms (Pike *et al.*, 1975; Alloway and Jackson, 1991) has similarly shown significant downwards movement of metals at these sites to the 50-100 cm layer of the soil profile, with slight increases in metal concentrations also being detected at lower depths (150-200 cm). From a survey of similar sites with long histories of sludge deposition (which may exceed 100 years in some cases), Stark and Carlton-Smith (1985) reported maximum soil concentrations of 1765 mg Cu kg^{-1}, 8400 mg Zn kg^{-1} and 314 mg Cd kg^{-1} reflecting large applications of sludge in the past. Organic matter contents of the soils may also be elevated (10-30%) and they can be particularly acidic (pH 4.5-6.5) (Smith, 1992; Smith *et al.*, 1992c) which would assist the leaching of heavy metals from the cultivation zone.

In general, however, the presence of PTEs in sludge-treated soil is confined to the cultivation zone with very little movement below that depth. Detailed analyses of soil profile samples from the long-term Woburn field experiment by McGrath and Lane (1989) showed that approximately 1% of the metals applied had moved 3.5 cm below the plough layer (0-23 cm) or less, but there was no evidence of accumulation of metals in deeper horizons down to 46 cm. Davis *et al.* (1988) measured movement of heavy metals from sludge surface-applied to grassland down to a depth of 10 cm, but most of the metal (60-100%, mean 87%) remained in the upper 5 cm of soil. They concluded, therefore, that sampling soil to a depth of 5 or 7.5 cm would be most suitable for monitoring long-term grassland treated with surface applications of sludge as currently recommended by the *Code of Practice for Agricultural Use of Sewage Sludge* (DoE, 1989a) (see Chapter six and Appendix).

Many reports indicate minimal or no movement of metals below the sludge-soil layer. The mass balance calculated by Emmerich *et al.* (1982c) showed that close to 100% of the metals added in sludge were recovered in the sludge-soil layer in soil columns following leaching treatment with 5.0 m of water over a 25-month period. In a field experiment, however, Chang *et al.* (1982a) only recovered 67% of the sludge-borne metals in the surface 30 cm of the soil profile four years following termination of sludge applications. Heavy metals had accumulated almost entirely in the 0-15 cm zone where sludges had been incorporated and there was apparently no evidence that metals had moved beyond the depth of the soil profile examined (0-60 cm). Dowdy *et al.* (1991) noted, however, that the silty clay soil in the study of Chang *et al.* (1982a) was subject to the occurrence of surface-connected desiccation cracks and seasonal water-logging at shallow depths which could partly explain the low recoveries. There is also evidence that the soil extraction procedure used by Chang *et al.* (1982a) did not recover fully the metal content from sludged soil and that changes in soil density could also be important (Chang *et al.*, 1984b). Cultivation displacement of soil and metals from the experimental plots could also partly explain the apparent low recovery (McGrath and Lane, 1989). On coarse and fine loamy soils Chang *et al.* (1984b) determined that over 90% of the heavy metals applied in sludge over a period of six years was found in the 0-15 cm layer corresponding to the depth of incorporation of the sludge. No significant increase in the concentrations of heavy metals was detected below the surface 30 cm of the soil profile. Williams *et al.* (1987) similarly incorporated sewage sludges continuously into the surface 20 cm of soil over a period of eight years, which amounted to a total addition of 1800 t ha^{-1} of sludge, but measured no significant movement of metals. Furthermore, no movement of metals occurred in the year following termination of sludge additions.

Laboratory studies using soil core microcosms with sewage sludges and metal salts have also demonstrated the minimal movement of metals by leaching processes. Sommers *et al.* (1979) found that the majority of the Zn, Cu, Cd, Ni and Pb added in sludge remained in the zone of sludge incorporation (0-7.5 cm) with minimal or no movement into the 7.5-15 cm soil depth. Khan and Frankland (1983) measured very little movement of Cd and Pb incorporated as chloride and oxide salts to give soil concentrations of 100 mg Cd kg^{-1} and 1000 mg Pb kg^{-1}. Interestingly the pH values of the soils used were low and in the range pH 4.4-5.4 and the total amount of water added was approximately equivalent to double the average rainfall in the UK. These data therefore support the view that soil pH value and quantity of rainfall generally have little influence on the extent of metal movement through sludge-treated soils. Hogg *et al.* (1978) similarly showed no movement of Hg below the 10-20 cm soil layer following incorporation of $HgCl_2$ and other mercuric compounds into the top 0-10 cm layer of a sand and clay loam soil at a soil concentration of 10 mg Hg kg^{-1} dry soil. Seven to 31% of the applied Hg was lost from the columns during the experiment presumably by volatilization, but the lack of movement of Hg was explained by the strong binding between Hg compounds and soil (see Chapters three, four and five).

On the basis of these reports it is apparent that environmental effects arising due to metal leaching from sludge-treated agricultural soils are unlikely. This is discussed in the next sections in relation to quantitative measurements of heavy metal concentrations in ground and surface waters.

Contamination of Groundwater by Sludge-borne PTEs

Extensive research conducted over the past 15 years of metal movement in sludge amended soils has shown that metals are highly spatially immobile even under conditions which raise bioavailability to crops. It is not surprising then that this survey of literature did not provide any reports showing that sludge application to agricultural land increases concentrations of heavy metals in groundwater to levels which may cause concern. Huylebroeck (1981) similarly noted that raised concentrations of heavy metals in groundwater had only been reported to a minor extent. In contrast, many examples can be quoted showing no effect of sludge spreading on the concentrations of toxic metals in groundwater supplies.

Sewage sludge was applied to a clay soil and a coarse-textured sandy soil by Melanen *et al.* (1985) in a series of field experiments conducted in Finland to study the movement of sludge constituents (nutrients, trace elements, organic matter and indicator micro-organisms). Metal concentrations in the groundwater were measured, but remained at the limit of detection of analysis (eg 0.0001 mg Cd l^{-1}). Melanen *et al.* (1985) concluded therefore that contamination of groundwater by heavy metals applied to soil in sludge was unlikely.

In Denmark, Grant and Olesen (1984) applied sewage sludge to mature spruce growing in acid sandy soil at a rate of about 45 t ha^{-1} (dry solids) which was more than twice that used by Melanen *et al.* (1985). No significant increases in the concentrations of heavy metals were detected in the groundwater except for Zn. Eighteen months after sludge treatment the concentration of Zn in the groundwater beneath the sludged plots was 500 μg Zn l^{-1} compared with <40 μg Zn l^{-1} for the control. However, this level of Zn in the groundwater is considerably smaller than the maximum guideline value set by WHO for aesthetic quality of drinking water of 5000 μg Zn l^{-1} (WHO, 1984) which has become the mandatory limit for water intended for human consumption in CEC and UK regulations (CEC, 1980b; SI, 1989b). The presence of Zn in the groundwater was probably explained because the sludge used was particularly contaminated with Zn (9600 mg Zn kg^{-1} dry solids) and supplied 407 kg ha^{-1} of Zn in a single dressing, which is nearly 30 times the maximum average annual rate of addition of Zn permitted by the UK Sludge Regulations (SI, 1989a).

Other Danish studies (Larsen, 1984) have measured metal leaching losses from sludges applied to an arable cropping rotation using lysimeters containing coarse textured soils. Annual applications of metal contaminated sludge up to a maximum rate of 16 t ha^{-1} (dry solids) over a period of seven years showed no detectable increase in the metal contents of leachates which remained below the limit of analytical detection (0.0001 mg Cd l^{-1}).

Effects of sludge application to soil on groundwater quality have been studied extensively in the US in both field and laboratory studies. Giordano and Mortvedt (1976) concluded from a series of laboratory experiments, using soil columns amended with sewage sludge and metal salts, that heavy metal contamination of groundwater was unlikely in heavy textured soils treated with sewage sludge. Concurrent application of inorganic nitrogen fertilizers did not affect the downward movement of Zn, Cd, Cr, Pb or Ni through soil. Duncomb *et al.* (1982) reported no detectable changes in the chemical composition of water from a 60 m well immediately adjacent to sludge storage lagoons used to supply the Rosemount sewage sludge watershed experiment in Minnesota. In other field experiments, Higgins (1984) applied annual dressings of 45 t ha^{-1} (dry solids) of sludge for three consecutive years. Groundwater quality was monitored for four years to establish the effects of high rates of sludge application, but no contamination of the groundwater by heavy metals was detected. Extensive monitoring surveys of sludge application field sites (Carroll and Ross, 1981; Berg *et al.*, 1987) similarly showed no apparent contamination of groundwater by metals applied in sludge.

Few if any studies have been conducted in the UK to assess effects of heavy metals applied to agricultural soils on groundwater quality. However, Baxter and Clark (1984) investigated the impact of sewage effluent disposal to land on groundwater quality at three sites over chalk and sandstone aquifers. Heavy metals contained in the effluents were largely removed during recharge through the soil and unsaturated zones such that only trace concentrations of metals potentially reached the groundwater by this method of sewage effluent treatment. It is emphasized, however, that effluent loading rates to such treatment sites are large relative to quantities of treated sludges applied to agricultural soils. For example, Baxter and Clark (1984) recommended an acceptable maximum loading would be 20 cm day^{-1}. To put this into context, a typical application of liquid sludge to agricultural land of 100 m^3 ha^{-1} represents a loading of only 1.0 cm and this would be applied only once during a growing season. Consequently there is little potential risk of metals percolating through to the groundwater from sewage sludge applied to soil at agronomic rates.

Contamination of Surface Water by Sludge-borne PTEs

Despite concern about possible contamination of surface waters by organic manures applied to agricultural soils (MAFF, 1991a) there are few examples in the literature demonstrating increased concentrations of metals in surface water supplies resulting from spreading sewage sludge on farmland. For example, Duncomb *et al.* (1982) detected no significant difference in the trace metal content of surface run-off from sludge-treated and control areas except for conditions of melting snow and shortly after sludge application to grass. However, Briggs and Courtney (1989) considered that the likelihood of run-off from grass was less than from cultivated soils (see Chapter nine), and Matthews *et al.* (1981) found no significant difference between the heavy metal concentrations in run-off water from sludge-amended and untreated grassland. These results are confirmed by Melanen *et al.* (1985) and Carroll and Ross (1981) who similarly reported no effects of sludge treatment on run-off water quality. Indeed, the total amounts of contaminants delivered to a waterway in run-off from surface applied sludge may actually be less than that from untreated areas if the sludge dries on the soil surface prior to a potential run-off event (Kladivko and Nelson, 1979). This is because the presence of sewage sludge stabilizes the soil surface thereby reducing the level of gross erosion which occurs. However, incorporation of sludge is generally considered the best management technique for minimizing run-off and surface water contamination from sewage sludge applied to agricultural land (Mostaghimi *et al.*, 1989; Gavaskar *et al.*, 1990). Effects of sludge application to farm land on heavy metal contamination of surface waters has not been studied in the UK, but research into the use of sludge in forestry (McPhail, 1984; Wolstenholme *et al.*, 1991) indicates that metal concentrations in surface water drainage are minimal and of no environmental concern. Forestry provides a potentially 'worst-case' model for surface water contamination with heavy metals due to the very low pH value (range 3.5-5.0) of forest soils, and available data from these catchments emphasize the minimal risk to potable surface water supplies from the agricultural use of sewage sludge.

Chapter eight:

Effects of PTEs on Soil Fertility and Natural Ecosystems

The Regulations and Soil Fertility

The current rules for controlling the use of sewage sludge on agricultural land have been criticized because they do not apparently consider the potential adverse effects of heavy metals in sewage sludge-treated soils on soil micro-organisms (McGrath *et al.*, 1994). This is not strictly true, however, because early researchers (Premi and Cornfield, 1969; Premi and Cornfield, 1971; Cornfield *et al.*, 1976) concluded that there was minimal risk to soil microbial processes from sludge applications using the experimental techniques which were available at the time for measuring C and N mineralization in soil. These processes are recognized as being essential to soil fertility and recent research has confirmed that they are relatively robust and unlikely to be affected adversely at the limit values for heavy metals in sludge-treated agricultural land (CEC, 1986a; SI, 1989a). This is because they are performed by a diverse range of microbes compared with other more specialized microbial functions, such as biological N_2-fixation and nitrification, which may be impaired (Smith, 1991; McGrath, 1994). The principal concerns at the outset were seen as the potential impacts of toxic metals on human health and effects on crop yields which formed the basis for the development of regulations to protect the environment when sludge is used in agriculture which are now enforced. The possibility that elevated concentrations of heavy metals in sludge-treated soil may be disruptive to soil microbial processes was discovered later (Brookes and McGrath, 1984) when new techniques for measuring the total soil microbial population (Jenkinson and Powlson, 1976) were applied to sludge-amended soil from the Woburn Market Garden Experiment. This opened the door to more specific studies on particular groups of soil microbes in the metal-contaminated soil at Woburn (McGrath *et al.*, 1988; Koomen *et al.*, 1990). Nevertheless it is reassuring that the scientific basis to the current regulations for PTEs in sludge-treated agricultural land was directed at protecting human health in the first instance. Studies on the effects of heavy metals on soil micro-organisms in sludge-treated farmland can proceed in the confident knowledge that the well-being of individuals consuming plant foods grown in sludged soil is not put at risk.

The UK Government commissioned an independent Scientific Committee (MAFF/DoE, 1993a) to review the implications for soil fertility of the current rules on sewage sludge applications to agricultural land. This Committee was organized in parallel with the scientific review of the implications for food safety as discussed previously (MAFF/DoE, 1993b). The Committee examining soil fertility aspects concluded that the experimental evidence on the effects of heavy metals from sludge on soil micro-organisms was inconsistent and incomplete. They also recognized that there were difficulties in assessing the soil limits required for individual metals to protect soil microbial processes due to the interacting effects of sludge, soil and environmental factors. Nevertheless, on the basis of very limited and often circumstantial evidence from field trials, which were arguably affected by unrealistic sludge quality, rates of application and the forms of metals supplied in sludge, the Committee were concerned that the soil limits for Zn (SI, 1989a) may not adequately protect soil fertility. In accordance with the 'precautionary principle' for soil protection, the Committee recommended reducing the maximum permissible concentration of Zn in sludge-treated agricultural land. However, there was no evidence to suggest that a lower limit for Zn would be required for soils of pH 5.0-6.0 than for those of pH 6.0-7.0. Therefore, a limit value of 200 mg Zn kg^{-1} was recommended for all soils of pH 5.0-7.0. For soils of pH >7.0, with a calcium carbonate content of >5% the Zn limit would be increased by 50% to 300 mg kg^{-1} as permitted in the Sludge Directive (CEC, 1986a). The current soil limits (sampling to 25 cm) for Ni, Cu, Cd, Pb, Hg and Cr were considered to adequately protect soil micro-organisms. Lead, Hg and Cr in particular have very low bioavailability in sludge-treated soil (Davis, 1984a) and are unlikely to have detrimental effects on soil quality.

Interestingly, the two UK Government Committees were inconsistent in the way recommendations to change the regulatory framework for agricultural utilization of sludge were approached. For example, MAFF/DoE (1993b) concluded that current evidence was insufficient to warrant a recommendation to change the soil limit values in relation to food safety and that the situation should be kept under review as results from research become available. The Committee reviewing soil fertility aspects (MAFF/DoE, 1993a), on the other hand, recommended a programme of research on the effects of heavy metals from sewage sludge on soil micro-organisms, which is now underway (Chambers *et al.*, 1994), as well as a reduction in the soil limits. Since it will take very long periods of time to reach the maximum permissible soil concentrations of heavy metals (Table 2.7), the long-term impacts on soil microbial processes can be thoroughly investigated and appropriate changes made to the regulations in the future, if necessary, without risk of compromising the fertility of agricultural land by sludge treatment. The problem facing the Water Undertakings is that once the soil limits have been reduced, there is no precedent within the EU for them to be increased again, which may constrain unnecessarily the economic, environmental and agronomic benefits gained from applying sewage sludge to agricultural land in the future. It also calls into question the requirement for the large investment, ultimately paid for by taxpayers and customers of the Water Undertakings, which is necessary to support the ongoing programme of research on soil fertility. The UK Government has recently announced (DoE, 1995) that it has decided to adopt the recommendations for changing the Zn limit as advisory values in a revised Code of Practice on sludge use in agriculture (see Appendix).

In developing the US regulations for land application of sewage sludge, the US EPA (1992a) considered soil organisms within the risk assessment procedure (Pathway 9) for deriving pollutant limits in sludge (Table A6). The criteria for the sludge → soil → soil organism pathway were set using data for the earthworm, *Eisenia foetida*, as the Highly Exposed Individual (HEI). Earthworms were not considered necessarily as the most sensitive species, but were selected because of the lack of data for other organisms. However, toxicity to earthworms was not identified as a limiting pathway to sludge application for any of the pollutants evaluated by the risk assessment. The apparent effects of heavy metals on soil microbial activity at the Woburn Market Garden Experiment were also considered (US EPA, 1992a), but the Agency concluded that further research was necessary to provide a meaningful interpretation of the observations in relation to setting limits for sludge-treated agricultural land.

The data on soil microbial activity obtained from sludge-amended field trials and from soils which have been treated with sludge on an operational basis in the past are described in the following sections particularly in relation to Zn and also Cd which may be important in soils treated with highly contaminated sludges in the past (MAFF/DoE, 1993a).

Micro-organisms

General

The effects of heavy metals on soil micro-organisms and microbial processes are complex and published data concerning the impacts of sludge applications on soil fertility are frequently contradictory (Smith, 1991). This is explained principally because of differences in reported experimental conditions, particularly in relation to timescales and the form and rate of applied metals. A further complication is that most studies generally make assessments based on measurements of total metal concentrations in soil. However, the bioavailability of metals and the toxicity to soil micro-organisms is strongly influenced by differences in soil physico-chemical properties such as pH value and texture. For example, metal toxicity may be reduced in soils with high clay and organic matter contents (Chander and Brookes, 1991a) due to the lower availability of metals to soil micro-organisms (Tyler, 1981; Bååth, 1989). Another possible source of contradiction is that the soil microbial population is highly adaptive to changing soil conditions and can develop resistance to toxic heavy metals (Tyler, 1981; Olson and Thornton, 1982; Doelman, 1986; Bååth, 1989).

Approaches to the Study of Microbial Effects

Sludge provides a source of organic matter and nutrients enhancing the growth and activities of soil micro-organisms. Therefore toxic effects of metals may only become apparent in the long-term once the readily available substrates have been mineralized, many years after the last application of sludge. Consequently, studies on the effects of heavy metals from sewage sludge on soil microbial processes necessarily require a long-term approach: short-term assessment in particular may provide contradictory information on microbial effects. Ironically, the improvement in sludge quality in recent years has made experimentation even more difficult inevitably extending the timescale of such studies.

To circumvent this problem researchers have frequently resorted to the application of metal salts, either directly to soil or in artificially spiked sludges, as a quick and convenient measure to raise metal concentrations of experimental soils within a short time-frame compared with the very slow increase in metal levels observed in practice. Chaudri *et al.* (1992a) argued that the addition of metal salts to Woburn soil would reflect the bioactivity prevailing after the decay of sludge organic matter. However, this is a doubtful supposition since at Woburn metal extractability remained constant with time independently of organic matter decomposition (Johnston *et al.*, 1989; McGrath and Cegarra, 1992). The application of sewage sludges spiked with metal salts was also considered to be acceptable by the Independent Scientific Committee reviewing soil fertility aspects of PTEs (MAFF/DoE, 1993a). They considered it appropriate to examine effects on the basis of total metals in soils regardless of their chemical form since there is no certainty over the availability of metals in soils in perpetuity. McGrath (1994), on the other hand, argued that the concentrations of metals in the soil solution which bathes plant roots and soil organisms may be a better indicator of exposure. However, this is not a new concept. Davis (1979) established there was considerable potential in measuring the concentrations of heavy metals in soil solution for assessing their environmental significance in sludge-treated soil. A refinement would be estimating the free metal-ion concentration since this is considered the principal factor exerting potentially a toxic effect on soil micro-organisms, although this can pose analytical difficulties due to the very small amounts present in the soil solution (McGrath, 1994).

There are many studies showing that the chemical form of applied metals can markedly alter toxicity responses and that metal salts have higher solubility increasing the severity of toxic effects on biological systems compared with the same amounts applied to the soil in sludge (Cunningham *et al.*, 1975; Giordano *et al.*, 1975; Street *et al.*, 1977; Bloomfield and McGrath, 1982; Korcak and Fanning, 1985). For this reason, artificial contamination with metal salts has been considered inappropriate for assessing the environmental impacts of sludge recycling to agricultural land (Logan and Chaney, 1983; Chaney *et al.*, 1987; Sommers *et al.*, 1987).

Alternatively, experimental trials have received very large and unrealistic doses of highly metal contaminated sludge to increase the soil metal content in a single treatment or in a limited number of applications (for example, Luddington and Lee Valley, see Chapter three). This approach has also over-estimated the availability and potential toxicity of heavy metals in soil compared with normal operational practice (Logan and Chaney, 1983), although it may be argued that, in time, the form in which the metals were originally applied is not important as the soil system approaches a new equilibrium condition. The question which remains unanswered, however, is whether the new equilibrium is a realistic model for sludge-treated soil in practice. An interesting observation is that metal availability to ryegrass was considerably smaller in historically sludge-treated soils compared with sludge-amended experimental soils (Carlton-Smith and Stark, 1987; Carlton-Smith, 1987) suggesting a lower bioactivity of heavy metals may occur in practice following small and repeated doses of sludge in the long-term.

Inconsistent results have also possibly arisen due to other artefacts of established long-term field experiments. For example, the patterns of metal concentrations in aged experimental trials, such as Woburn, are incompatible with current and future sludge quality (Smith, 1991). The large Cd concentration of the sludge-treated soil at Woburn is a legacy of uncontrolled applications of highly metal-contaminated sludge from industrial inputs in the past which would be illegal under the current regulations. Consequently, it may be inappropriate to extrapolate the results obtained from these highly metal contaminated experimental soils to sludge-treated agricultural land in practice. Indeed, MAFF/DoE (1993a) repeatedly noted the possible limitation of the information on microbial effects obtained from Woburn due to the excessive amount of Cd in sludge-treated soil at this site.

To obtain information on the extent of potentially adverse and long-term effects of heavy metals on soil micro-organisms in sludge-treated agricultural land, Smith and co-workers (Smith and Giller, 1992; Smith *et al.*, 1992e; Smith, 1994c,d) have measured microbial properties in soils from operational field sites with long histories of sludge treatment (10-100 years), but which have had no sludge applied in the past 5-30 years. A large number of sites is necessary to give a range of contrasting soil physico-chemical properties. These sites were carefully selected for study using the detailed soil registry information on heavy metals collected and held by the Water Utilities in accordance with the Sludge Regulations (SI, 1989a). The significance of increasing heavy metal concentrations in the different soils was quantified by measuring the total concentrations of metals in plant leaves from a five-leaf-stage barley bioassay and in displaced soil solutions. Whilst this approach does not measure directly the bioavailability of metals to soil microbes, it can be used routinely and provides a superior basis for assessing effects of metals on the presence and activities of micro-organisms under different soil conditions compared with estimates of total metals. Furthermore, it offers a more practical way of assessing the relative bioavailability of metals and the potential toxicity to micro-organisms in soils with different physico-chemical properties compared with the preferred, but difficult and laborious measurement of metal species in soil solution (McGrath, 1994). The examination of biological and chemical properties of such historic site soils provides a complementary approach to conventional experimental field trials. In contrast to field experimentation, however, the time frame and cost of field monitoring studies is considerably reduced.

Microbial Resistance to Heavy Metals

The necessity for soil micro-organisms to tolerate the presence of heavy metals in the environment is not a recent phenomenon associated with industrial development of the 19th and 20th centuries. This is put into perspective by the account of Silver *et al.* (1989) concerning the possible route of development of bacterial resistances to toxic heavy metals. Bacterial cells have co-existed with toxic heavy metals since the origin of life, perhaps 3 or 4×10^9 years ago. In the early stages of the evolution of life, when volcanic activity and other sources of heavy metals were ubiquitous, it was essential to develop mechanisms to cope with the toxic metals that were abundant in the environment. As prokaryotic evolution progressed, the genes for resistance to toxic heavy metals were less essential, and it became important to minimize the genetic burden of carrying these genes which were periodically essential, but were sometimes burdensome. It seems likely that the genes for metal resistances were 'packaged' into bacterial plasmids and transposons. These genes could be lost from particular lines of bacterial cells easily without affecting the rest of the genetic coding. Even more importantly they could move from cell to cell by facile intercellular gene transfer, thus facilitating the spread of resistance among a population of bacteria including more than a single strain or species, when stress from toxic heavy metal pollution made this resistance an asset for survival.

Consequently, Silver *et al.* (1989) considered that the resistance processes to toxic metals should be quite ancient, and that the genetic and biochemical mechanisms involved are likely to show common properties with those for essential cellular roles, such as growth, metabolism of energy and carbon sources, and biosynthesis of essential nutrients. However, each of the mechanisms for microbial resistance to toxicity requires inputs of cellular energy. The possible mechanisms of resistance to toxic heavy metals exhibited by soil micro-organisms were listed by Doelman (1986) as follows:

1. An energy-dependent efflux system that keeps the intracellular metal concentration low.
2. Oxidation to less toxic compounds.
3. Biosynthesis of intracellular polymers that serve as traps.
4. Binding of metals to the cell surface.
5. Precipitation at the cell surface.
6. Biomethylation.

Although these phenomena have been identified in natural populations of micro-organisms in metal contaminated soils (Doelman, 1986), it has been argued (McGrath, 1994) that detrimental effects on the size and diversity of the microbial biomass, and on certain microbial processes, apparently occur below the maximum soil limits for heavy metals in sludge-treated agricultural land (CEC, 1986a; SI, 1989a).

Microbial Biomass and Activity

The soil microbial biomass mediates the biochemical processes occurring in soils and also acts as a reservoir of labile plant nutrients (McGill *et al.*, 1975; Marumoto *et al.*, 1982; Brookes *et al.*, 1984; Jenkinson, 1990a). Consequently the function of the microbial biomass is essential to the intrinsic fertility of agricultural soils by mineralizing nutrients from the residual soil organic matter to support plant growth. Although of critical importance to soil fertility, the microbial biomass generally accounts for as little as 1-3% of the soil organic carbon (Jenkinson and Ladd, 1981). In addition, there is normally an approximately linear relationship apparent between the amounts of microbial biomass and organic carbon present in soil, although there are marked seasonal fluctuations in biomass size, for example, due to differences in soil moisture status (van Gestel *et al.*, 1992) and cropping (Flieβbach and Reber, 1992), which may also influence the apparent toxicity of heavy metals to the microbial biomass. There are also gross differences in the microbial biomass content of soils according to the type of agricultural management (Insam *et al.*, 1989; Zelles *et al.*, 1994). Whilst measurements of the size of the microbial biomass may provide a sensitive indication of changes in soil management practices, the importance of the amount of microbial biomass in relation to nutrient cycling and decomposition processes in soil is poorly understood (Witter, 1989). Nevertheless, the microbial biomass has been considered a sensitive and useful indicator of metal pollution (McGrath, 1994).

Brookes and McGrath (1984) reported that the microbial biomass was halved in sludge-treated soil at Woburn compared with soil receiving FYM for a similar period of time. This effect was attributed to the elevated concentrations of potentially toxic heavy metals in the sludge-amended plots which generally approximated to the current UK statutory limits for the agricultural use of sewage sludge, except for Cd which was more than five times the limit value for this element in soil (McGrath *et al.*, 1988; SI, 1989a). The minimum concentrations of Cd and Zn in soil which negatively affected the soil microbial biomass at Woburn were 6.0 mg Cd kg^{-1} and 180 mg Zn kg^{-1} (McGrath *et al.*, 1994). However, the rate of soil respiration was similar for both high- and low-metal treatments indicating that the metabolic activity of the biomass was increased in the metal-contaminated soil. This suggested a less efficient conversion of substrate carbon into microbial carbon in the high metal soil (Chander and Brookes, 1991a,b,c). Such increases in the specific activity, or metabolic quotient (qCO_2), of the microbial biomass is significant because it indicates stress to the soil microbial population since disturbance to an ecosystem causes diversion of energy from growth to maintenance (Odum, 1985). Reductions in the size of the soil microbial biomass have also been measured in other sludge-treated soils rich in Cd. For example, Stark and Lee (1988) reported that the soil microbial biomass content in certain sludge-amended pasture plots at the Cassington field experiment was smaller compared with an untreated control for soils sampled in the summer. Cadmium concentrations in soil at Cassington were raised to more than three times the maximum limit value for this element. Unlike Woburn, however, this increase was planned as part of a study into the potential transfer of Cd to the human food chain from sludge-treated agricultural land (Carlton-Smith, 1987). There was also evidence of a significant decrease in the microbial biomass content in the pasture soil at 210 mg Zn kg^{-1} and containing approximately 1.7 mg Cd kg^{-1} (see Carlton-Smith (1987) for estimated soil Cd based on 15 cm depth of sludge incorporation), albeit at a marginally lower pH value (pH 5.6), compared to the untreated control (pH 5.9). However, no significant effects of heavy metals were observed when pasture soil was sampled in the spring, autumn or winter period and no significant reductions in microbial biomass content occurred in sludge-amended arable soil compared with the control for any of the different sampling times.

An extensive analysis of the microbial biomass contents of historically sludge-treated field sites has been reported by Smith (1994c). Microbial biomass content and activity were measured in 13 sludge- and waste-treated soils. The sites were selected for either high Zn or Cd contents compared with the other elements present in sludge-treated soil and in relation to the limit values (SI, 1989a) to assess their potential effects on soil fertility. The largest concentrations of these metals measured in the soils were 540 mg Cu kg^{-1}, 3000 mg Zn kg^{-1} and 16.9 mg Cd kg^{-1}. It is emphasized that the sites represented potentially a 'worst-case' of soil contamination with heavy metals due to sewage sludge application because they had received long-term treatment for over 100 years in some cases and because there had been no sludge applied in the past 5-30 years. Untreated control sites were also sampled with more or less background concentrations of heavy metals.

As expected, the soil microbial biomass content was raised with increasing soil organic matter. However, the microbial biomass declined as soil Zn increased, although an upper critical concentration for Zn was apparent on examination of the pooled data in a regular grid format (Fig. 8.1). A statistical model of the relationship between the biomass content in soil and the total soil concentrations of Zn and organic C estimated the upper critical concentration for Zn was 500 mg kg^{-1} in a potentially vulnerable sandy soil with pH in the range 4.9-6.4 (mean = 5.8) and organic C content of 1.2%-8.2% (mean = 1.8%). This amount of Zn was associated with concentrations of Cd and Cu in the soil of 2.6 and 79 mg kg^{-1}, respectively. Concentrations of metals in 5-leaf-stage barley plants grown in sub-samples of the same soils, and in displaced soil solutions, indicated Zn was potentially in a highly bioavailable form in the soil.

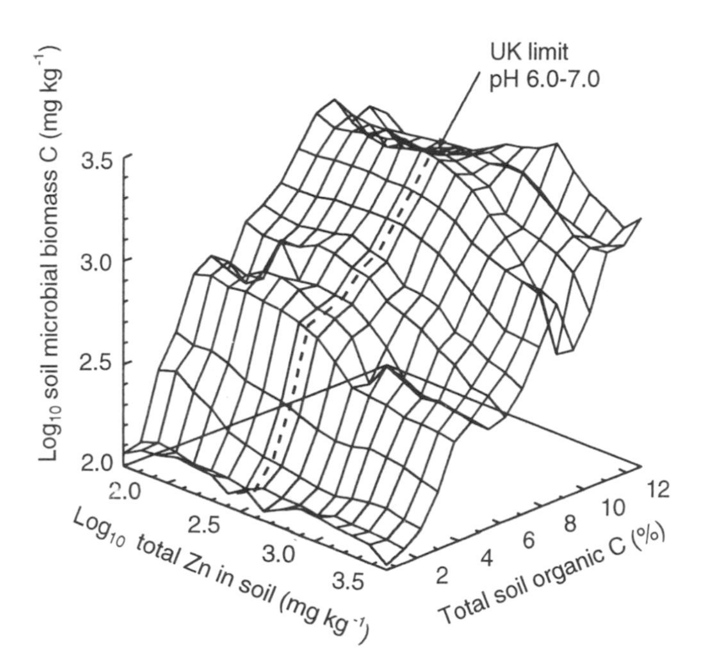

Fig. 8.1. Three-dimensional grid showing the relationship between the microbial biomass carbon (C_{mic}) content and the total concentrations of zinc and organic carbon in soils following long-term treatment with sewage sludge (Smith, 1994c).

Three arable soils were selected for study in the survey of historic sites because they contained proportionally higher concentrations of Cd compared with Zn, and Cu in relation to the soil limits. The soils were silt or silty loams with maximum concentrations of 350 mg Zn kg^{-1}, 9.8 mg Cd kg^{-1} and 240 mg Cu kg^{-1}. Soil pH value and total organic C content were typical of intensively cultivated arable soil and in the range 6.9-7.7 and 0.6-3.6%, respectively. Interestingly, the sites had been treated in the past with sludge obtained from the same treatment works which had supplied the Woburn experiment. Consequently, the patterns of heavy metal concentrations were broadly similar to those present in the sludged soil at Woburn (McGrath, 1984). As expected, microbial biomass increased with total soil organic C. However, a multiple regression model indicated that there was also an overall reduction in biomass content with increasing Cd concentration in soil (Table 8.1). The upper critical concentration of Cd was close to the maximum permissible limit value for sludge-treated agricultural land and probably in the range 3-4 mg kg^{-1}. In contrast, no significant effects of Zn or Cu were apparent (Table 8.1) despite exceeding the soil limit values (SI, 1989a) and being in potentially bioavailable form. Interestingly, the statistical analysis indicated that differences between the sites were mainly attributed to the overall differences in Cd concentrations between the soils since the significance of this term was markedly smaller when Cd was included in the regression model compared with either Zn or Cu. These results suggest that the adverse effects on the soil microbial biomass observed at Woburn and Cassington could be explained by the high Cd content of those sludge-treated experimental soils. Interestingly, Cd had no effect on the microbial biomass at 6 mg kg^{-1} (twice the soil limit) in soil at the Lee Valley experimental site in the UK (Chander and Brookes, 1991a). However, the minimum concentrations of Zn and Cu, which negatively affected the microbial biomass in this soil at pH 5.6-5.9, were also relatively large at 857 mg Zn kg^{-1} and 384 mg Cu kg^{-1}.

In another soil, contaminated uniquely with Cu and containing up to 460 mg Cu kg^{-1} from previous treatment with whisky distillery waste (pot-ale), there was no significant correlation between Cu concentration in soil and the microbial biomass content. Soil solution and plant tissue analysis indicated the Cu was in potentially highly bioavailable form. Information from this site, and a sludge-treated soil containing a similar amount of Cu (29-460 mg kg^{-1}), but comparatively little Zn (105-330 mg kg^{-1}) and Cd ($\leq$1.0 mg kg^{-1}), indicated that the current soil limits for Cu in sludge-treated agricultural land may provide a high margin of protection to the microbial biomass. In contrast, Chander (1991) suggested that Zn and Cu were additive in their adverse effects on the soil microbial biomass close to, albeit higher than, the limit values in soil treated with sludge prepared by spiking raw sewage with metal salts.

The metabolic quotient (qCO_2) of the microbial biomass, and the ratio between the biomass and organic C (C_{mic}:C_{org}) contents of soil, were estimated by Smith (1994c) to assess the efficiency of biosynthesis in the metal contaminated soils. The results showed that qCO_2 and C_{mic}:C_{org} were inversely correlated. Diverse microbial communities (high C_{mic}:C_{org}) are generally associated with a low energy requirement (low qCO_2) for maintenance and growth (Insam and Haselwandter, 1989; Insam *et al.*, 1991). Reductions in C_{mic}:C_{org} and increases in qCO_2 were apparent with increasing concentrations of Zn in soil as C was diverted to maintenance needs and the community structure of the microbial population was simplified favouring more metal-tolerant forms. In this diverse range of soils, however, the disruptive effects of Zn on the soil microbial biomass occurred at soil concentrations exceeding the maximum permissible limit value at pH 6.0-7.0.

Table 8.1. Multiple regression models showing the effects of heavy metals on the microbial biomass carbon content of arable soils treated in the past with sewage sludge.

Source of variation	+/-	F-ratio	Significance level (P)
Total copper (r^2=0.65, P<0.001)			
Total organic C	+	33.8	<0.001
Total Cu in soil	-	1.5	0.226 ns
pH	-	11.1	0.002
Site	na	7.9	0.002
Total zinc (r^2=0.65, P<0.001)			
Total organic C	+	34.2	<0.001
Total Zn in soil	-	0.8	0.382 ns
pH	-	11.9	0.002
Site	na	8.2	0.001
Total cadmium (r^2=0.72, P<0.001)			
Total organic C	+	42.7	<0.001
Total Cd in soil	-	22.0	<0.001
pH	-	12.5	0.001
Site	na	5.1	0.012

Regression df = 5, error df = 34
na - not applicable
ns - not significant
Source: Smith (1994c)

Soil pH value can also have an important influence on the size and activity of the soil microbial biomass (Adams and Adams, 1983; Anderson and Domsch, 1993; Carter, 1986). This is partly explained because soil acidifcation reduces the availability of soil organic C to micro-organisms (Persson *et al.*,

1989). Smith (1994c) showed that the qCO_2 of the microbial biomass increased in response to both low and high soil pH values about the minimum optimum range of 5.5-6.0. This observation is particularly important in relation to the Sludge Regulations in the UK (SI, 1989a) because different soil limit values for Zn, Cu and Ni are allowed according to banded ranges of soil pH value. Indeed, the reduced microbial biomass content reported by Witter *et al.* (1993) for sludge-treated soil compared with an FYM-treated control in a long-term field experiment at Ultuna in Sweden may be largely attributed to differences in soil pH value as an artefact of the trial. Here, the pH value of the sludged soil was 5.3 whereas the FYM-treated soil had a pH of 6.6. The problem is that these results can be taken at face value, implying that an adverse effect is due to increased concentrations of heavy metals in the sludge-amended soil (McGrath *et al.*, 1994). Interestingly, the concentrations of Zn and Cu in the sludged soil at Ultuna were 230 mg Zn kg^{-1} and 125 mg Cu kg^{-1}, which exceed the permitted UK maxima by 15% and 56% for these elements, respectively, for soil of pH 5.0-5.5 (Table A2).

Two field experiments at Braunschweig in Germany were established in 1980 on an old woodland site (pH 5.3-5.7) and an adjacent arable area (pH 6.0-6.6) with identical soil texture (silty loam) and cation exchange capacity (Flieβbach *et al.*, 1994). Low and high metal sludges were applied annually for 10 years at rates of 100 and 300 m^3 ha^{-1}. The unsludged control plot received inorganic fertilizer only. After the first year, the high metal sludge was enriched with metal salts by artificial spiking followed by anaerobic incubation for six weeks. The original objective of the trials was to provide information to the German Government in setting sludge and soil quality limits based on crop uptake data. Chemical properties of the soils are presented in Table 8.2. Compared with Woburn and Cassington, the concentrations of metals in the sludge-treated soils are more representative of current and future sludge quality containing proportionally more Zn and also Cu than they do Cd in relation to the current maximum soil limit values (CEC, 1986a; SI, 1989a). The application of sewage sludge at Braunschweig increased the total soil microbial biomass compared with the unsludged control although this beneficial effect was less pronounced with the high metal sludge (Flieβbach *et al.*, 1994). On average, a reduction in microbial biomass of about 26% was measured for soil treated with the high-metal sludge at 300 m^3 ha^{-1} compared with the same rate of low-metal sludge. Similar effects were apparent after only seven years of addition of the metal salt enriched sludge, even at the lowest rate of treatment (Flieβbach and Reber, 1992). It could be argued, however, that these effects were associated primarily with the artificially spiked sludge and were therefore more severe due to the higher solubility and toxicity of the added metal salts.

Flieβbach *et al.* (1994) also reported a reduction in the efficiency of biomass synthesis of the total C applied in sewage sludge with increasing metal concentration in soil at Braunschweig. These effects were most pronounced in the low pH soil. Indeed, the increase in total organic C content of the low pH soil at the highest contamination level (Table 8.2) was attributed to the lower synthesis of microbial biomass reducing the extent of mineralization of the organic matter. Clearly, the accumulation of organic material in soil reflects a severe impairment of soil function. However, no adverse effects on C mineralization in sludge-treated soils have been detected at and above the maximum permissible limit values for heavy metals (CEC, 1986a; SI, 1989a) in other long-term field experiments (Johnston *et al.*, 1989). Decreases in microbial biomass in soil treated with the high-metal sludge were accompanied by an increase of qCO_2. The largest increases in qCO_2 were measured in the former woodland soil compared with the old arable soil. These responses were attributed either to a higher availability of metals in the former woodland soil since total metals were similar in both soils, or to an additional stress, exerted by the low pH value.

Table 8.2. Some chemical properties of soils at the Braunschweig experimental site in Germany.

Sludge addition (rate ha^{-1})	Total concentration ($mg\ kg^{-1}$) Cu	Zn	Cd	Total organic C (%)	pH (0.1 M KCl)
Arable soil					
Unsludged	15.2	48.2	0.23	0.79	6.6
100 m^3 low metal	23.1	91.9	0.43	0.98	6.5
100 m^3 high metal	43.1	155.0	0.96	0.99	6.5
300 m^3 low metal	40.6	204.3	0.83	1.37	6.2
300 m^3 high metal	99.5	**345.2**	2.64	1.38	6.0
Former woodland soil					
Unsludged	8.8	41.0	0.17	1.62	5.7
100 m^3 low metal	16.3	88.3	0.43	1.75	5.7
100 m^3 high metal	40.5	162.3	1.03	1.79	5.5
300 m^3 low metal	32.5	190.9	0.83	2.19	5.6
300 m^3 high metal	**107.8**	**404.1**	**3.19**	2.40	5.3

Values in bold exceed UK soil limit values (SI, 1989a)
Source: Fließbach and Reber (1994)

The total Zn concentration in soil from the ex-woodland site (Table 8.2), treated with the highest rate of high-metal sludge, was more than twice the permitted UK maximum for soil of pH 5.0-5.5 and the Cu concentration was 34% higher than the UK limit. There was no significant difference in qCO_2 of the microbial biomass for soil treated with low-metal sludge and the unsludged control. Furthermore, the qCO_2 was only marginally increased in the old arable soil treated with the high rate of artificially contaminated sludge compared with the unsludged control. However, the C_{mic}:C_{org} ratio was reduced in the arable soil treated with the highest rate of low-metal sludge at a metal concentration below the UK soil limits for soil of pH 6.0-7.0 (SI, 1989a).

Interestingly, the minimum concentrations of Zn listed by McGrath *et al.* (1994) (quoting Fließbach and Reber, 1991) which negatively affected the soil microbial biomass at Braunschweig were 360 mg kg^{-1} and 386 mg kg^{-1} for the high and low pH trials, respectively. Both values exceed the permitted maximum limits for Zn stipulated in the UK regulations (SI, 1989a).

The field trials at Braunschweig are of particular interest to experimentalists because controlled applications of sludge have been supplied to give a range of concentration levels of metals under contrasting soil pH conditions. However, the use of sludges spiked with metal salts necessarily requires a cautious approach in extrapolating the results to sludge applications in the field in practice. For example, potentially disruptive effects on soil microbial function and diversity were apparent at much smaller concentrations of metals in soil (albeit generally above the UK soil limits) compared with those found in operationally sludge-treated field soils (Smith, 1994c).

Most of the research effort on soil fertility aspects of heavy metals has focused on sewage sludge, mainly because agricultural recycling of sludge has been a highly emotive issue and concerns are often perceived about disposing of the material. Comparatively little attention has been given to the effects of heavy metals applied in the wastes of farm animals, yet at the local as well as the global level, additions to soil from these wastes of the important elements Zn and Cu significantly outweigh those from sludge (Table 2.4). In a long-term field experiment with animal slurries, Christie and Beattie (1989) measured a significant reduction in microbial biomass content and increased soil organic C content in the top 5 cm of a grassland soil treated with pig slurry and containing 110.8 mg kg^{-1} of both Zn and Cu at pH 5.1. Huysman *et al.* (1994) detected changes in the microbial communities of soils treated with pig manure at concentrations of Cu as low as 20 mg kg^{-1}, although Bååth (1992) observed no effect on thymidine incorporation into soil bacteria DNA following the addition of 120 mg Cu kg^{-1} as copper sulphate to a sandy loam. Whilst these results may also be relevant to recycling sewage sludge on agricultural land, it is ironical that farm wastes are not subject to any soil protection legislation, yet they may prove to have a greater and wider adverse impact on soil quality in the long-term compared with sewage sludge. However, MAFF/DoE (1993a) recommended that the impacts of other organic wastes including animal manures applied to land should be evaluated for effects on soil fertility.

In summary, the microbial biomass can be disrupted by heavy metals in sludge-treated soil, but apparently at concentrations which generally exceed the permitted maxima in sludge-amended agricultural land (MAFF/DoE,1993a). Indeed, there are data (Jones *et al.*, 1987) which show no adverse effects on the microbial biomass content of soil low in Cd (1.29 mg kg^{-1}) but at five times the

UK limit for Zn which was identified by MAFF/DoE (1993a) as the principal element of concern. Problems may have arisen in experimental studies at smaller concentrations of Zn than the soil limits possibly due to differences in soil pH value between sludge-treated and control plots, the application of sludges spiked with metal salts, or because the Cd concentration was potentially toxic and exceeded the limit value for this element. Following the initial concerns, it is reassuring that the current UK soil limits for heavy metals appear to protect the soil microbial biomass in sludge-amended agricultural land. Of course, it is possible to argue that subtle changes in the microbial community may still occur which cannot be detected by simply measuring the size of the soil microbial biomass as a whole. It is interesting to speculate whether, in fact, this will be the case.

Nitrification

In contrast to mineralization processes, more specialized groups of soil microbes may be less tolerant of the potentially toxic effects of metals. For example, nitrification processes, which are performed by specialized groups of soil bacteria, were suppressed in sludge-amended soil at Woburn compared with the FYM control (Brookes *et al.*, 1984). However, Leung *et al.* (1993) isolated both ammonia and nitrite-oxidizing organisms from a range of metal contaminated soils without exception, and other studies at Cassington and at old sewage farms (Kwan and Smith, 1990; Smith *et al.*, 1990) have indicated no detrimental effects on nitrification of metal contents up to 1250 mg Cu kg^{-1} and 6900 mg Zn kg^{-1} in soil. The increasing rate of nitrate production in these soils was apparently explained only as a simple function of increased soil organic matter content. Tolerance and adaptation of nitrifying micro-organisms to high metal concentrations in soil has also been reported (Rother *et al.*, 1982a).

More recently, Smith (1994d) measured the nitrification rate in selected soils contaminated specifically with Cu and Zn (the principal elements limiting sludge recycling; Table 2.7), resulting from past applications of sewage sludge, or containing only increased amounts of Cu from land treatment with whisky distillery waste. Nitrification rate increased with increasing soil pH value and organic matter content. The rate of nitrification also differed significantly between sites presumably due to localized variations in site conditions and in the activities of different strains of nitrifying bacteria present. However, there was no effect of Cu on nitrification rate at the maximum concentration measured of 460 mg Cu kg^{-1} (in soils treated either with sewage sludge or with 'pot-ale'). This is strongly supported further because other evidence indicates that Cu in soils treated with 'pot-ale' is in more bioavailable form compared with Cu in sludge-amended soils (Berrow *et al.*, 1990). An upper critical level of Zn was detected for nitrification in the vulnerable sandy soil described by Smith (1994c) in the previous section in connection with studies on the microbial biomass content of historically sludge-treated farmland. This was equivalent to a plant tissue concentration of 500 mg Zn kg^{-1} determined by a 5-leaf-stage barley bioassay. The critical Zn concentration in soil was obtained from a statistical model of the relationship between plant uptake of Zn and the total concentration of Zn in soil and pH and is presented in Table 8.3 for different soil pH values.

The results demonstrated that the current UK soil limit values set for Cu and Zn (SI, 1989a) protect sensitive nitrification processes in sludge-treated soils and generally with very considerable margins of safety.

Table 8.3. Estimated upper critical concentrations of Zn for nitrification in relation to soil pH value.

Soil pH value	Zn concentration in soil (mg kg^{-1})
5.0	230
5.5	390
6.0	660
7.0	1900

Source: Smith (1994d)

Nitrogen-fixation by Free-living Micro-organisms

Biological N_2-fixation may also be a sensitive indicator of soil metal pollution (Smith, 1991) and apparently adverse effects of metals applied to soil in sewage sludges on the N_2 fixing activity of free-living blue-green algae and heterotrophic bacteria have been reported (Brookes *et al.*, 1984; Brookes *et al.*, 1986; Mårtensson and Witter, 1990). However, the contribution made by free-living N_2-fixing organisms to the N budget of agricultural systems is small and probably of little agronomic significance (Giller and Day, 1985). For example, inputs of N deposited on land from the atmosphere are as high as 40 kg N ha^{-1} y^{-1} (Goulding, 1990) and significantly outweigh those from biological fixation by free-living organisms (Powlson *et al.*, 1989). Furthermore, not all soils apparently support free-living N_2-fixation. Lorenz *et al.* (1992) measured N_2-fixation activity by free-living blue-green algae and heterotrophic bacteria in some soils, but little or no activity was detected in others at background soil concentrations of heavy metals. No apparent explanation could be given for the absence of fixation activity in certain uncontaminated soils and both methods were therefore rejected as biological indicators of metal toxicity. For metal mine spoil contaminated soils, Rother *et al.* (1982b) reported that there were no consistent effects on N_2-fixation activity of non-symbiotic, heterotrophic bacteria in soils containing up to 200 mg Cd kg^{-1} and 26 000 mg Zn kg^{-1}.

Symbiotic Nitrogen-fixation by Clover

In contrast to N_2-fixation by free-living organisms, symbiotic fixation can contribute an important source of N to agricultural soils. In particular, the white clover-*Rhizobium* symbiosis can supply as much as 200 kg N ha^{-1} y^{-1} (Robson *et al.*, 1989) and much research has therefore focused on the effects of heavy metals in sludge-treated soils on this process.

Nitrogen fixation by white clover (*Trifolium repens* L.) was completely absent on the sludge-treated plots at Woburn (McGrath *et al.*, 1988) and yields of white clover were reduced by up to 60% compared with FYM plots (MAFF/DoE, 1993a) apparently due to toxic effects of metals on *Rhizobium leguminosarum* biovar *trifolii*. *Rhizobium* isolated from the metal-contaminated soil were shown to be ineffective in N_2-fixation with white clover (Giller *et al.*, 1989; Chaudri *et al.*, 1992a). The ineffective isolates formed numerous, small, white nodules on clover roots compared to the fewer large and pink nodules characteristic of effective strains from the FYM plots. Plasmid conformity and reduced genetic diversity of ineffective rhizobia contrasted with the normal patterns of diversity found in FYM isolates (Giller *et al.*, 1989; Hirsch *et al.*, 1993). This suggested that the ineffective isolates of *Rhizobium* from nodules of white clover grown on the metal-contaminated soil represented a single strain. Inoculation of the contaminated soil with an effective strain of *Rhizobium* resulted in effective N_2-fixation, but this ability was lost over a two-month incubation period unless very large inoculum densities were used (Giller *et al.*, 1989). This indicated that effective *Rhizobium* strains were unable to survive, or at least remain effective, in the presence of sludge-derived heavy metals.

However, the extrapolation of the results from Woburn as a universal model of the potential effects of current and future sludge applications on soil micro-organisms has been questioned (Smith, 1991). This is because the metal concentration profile at Woburn is representative of soils which have received highly contaminated sludges in the past. It is emphasized that such soils contain significantly more Cd than the current soil limit for this element which may be toxic to the microbial biomass and symbiotic N_2-fixation compared with the concentrations of other PTEs present in the soil (Chaudri *et al.*, 1992a; Smith and Giller, 1992; Obbard and Jones, 1993). This is explained because sludge applied at Woburn in the late 1950s, for example, contained up to 160 mg Cd kg^{-1} (McGrath, 1984). However, the concentrations in sludge have declined significantly since then: the median value reported for Cd in sludge from a recent survey was 3.2 mg Cd kg^{-1} (CES, 1993).

The toxicity of Cd to N_2-fixation has been widely reported. McIlveen and Cole (1974) ranked the toxic effects of metals to nodule formation by red clover in the order Cd>Co>Cu>Zn. The order of decreasing toxicity to symbiotic

N_2-fixation is similar and has been ranked as Cd>Ni>Cu>Zn>Pb (McGrath *et al.*, 1988). Chaudri *et al.* (1992b) ranked the toxicity to *R. leguminosarum* bv. *trifolii* isolated from sludge-amended plots at Woburn as Cu>Cd>Ni>Zn in simple solution tests. In metal salt-spiked soil, however, the order was Cd>Zn>Cu due to differences in the bioactivity of metals in soils compared with solutions (Chaudri *et al.*, 1992a). No rhizobia survived at Cd concentrations $\geq$7.1 mg kg^{-1} in the spiked soil. It is unlikely that such metal salt additions reflect bioavailability in sludge-treated soils. However, the apparent 50% reduction in N_2-fixation activity reported earlier by McGrath *et al.* (1988) at soil concentrations of Zn and Cu approximating to their limit values, could be explained by the large Cd content of the soil of 10 mg Cd kg^{-1} which occurred in the 'cocktail' of metals present. The loss of effective rhizobia from metal-contaminated soil at Woburn, reported by Chaudri *et al.* (1992a), could similarly be explained by the large Cd content of the soil which was four times the limit value in this case.

Other data also implicate Cd toxicity in the absence of *Rhizobium* from sludge-treated soils. For example, the soil inoculation experiments described by Turner *et al.* (1993) demonstrated that metals had no effect on rhizobial populations in a sludged soil containing 383 mg Zn kg^{-1} and 1.3 mg Cd kg^{-1} compared with an untreated control soil, whereas the number of effective rhizobia surviving decreased in another soil with 306 mg Zn kg^{-1} and 9.8 mg Cd kg^{-1}. Other work by Giller *et al.* (1993) showed that a reduction in the numbers of *R. leguminosarum* bv. *trifolii* occurred in mixtures of sludge and FYM treated soil from Woburn containing as little as 3.4 mg Cd kg^{-1} after laboratory incubation for 171 days. This value is very close to the toxic threshold concentration of 4 mg Cd kg^{-1} reported by Chaudri *et al.* (1992a) for this element in soil from the FYM-treated plots at Woburn spiked with metal salts. It would appear unlikely that Zn was causing an effect because the Zn content of the soil mixture was only 127 mg kg^{-1} which is well below the level causing a reduction in rhizobial numbers in soil spiked with metal salts (Chaudri *et al.* 1992a) and in a sludge-treated field experiment low in Cd (Chaudri *et al.*, 1993). Soil from the FYM-treated plots at Woburn contained 96 mg Zn kg^{-1} and 2.4 mg Cd kg^{-1}.

Obbard and Jones (1993) did not detect rhizobia using an enclosed tube method for nodulation by white clover (Vincent, 1970) in sludge-amended arable soils containing Zn, Cu and Ni within their respective limit values, but with up to 4.5 times the maximum permissible concentration of Cd. In another more detailed survey of operationally sludged soils, Smith and Giller (1992) showed that only 50% of inoculations by the enclosed tube test were successful (indicating the presence of effective rhizobia) for an arable soil with large concentrations of Cd up to 7.6 mg Cd kg^{-1}, but containing less than the permitted levels of the other elements. This indicated that the population was probably reduced, but not completely eliminated as suggested by Obbard and Jones (1993). However, the available data indicate that detrimental effects of Cd on symbiotic N_2-fixation are probably unlikely to occur at the maximum permissible limit value of 3 mg Cd kg^{-1} in soil.

There is an increasing body of evidence which demonstrates that the apparent selection of an ineffective strain of *R. leguminosarum* bv. *trifolii* in sludge-treated plots at Woburn is probably unique to that specific site. *Rhizobium* isolated from a range of other sludge-amended soils all exhibit effectiveness in N_2-fixation with white clover (Smith *et al.*, 1990; Leung and Miles, 1991; Obbard *et al.*, 1992a; Smith and Giller, 1992; Smith *et al.*, 1992e; Leung *et al.*, 1993). Furthermore, the relative effectiveness of strains isolated from operationally sludged field soils is not influenced by soil metal content (Leung *et al.*, 1993). However, natural populations of *Rhizobium* show considerable variability in effectiveness in N_2-fixation (Gibson *et al.*, 1975), and ineffective or only partially effective strains normally occur in the environment. For example, Nutman and Ross (1970) estimated that 15% of the clover bacteria they isolated from Barnfield at Rothamsted were poorly effective or ineffective in N_2-fixation. Ineffectiveness is often associated with acid soil conditions (Holding and King, 1963; Fåhraeus and Ljunggren, 1968). The better survival of these strains in metal-contaminated soil compared with effective ones is thought to be due to their relatively high tolerance to toxic metals (Fåhraeus and Ljunggren, 1968). This may offer a possible explanation for the observations made by McGrath and co-workers at Woburn (McGrath *et al.*, 1988; Giller *et al.*, 1989; Chaudri *et al.*, 1992a,b). Interestingly, Hirsch *et al.* (1993) showed that the isolates of *R. leguminosarum* bv. *trifolii* from the sludge-treated plots, which were ineffective in N_2-fixation with white and red clover, could effectively nodulate subterranean clover (*Trifolium subterraneum*). It was not possible to determine whether strains resembling these isolates were present in the FYM-treated soil due to the dominance of the strains which formed effective nodules with white clover in this soil. One suggestion is that the characteristics of the surviving strain could have been acquired by mutation as heavy metals accumulated in the soil. Alternatively, these characteristics could be intrinsic to that rhizobial population possibly induced by some past influence of soil conditions at the site, such as acidification processes, which is conceivable given the potential vulnerability of the coarse-textured soil at Woburn. Consequently, the ineffective strain could have already been present in the soil at Woburn long before sewage sludge applications commenced.

Site specificity of the potentially detrimental effects of metals on *Rhizobium* was demonstrated by Smith and Giller (1992) from a monitoring exercise of the presence of *Rhizobium* in soils sampled from historic operationally sludge-treated fields without indigenous clover. Critical concentrations of metals which resulted in the complete absence of nodulation by white clover could not be determined across the range of metal levels measured at each site examined. Differences in the patterns of nodulation were only observed between the sites studied. The presence of *Rhizobium* in soils containing metals above the maximum permissible levels for sludge-treated soils suggested that tolerance to elevated metal conditions could develop without losing the ability to fix N_2 with white clover, possibly indicating the potential for adaptation to local site-specific conditions. From a study of the metal tolerance characteristics of *R. meliloti*, El-Aziz *et al.* (1991) suggested that the adaptation of isolates of *Rhizobium* to local conditions could occur resulting in divergent lines of the same species and put this forward as a possible explanation of the effects reported at Woburn. Indeed, different strains of *R. leguminosarum* bv. *trifolii*, effective in N_2-fixation can exhibit considerable differences in sensitivity to heavy metals. For example, the strains of *R. leguminosarum* bv. *trifolii* from diverse geographical origin tested by Angle *et al.* (1993) were all more resistant to Zn than isolates of the same species from FYM-amended soil at Woburn. Angle *et al.* (1993) suggested that the isolates from Woburn soil apparently represented a Zn-sensitive population of *R. leguminosarum* bv. *trifolii*, that may be responsible for the metal toxicity observed from the sludge-amended soils at this site.

Tolerance to toxic metals by *R. leguminosarum* bv. *trifolii*, effective in N_2-fixation with white clover, was shown by Leung *et al.* (1993) for 44 isolates from different sludge-amended and untreated control soils. The majority of the field isolates had an apparent higher tolerance to Cu than a standard laboratory strain (TA1) used for comparison. Three strains were more tolerant to Cd and 22 strains were more tolerant to Zn. A two- to three-fold increase in tolerance to metals was measured overall compared with the laboratory strain. However, there was no correlation between the heavy metal concentration in soil and the presence of metal-tolerant strains, except for one strain showing high tolerance to Cu which was isolated from soil containing increased amounts of Cu resulting from past applications of 'pot-ale'. Therefore, it appears that heavy metal concentration may be relatively unimportant compared to other physical, biological and chemical properties of soils which may indirectly confer tolerance to heavy metals and which select strains of *Rhizobium* best adapted to survival in a particular soil.

Obbard *et al.* (1992a) measured the N_2-fixation activity of clover by ^{15}N-dilution techniques in a mixed sward with ryegrass grown on soil from the Luddington site. The severity in adverse effects of heavy metals on N_2-fixation decreased in the order Zn>Cu>Ni although Zn also generally exceeded the soil limits in the Cu and Ni plots (Table 8.4). No significant difference in N_2-fixation was detected between the sludge-treated control and soil from the low Cu treatment which contained 254 mg Zn kg^{-1} and 147 mg Cu kg^{-1}. Both soil concentrations exceeded the UK limits for soil at pH 5.0-5.5. The rate of N_2-fixation was actually increased relative to the control in soil exceeding the limit value for Ni in the low Ni treatment at pH 5.2, and in the Cr contaminated soil. In the same soils, Smith *et al.* (1990) found that the high rate of Zn and Cu treatments markedly reduced the size of the rhizobial population more than Ni, and Cr had no effect compared to the sludged control. This suggested that Zn and Cu exerted a toxic effect at these extreme soil concentration values on the free-living bacteria released from root nodules in soil. The results confirmed the minimal toxicity of Cr due to the low bioavailability of this element in sludge-treated soil. However, it is arguable how useful data from the Luddington (and Lee Valley) sites are in estimating appropriate soil limits for the other elements of concern in sludge-treated agricultural land. This is because only two rates of heavy metals were supplied to the plots thus preventing the determination of the minimum concentrations at which metals may adversely affect soil micro-organisms (MAFF/DoE, 1993a). Even more importantly, however, the concentrations of Zn, Cu and Ni exceeded the respective limit values in soil at Luddington for both the low- and high-metal treatments due to the low pH value of the soil (Table 8.4). These combinations of metal concentrations, low soil pH and experimental conditions (see Chapter three) constituted an extreme example of soil contamination with heavy metals from sewage sludge application. The *Code of Practice for Agricultural Use of Sewage Sludge* (DoE, 1989a) recommends extreme caution in using sewage sludge on soil at pH<5.2 and that sludge applications should be made only in accordance with specialist agronomic advice under those circumstances. Marks *et al.* (1980) alluded to the vulnerability of Luddington soil to acidification. Presumably only minimal management inputs to the trial had occurred in the years preceding the work reported by Obbard *et al.* (1992a) and Smith *et al.* (1990) due to the impending closure of the site shortly afterwards. Indeed, in another batch of soil samples (Obbard *et al.* 1992a) the soil pH had declined still further with a mean value of 4.9 and range of 4.4-5.2. Consequently, the 'idea' (MAFF/DoE, 1993a; McGrath, 1994) that rhizobia may be continually killed in sludge-treated soil in practice on the basis of the data obtained from the Luddington experiment is purely conjecture.

Severe reductions in the yield of red clover at Luddington were reported previously on the high Zn plots at 455 and 511 mg Zn kg^{-1} in soil (Jackson, 1985) and there have been attempts to link this yield decline with the toxic effects of metals on rhizobial populations in the soil (McGrath, 1994; McGrath *et al.*, 1994). However, at these concentrations in sludge-treated soil, Zn was phytotoxic to clover grown in pots in fertilized soils at pH ≥6.0 (Carlton-Smith *et al.*, 1985; Carlton-Smith *et al.*, 1987; Sanders *et al.*, 1987). In any case, the maximum permissible concentration of Zn in sludge-treated agricultural land at pH 6.0-7.0 (SI, 1989a) was exceeded at Luddington. Interestingly, there was no decrease in yield measured at 238 mg Zn kg^{-1} in Luddington soil (Jackson, 1985). The higher availability and toxicity to crops of heavy metals in the Luddington experimental soil has been demonstrated (Carlton-Smith *et al.*, 1987) compared with soils treated in the long-term with sludge in practice.

The leaf Zn concentration in red clover grown on the high Zn plots at Luddington was 301 mg Zn kg^{-1} (Jackson, 1985). This is at the top end of the upper critical concentration range for Zn of 250-300 mg kg^{-1} in plant leaves considered appropriate for clover (Boawn and Rasmussen, 1971; Macnicol and Beckett, 1985) suggesting that phytotoxicity could explain the apparent effects

Table 8.4. Some chemical properties of soils at the Luddington experimental site in the UK.

Treatment	pH	Organic matter (%)	Total concentration (mg kg^{-1}) Zn	Cu	Ni	Cr
Control (unsludged)	5.5	3.4	92	27	26	61
Control (sludged)	5.3	4.3	112	33	48	78
Low Zn	5.0	5.4	**297**	32	31	81
High Zn	5.0	4.5	**542**	34	48	117
Low Cu	5.1	5.0	**254**	**147**	30	135
High Cu	5.0	6.7	**249**	**235**	35	160
Low Ni	5.2	5.9	121	30	**63**	144
High Ni	5.4	5.3	**230**	35	**153**	168
Low Cr	5.7	3.8	99	32	25	230
High Cr	5.6	3.6	143	74	31	331

Values in bold exceed UK soil limit values (SI, 1989a)
Source: Smith *et al.* (1990)

of Zn on clover yield at Luddington. However, Sanders *et al.* (1987) reported that the critical leaf Zn concentration for clover was 400 mg Zn kg^{-1} (Chapter three). Interestingly, the corresponding total soil concentration for this element was 250 mg Zn kg^{-1} in their pot trial. Normally, the metal content in tissues of forage and arable crop plants is considerably smaller than the total concentration present in sludge-treated soil (Cottenie *et al.*, 1984; Davis and Carlton-Smith, 1984; Carlton-Smith, 1987).

For example, Davis and Carlton-Smith (1984) estimated that the upper critical concentration of Zn in ryegrass was 140 mg Zn kg^{-1} whereas the corresponding total concentration in soil was 319 mg Zn kg^{-1}. The reason why this fundamental relationship was apparently inverted for a range of crops grown by Sanders *et al.* (1987) was probably due to the problems associated with glasshouse experimentation with small pots (Logan and Chaney, 1983). This would suggest that the critical concentration estimated for Zn in clover may be atypically large. Indeed, the threshold level of Zn measured in red beet was as high as 900 mg Zn kg^{-1} in the same trial, which would appear to be an excessive amount of Zn in the plant, given that red beet is particularly sensitive to Zn accumulation (Davis and Carlton-Smith, 1980) even taking a liberal upper critical tissue concentration (Table 3.3). It is questionable, therefore, whether the critical value for clover is also appropriate for extrapolation to field-grown crops.

In another field trial at Gleadthorpe Experimental Husbandry Farm in the UK, remarkable reductions in the yield of clover have occurred in soil amended with sludges prepared from sewage spiked with metal salts (Royle *et al.*, 1989). The sludges were applied in 1982. However, recoveries of the added metals were very poor (35-60% an average to 90 cm for Zn in selected plots) in comparison with the target concentrations in soil and no plausible explanation could be found for the apparent losses (Unwin *et al.*, 1989). Indeed it was necessary to supplement a number of the plots with further applications of sludge, highly contaminated with Zn and Cu (no data were given on sludge concentrations), to make up the difference in 1986. White clover was sown in 1988 after a barley crop. Clover yield declined in soil containing more than 170 mg Zn kg^{-1} and 100 mg Cu kg^{-1} and the crop failed altogether on two plots with soil Zn concentrations of 214 and 378 mg kg^{-1}. The concentration of Zn in herbage (220 mg Zn kg^{-1}) was approaching the potentially phytotoxic range for a soil containing 325 mg Zn kg^{-1}. Interestingly, crop failure only occurred on plots treated with the second application of high Zn sludge suggesting that the sludge itself was potentially phytotoxic rather than having an indirect effect on yield due to impacts on N_2-fixation. Royle *et al.* (1989) also suggested that the application of the fresh sludge might in fact explain the apparently anomalous result. Indeed, sufficient residual N to support clover growth would still be anticipated to remain in the soil (Johnston *et al.*, 1983) on the basis of the normal diminishing return of available N residues following the application of sewage sludge to agricultural land (Hall, 1985) even if N_2-fixation were deficient (McGrath *et al.*, 1988). In addition, phytotoxicity cannot be fully discounted for the failed crop because there was no plant material available for chemical analysis. The absence of an apparent yield effect on barley could be explained because this crop is more tolerant to heavy metals in sludged soil compared with clover (Carlton-Smith *et al.*, 1987; Sanders *et al.*, 1987).

On balance, the suggestion by McGrath *et al.* (1994) that metal effects on clover yields at Luddington and Gleadthorpe are due to an effect on N_2-fixation, rather than phytotoxicity, because there was no phytotoxicity in clover apparent at similar total concentrations of metals in the sludge-amended soil at Woburn, would appear to be highly speculative. This is because comparisons based on total metal concentrations do not account for actual differences in the chemical forms of metals in soil which influence their potential toxicity to plants and soil organisms (McGrath, 1994). To the contrary, there would appear to be a reasonable amount of evidence suggesting that phytotoxicity was a strong possibility in Zn contaminated soils at these experimental sites.

Stark and Lee (1988) used an *in situ* acetylene reduction technique to measure N_2-fixation by white clover growing in the sludge-treated plots at the Cassington experimental site. A 15% reduction in activity was detected for soil containing less than the permitted maximum amounts of heavy metals in sludge-treated agricultural land in the UK (SI, 1989a) compared with an untreated control. The rate of N_2-fixation decreased by 50% in soil containing 300 mg Zn kg^{-1} and 6.0 mg Cd kg^{-1} (Cd concentration estimated from Carlton-Smith, 1987) at pH 5.7. The UK limit for Zn in soil at pH 5.5-6.0 is 250 mg Zn kg^{-1}. In contrast, there was no effect on N_2-fixation at 343 mg Zn kg^{-1} in a historically sludge-treated soil monitored in the same study.

The increased severity of toxic effects of metals on N_2-fixation at the experimental site compared with the operationally sludge-treated soil may be attributed to differences in soil type. It may also be due to the higher availability of metals in soils receiving large doses of sludge on an experimental basis compared with the long-term treatment of soil with sludge in practice (Carlton-Smith and Stark, 1987). However, considerable caution is necessary in interpreting N_2-fixation rates measured by the acetylene reduction assay due to the possibility of other interfering factors influencing the usefulness of the results (Witty and Minchin, 1988).

The Independent Scientific Committee reviewing soil fertility aspects of heavy metals (MAFF/DoE, 1993a) concluded that the soil limits for Zn needed reviewing in light of the possible effects on soil microbial processes and particularly with respect to symbiotic N_2-fixation. The strongest evidence which is available indicating potentially a toxic effect of Zn at the UK and EU upper limits (CEC, 1986a; SI, 1989a) on *R. leguminosarum* bv. *trifolii* is from the sludge experiment at Braunschweig in Germany (Chaudri *et al.*, 1993; Obbard *et al.*, 1993a). Even with 300 m^3 y^{-1} of unamended sludge, rhizobial numbers decreased by several orders of magnitude compared to the unsludged control soil. The minimum concentrations of Zn in the low and high pH soils which reduced the population of *R. leguminosarum* bv. *trifolii* were 130 mg kg^{-1} and 200 mg kg^{-1}, respectively (McGrath, *et al.*, 1994). However, it could be argued that the very large rates of sludge application may also have contributed to the observed deleterious effects on soil micro-organisms (MAFF/DoE, 1993a). Furthermore, there are no published data available for the Braunschweig trial dealing with the basic aspects of metal uptake by crops or crop yields. It would be interesting to examine these parameters, in relation to the total concentrations of heavy metals present in the soils, to determine if the general behaviour of the metals in the sludge-amended experimental soils at Braunschweig compare with other well characterized sludge-treated soil systems.

In an overall assessment, the effects of Zn on *Rhizobium* in particular may require further scrutiny. On the other hand, the maximum concentrations permitted for the other elements which occur in elevated amounts in sludge-treated soil may adequately protect symbiotic N_2-fixation by white clover in sludge-treated agricultural soil. However, the widely reported presence of effective *Rhizobium* even in soils contaminated well above statutory metal limits (Obbard *et al.*, 1992b; Smith, 1992; Smith *et al.*, 1992e; Leung *et al.*, 1993) suggests that nodulation and N_2-fixation would probably occur on introducing white clover to sludge-treated fields at the soil limit for Zn. Once inside the root nodule, however, rhizobial bacteroids are largely protected by host plant tissue from the outside environment and the potential toxic effects of metals (Giller *et al.*, 1989), allowing the full potential of the symbiotic relationship to develop. Thus, the agronomic significance of potentially detrimental effects of sludge application to farmland on soil rhizobial populations, and concomitantly on symbiotic N_2-fixation with white clover, has not yet been established in practice. Interestingly, there have been no reports by the farming community that yields of clover have been compromised due to sewage sludge application. This is in spite of the fact that organized sludge recycling to agriculture has been operating since the late 1950s, although sludge was used on farmland long before this date (Davis, 1989). However, seed inoculation with *Rhizobium* offers an inexpensive means of ensuring successful establishment of N_2-fixation activity (Crush, 1987), but should not be regarded as a universal solution to the problem if innate aspects of soil fertility are being impaired irrevocably by metal additions to soil in sludge.

Symbiotic Nitrogen-fixation by Other Legume Crops

Compared with clover there are few experimental data reported of the potential effects of heavy metals in sludge-treated soils on other important legume crops such as peas or beans. Obbard *et al.* (1992a) assessed nodulation, N_2-fixation activity and yield of peas (*Pisum sativum*) and beans (*Vicia faba*) grown in the Luddington soils, and the nodulation and yield potential of beans grown in soil from the Braunschweig experimental site in a controlled environment study. These crops had not been grown previously at the experimental sites and effective nodulation depended on indigenous populations of the symbiont *Rhizobium leguminosarum* bv. *viciae* occurring in the soils. Nodulation declined in Luddington soils exceeding the permitted metal concentration values compared with control soils, but there was no significant correlation found between plant yield and metal content in soil. Field beans and peas showed active symbiotic N_2-fixation even in acid soils (pH 4.9) containing 442 mg Zn kg^{-1}. It is emphasized that sewage sludge should not be applied to soil with pH<5.0 and the Luddington samples tested represent extremes of soil contamination and pH value (which was as low as pH 4.4 in one case) not permitted under the regulations (SI, 1989a). In the Braunschweig soils, nodulation increased as soil pH value was raised. However, no significant correlation between plant yield or soil chemical properties was found, although nodule number was reduced with increasing metal content in soil. Consequently, the apparent effects of metals in sludge-treated soils on the extent of nodulation may be of little overall agronomic significance. There is other evidence, for example, that decreased nodule number on roots of legume crops in the presence of heavy metals is compensated for by an increase in nodule size and higher leghaemoglobin content (Leung and Chant, 1990). Thus, Obbard *et al.* (1992a) detected only small changes in N_2-fixation activity of field beans grown in Luddington soils despite the extreme contamination of these soils with Zn and Cu at low pH. This also supports the suggestion (McGrath, 1994) that the *R. leguminosarum* bv. *viciae* may be less sensitive to metals in sludge-treated soil than the clover *Rhizobium*.

Different species of symbiotic N_2-fixation organisms vary widely in their sensitivity to heavy metal contamination in soil. For example, Giller *et al.* (1993) have reported the greater tolerance of *Rhizobium meliloti* (host plant: *Medicago sativa* (lucerne)) to heavy metals compared with *R. leguminosarum* bv. *trifolii* and *R. loti* (host plant: *Lotus corniculatus*) which were equally sensitive to metal contamination when inoculated into sludge-treated soil from Woburn. Indeed, El Aziz *et al.* (1991) found little relationship between the heavy metal concentration and growth and survival of *R. meliloti* in soil adjacent to a Zn smelter (containing up to 1540 mg Zn kg^{-1}, 31.8 mg Cu kg^{-1} and 32.4 mg Cd kg^{-1}) due to the intrinsic level of metal tolerance exhibited by this species of rhizobia. *Bradyrhizobium japonicum* (host plant: soybean) tolerated five times the amount of Zn in an artificial growth medium compared with strains of *R. leguminosarum* bv. *trifolii* (Angle *et al.*, 1993). The isolates of *R. leguminosarum* bv. *phaseoli* (host plant: *Phaseolus* beans) and *R. leguminosarum* bv. *viciae* (host plants: peas and *Vicia* beans) examined by Angle *et al.* (1993) were broadly comparable with *R. leguminosarum* bv. *trifolii* in sensitivity to heavy metals in artificial media.

Reddy *et al.* (1983) observed a decline in the population of *B. japonicum* in soils freshly amended with large rates of sludge and suggested that sludge-borne heavy metals may be the cause. However, it appears that soluble salts applied to soil in sludge, and not heavy metals, were responsible for the reduction in bradyrhizobial populations (Madariaga and Angle, 1992). The application of very high rates of sludge to soil may reduce the population of *B. japonicum* due to the adverse effects of soluble salts (Singleton *et al.*, 1982), although the effects are only transient under field conditions.

Vesicular-arbuscular Mycorrhizae

Vesicular-arbuscular mycorrhizae (VAM) are fungus-root associations formed between fungal root endophytes and plants including most agricultural crops and are responsible for enhancing the overall phosphorus (P) nutrition of the host plant (Smith, 1991). Improved P status of legumes through root mycorrhizal infection increases growth, nodulation and N_2-fixation by the crop (Barea and Azcon-Agiular, 1983). Mycorrhizae also increase crop yields due to enhanced uptake of essential trace elements when these are deficient in soil, improved plant resistance to soil-moisture stress, modifications in plant hormonal balance, and greater resistance to plant pathogens (Lambert *et al.*, 1979; Dehne, 1982; Hayman, 1983).

Exposure of VAM to elevated heavy metal concentrations in soil can significantly reduce the level of host plant infection and P uptake (Gildon and Tinker, 1983a,b). Adverse effects can be exacerbated by low soil pH conditions which increase the availability of most metals in soil (Angle and Heckman, 1986). However, Koomen *et al.* (1990) demonstrated that metal-tolerant strains of VAM persisted in sludge-treated soil at Woburn, which formed normal associations with clover roots, although the rate of root infection was reduced compared with FYM plots. Obbard *et al.* (1993b) detected VAM infection of an *in situ* wheat crop in a highly contaminated soil from a historically sludge-amended site containing over 5000 mg Zn kg^{-1}, 750 mg Cu kg^{-1}, 360 mg Ni kg^{-1} and 20 mg Cd kg^{-1} and concluded that VAM fungi were tolerant to the presence of heavy metals. Furthermore, VAM infection was not correlated in general with any soil or plant chemical properties of samples taken from the sludge-amended field, although a lower level of infection was noted overall compared with untreated soil. Other studies by Obbard *et al.* (1993b) with soils from the Braunschweig experiment, using onions as a test species (onions are particularly sensitive to root infection by VAM), showed that the extent of infection was very low in both sludge-amended and untreated soils over the ten-week duration of the investigation. In contrast, Koomen *et al.* (1990) reported significant levels of VAM infection of clover roots after six-eight weeks growth of clover. Soil pH value alone had an important effect on VAM infection in Braunschweig soil (Obbard *et al.*, 1993b), and onions grown in untreated high-pH soil (pH 6.8) had a higher incidence of infection than for the equivalent low-pH treatment (pH 5.8). However, the direct toxicity of metals on the VAM symbiosis could not be clearly demonstrated because the patterns of VAM infection reflected plant yield which were influenced indirectly by the possible phytotoxic effects of metals on the host plant. In addition, sludged soils often contain much available phosphate which is known to inhibit or reduce VAM infection (Pera *et al.*, 1982) such that potentially reduced infection, resulting from elevated metal levels, is unlikely to affect plant growth (Koomen *et al.*, 1990).

Recent work examining VAM infection in grassland following long-term slurry application showed that mycorrhizal infection decreased with increasing Zn concentration in soil and decreasing soil pH value (Christie and Kilpatrick, 1992). The concentration of Zn in animal slurries may be comparable or higher than the levels of Zn present in sewage sludge (see Chapter two). This would suggest that regulations designed to protect the environment from potentially harmful effects of heavy metal contamination of soil should also encompass the application of farm wastes to agricultural land.

Invertebrate Macrofauna

Soil animals play an important role in breaking down plant remains and the larger animals, particularly earthworms, are effective in mixing the surface organic matter into the soil. Material which has passed through the gut of soil animals is more readily colonized and decomposed by the soil microbial community and the rate of mineralization is increased (Satchell *et al.*, 1984; Newman, 1988; Parkin and Berry, 1994; Siepel and Maaskamp, 1994). Soil fauna also assist in the preservation of soil physical characteristics. The activity of the soil fauna is therefore essential to the overall fertility of agricultural soils.

Elimination of detrivorous invertebrates due to metal toxicity could ultimately reduce the rate of decomposition of plant residues returned to the soil (Hopkin, 1989). Furthermore, soil animals form an important link in the food chain of natural ecosystems discussed in the following section. In contrast to soil micro-organisms, comparatively few studies have investigated the potential effects of heavy metals applied to soil in sewage sludge on invertebrate macrofauna.

The action of earthworms is particularly important in the maintenance of soil fertility by digesting organic matter into forms more available to soil micro-organisms and through improving soil physical characteristics due to mixing and aerating the soil as well as improving water permeability (Ash and Lee, 1980; Newman, 1988). Interestingly, however, a recent review (Palzenberger, 1995) has suggested that it is not possible to draw any firm conclusions concerning metal toxicity to earthworms in soil from the available data-base of information.

Earthworms are classified into three types based on life-style and burrowing habit (Bouche, 1972). Epigeal forms (e.g. *Lumbricus rubellus* and *Eisenia foetida*) hardly burrow in soil at all, but inhabit decaying organic matter on the surface including manure or compost heaps. Indeed, *E. foetida* has potential for use in the management of sewage sludges (Hartenstein *et al.*, 1981). Endogenous species (e.g. *Allolobophora chlorotica* and *A. caliginosa*) produce shallow branching burrows in the organo-mineral layers of the soil. Anectic forms (e.g. *Lumbricus terrestris, Allolobophora longa, A. nocturna* and *Octolasion cyaneum*) are deep burrowing species, producing channels to a depth of one metre or more.

No observed effect concentration (NOEC) values for effects of Cd and Zn on the growth and reproduction of *Eisenia andrei* were <10 mg kg^{-1} and 320 mg kg^{-1} for these elements, respectively, following a recommended test protocol (OECD, 1984) using an artificial soil spiked with metal salts (van Gestel *et al.*, 1993). Adopting the same procedure, Spurgeon *et al.* (1994) measured LC_{50}, EC_{50} and NOECs for *E. foetida* exposed to heavy metals. The estimated NOEC values were 39.2 mg Cd kg^{-1}, 32 mg Cu kg^{-1}, 1810 mg Pb kg^{-1} and 199 mg Zn kg^{-1} in this study. The NOECs for Cu and Zn were less than the soil limits for these metals in sludge-treated agricultural land (SI, 1989a). However, the OECD standard test was shown to significantly overestimate the potential effects of metals on earthworm populations, since earthworms were found in metal contaminated soils close by (1 km) to a smelting works study site. The concentrations of metals in the contaminated field soils were 300 mg Cd kg^{-1}, >683 mg Cu kg^{-1}, >4480 mg Pb kg^{-1} and >1010 mg Zn kg^{-1}. The discrepancy between the test and field observations was probably explained due to the greater availability of the metals in the artificial soil.

The concentrations of heavy metals in salt-spiked sludges reported by Hartenstein *et al.* (1981), that were potentially inhibitory to growth of *E. foetida* colonizing the sludge, were also considerably larger than those obtained from the standard soil test (Table 8.5). Manganese, Cr and Pb had no toxic effect even at the highest concentrations of 22 000 mg Mn kg^{-1}, 46 000 mg Cr kg^{-1} and 52 000 mg Pb kg^{-1} added. However, this study is also potentially open to criticism along with others where earthworms have been exposed to metal-contaminated food material such as sewage sludge overlaying a pollutant-free soil. This is because earthworms are able to avoid polluted soil, so it seems likely in these studies that the earthworms were able to escape from the metal contaminated food into the underlying clean soil, and hence were not exposed continuously to elevated concentrations of metals (Spurgeon *et al.*, 1994). The toxic threshold concentrations measured for both Cu and Zn by Hartenstein *et al.* (1981) exceeded the maximum UK limits for these elements in sludge-treated soil.

Table 8.5. Minimum toxicity threshold concentrations of heavy metals which may inhibit growth of *E. foetida* in sewage sludge.

Element	Toxic threshold (mg kg^{-1} dry solids)
Co	300
Hg	480
Cu	1100
Ni	1200
Zn	1300
Cd	1800

Source: Hartenstein *et al.* (1981)

Mitchell *et al.* (1978b) found that introduced *E. foetida* thrived in sludge-treated soils containing up to 370 mg Cu kg^{-1} and 1101 mg Zn kg^{-1} in the surface layers. Differences in the biomass of worms in the soils depended on organic matter content: low biomass was associated with low soil organic content. Andersen (1979) also reported an increase in the number and biomass of the species *A. longa* and *L. terrestris* in response to the addition of sludge organic matter. Other species were suppressed in the sludge-treated plots compared with FYM and slurry and this was attributed to the effects of metals also applied in the sludge. However, the numbers of worms (all species) isolated from control plots receiving only K and NPK fertilizers were actually smaller than from sludge-amended soil indicating that factors other than metals were probably responsible in this case. Wade *et al.* (1982) estimated that *L. terrestris* were about 25 times more plentiful and were generally larger in control soil than in soil which had received an application of 224 t ha^{-1} of metal-contaminated sludge (dry solids). It was considered that this could be due to metal toxicity or to the presence of organic constituents in sludge, such as factory oils, which may repel earthworms.

Despite the apparent inability of earthworms to accumulate Cu compared with other elements such as Cd, the metal is considerably more toxic (Hopkin, 1989). Thus, Cu concentrations of about 100 mg kg^{-1} in soils treated with pig slurry inhibited reproduction in earthworms in grassland (Van Rhee, 1975; Curry and Cotton, 1980). Nielson (1951) observed from an accidental spill of copper sulphate to grassland that earthworm populations were reduced by half at a concentration of 150 mg Cu kg^{-1} in the surface soil, with an almost total elimination at 260 mg Cu kg^{-1}. Van Rhee (1967) reported a gradual decline of earthworm populations in an orchard soil containing 95 mg Cu kg^{-1} accumulated in the upper soil layer as a consequence of past spraying with Cu-based fungicides. Van Rhee (1969) later indicated that 50 mg Cu kg^{-1} applied in salt form was sufficient to interfere with reproductive activity in earthworms. Ma (1982) showed that both Cu and Cd, applied to soil as metal salts, were much more toxic to *L. rubellus* than Pb or Ni in a sandy loam soil (Table 8.6). However, Cu and Cd only became toxic at soil concentrations exceeding the current maximum limits for sludge-treated soil (SI, 1989a).

Table 8.6. Toxicity of metals to *Lumbricus rubellus* in a sandy loam soil after 12 weeks.

Rate (mg kg^{-1} dry soil)	% Mortality			
	Cd	Cu	Pb	Ni
Control[(1)]	12	12	13	13
20	12	8	30	10
150	3	8	13	23
1000	100	100	17	40
3000	100	100	20	100

[(1)] Uncontaminated soil contained 0.5 mg Cd kg^{-1}, 12 mg Cu kg^{-1}, 26 mg Pb kg^{-1} and 17 mg Ni kg^{-1}
Source: Ma (1982)

Not surprisingly, Cr(VI) is even more toxic to earthworms than Cu. A concentration of only 10 mg Cr kg^{-1} in soil contaminated by effluent containing Cr(VI) from a tannery was fatal to *Pheretima posthuma* and other species (Soni and Abbasi, 1981; Abbasi and Soni, 1983). The critical concentration value for Cr(III) of 32 mg kg^{-1} reported by van Gestel *et al.* (1993) for *E. andrei* in an artificial soil spiked with metal salts contrasts with the earlier work of

Hartenstein *et al.* (1981) which effectively showed there was no adverse effect of Cr on *E. foetida* in metal salt-amended sludges. This may reflect the differences in sensitivity to metals between species of earthworm (Hopkin, 1989), although the higher bioavailability of Cr in artificial media compared with sewage sludges and sludge-treated soils is also a strong possibility. Chromium occurs only as Cr(III) in sewage sludges and sludge-treated soil and is tightly bound in residual, unavailable forms (see Chapter three) limiting its toxicity to soil organisms and plants.

Soil factors also influence the extent of metal toxicity to earthworms. For example, Ma (1984) reported significant decreases in cocoon production by *L. rubellus* at two to six weeks at sublethal concentrations of Cu in the range 100 to 150 mg Cu kg^{-1} in an acid sandy soil (pH 4.8 [KCl]). However, raising soil pH value by lime addition reduced the toxicity of Cu and no significant effect of Cu on cocoon production, litter breakdown or body weight gain was detected in soil containing 148 mg Cu kg^{-1} at pH 6.0 (KCl).

Relatively few studies have assessed effects of metals in sludge-treated soils on other soil animals. However, a comprehensive review of the general ecophysiology of metals in terrestrial invertebrates has been published (Hopkin, 1989). Mitchell *et al.* (1978b) found that mite and collembola populations in sludge-amended soil were low, but similar to populations in arable soils. Soils which were initially low in bacterial feeder nematodes showed a rapid increase in these worms following sludge addition in response to the increased bacterial population. Rhabditidae became the most abundant family. Enchytraeid worms also increased in sludge-amended soils, but at a slower rate than the nematodes. The status of the plots converged over time indicating that the populations of soil animals were not impacted detrimentally by the application of metal-contaminated sludge.

Nematodes have also been studied in detail by Weiss and Larink (1991) in soils which had received 12 t ha^{-1} y^{-1} of metal salt-spiked and unspiked sludges for nine years. Soil treated with spiked sludge contained 102 mg Cu kg^{-1}, 360 mg Zn kg^{-1} and 2.75 mg Cd kg^{-1} and thus approximated to the UK limits (SI, 1989a). Total nematode numbers were highest in soil treated with the spiked sludge and were lowest in the untreated control indicating that nematodes generally tolerate elevated soil metal conditions. However, the population composition changed due to the increased availability of nutrients following sludge amendment. The plant-feeding nematode genera showed different patterns of abundance depending on the sludge treatment and heavy metal content of the soil. For the mycophagic and bacteriophagic nematodes, numbers increased with the amount of sludge, especially in the sites with higher heavy metal content. Weiss and Larink (1991) explained that this was probably caused by the increased microbial biomass in the sludge-treated soil. The Rhabditidae became the most numerous group in the high-metal soil because they are insensitive to heavy metals and are well adapted to saprobic conditions in sludged soil, which is in agreement with the earlier work of Mitchell *et al.*

(1978b). In contrast, omnivorous nematodes were very rare in the sludge-treated plots and were completely absent from the high-metal soil. This could be explained by the possible reduction in algae and cyanobacteria in sludge-treated soil which may be an obligate requirement in the diet of these nematodes. Alternatively, the lack of omnivores may be because they were deterred by the changes in soil physical characteristics which occurred following sludge addition. Omnivores are particularly sensitive to heavy metals and Weiss and Larink (1991) suggested that the distribution of the genus *Ecumenicus* could be used as a bio-indicator of metal toxicity in contaminated soils. Certain predatory nematodes are highly resistant to metals and were only isolated from sludge-treated soil.

Glockemann and Larink (1989) reported that the population size of soil mites was raised following the application of sewage sludge to soil. An application rate of 5 t ha^{-1} (dry solids) of sludge also increased species diversity, whereas a larger rate of 15 t ha^{-1} reduced diversity compared to an untreated control. Not surprisingly, sludges spiked with metal salts reduced the number of species present, although population size was larger (23 900 mites m^{-2}) in high-metal soil containing 353 mg Zn kg^{-1}, 92.7 mg Cu kg^{-1} and 2.08 mg Cd kg^{-1} compared with control soil (5700 mites m^{-2}) at background metal levels.

Bengtsson and Tranvik (1989) suggested that the maximum allowable concentrations of heavy metals to protect forest soil invertebrates were 100-200 mg Pb kg^{-1}, <100 mg Cu kg^{-1}, <500 mg Zn kg^{-1} and 10-50 mg Cd kg^{-1} in soil. These critical levels considered the potential effects of soil contamination on abundance, diversity and life history parameters of invertebrate animals in forest soils and were regarded as conservative and tentative.

The benefit of sewage sludge application to soil on soil fauna was demonstrated by Ducommun and Matthey (1989). Sludge application (50 m^3 ha^{-1} y^{-1} for eight years) increased the number of decomposers in soil compared with a mineral fertilizer control. The Diptera particularly benefited from the sludge treatment: these organisms are actively involved in incorporating sludge into the soil. Predator and parasitic groups consequently increased in numbers in response to the larger population size of decomposers, thus maintaining the species balance in the soil. Ducommun and Matthey (1989) concluded, therefore, that in the medium term (5-10 years), at least, sludge does not have any toxic effects on soil macroinvertebrates.

Available data suggest that soil invertebrates are generally more tolerant to soil contamination with heavy metals compared with certain soil microbial processes. For example, earthworms would appear to be more seriously impacted by soil cultivation practices than the presence of heavy metals in soil and Newman (1988) considered that the problems of soil contamination with metals from spreading sewage sludge on farmland would arise more through effects on crops rather than from effects on earthworms. As with soil micro-organisms, many groups of soil invertebrate macrofauna remain unstudied in sludge-treated soils, but the predominance of metal-tolerant forms is likely to ensure that soil function and fertility is maintained. However, soil animals form an important link in the food chain of natural ecosystems and the assimilation by them of toxic elements, and the potential impacts on the environment of transferring heavy metals to higher trophic levels, are considered in the next section.

Natural Ecosystems

Terrestrial ecosystems may be impacted by the potential transfer and bioconcentration of PTEs from contaminated soils through food chains (Roberts and Johnson, 1978). Many reports demonstrate that earthworms readily accumulate heavy metals into their body tissues (Andersen, 1979; Helmke *et al.*, 1979; Hartenstein *et al.*, 1980; Beyer *et al.*, 1982; Ma, 1982, 1987) and their use as indicators for monitoring the effects of sludge disposal on terrestrial communities during secondary succession has been recommended (Kruse and Barrett, 1985; Levine *et al.*, 1989).

Earthworms have a particularly high affinity for accumulating Cd. For example, Beyer *et al.* (1982) reported concentrations of Cd as high as 100 mg kg^{-1} (dry matter) in earthworms from sludged soil containing 2 mg Cd kg^{-1} compared with 8.6 mg Cd kg^{-1} in earthworms from an untreated control plot (0.14 mg Cd kg^{-1} in soil). However, the mean concentration of Cd in earthworm tissue determined by Kruse and Barrett (1985) was larger than this value at 135 mg Cd kg^{-1} for sludge-amended plots containing only 1.3 mg Cd kg^{-1} on average in soil. Cadmium concentrations were 9-11 times greater in earthworms from sludge-treated plots than those from control or fertilizer only plots in these studies, demonstrating the significant bioconcentration of Cd by earthworms which may potentially be hazardous to wildlife that eat worms.

According to Beyer *et al.* (1982), Zn is also accumulated by earthworms relative to soil concentrations, whereas Cu, Pb and Ni are not. Helmke *et al.* (1979) showed that concentrations of Cd, Cu and Zn in earthworms increased, whereas Se content decreased, with increasing rate of sludge application. The effect on Se may be explained due to antagonism between Se and heavy metals. In contrast, Hg and Cr were not bioavailable to earthworms. Levine *et al.* (1989) reported the bioconcentration of Cd, Cu and Zn at the detritivore trophic level (earthworms) in an established field community which had been treated with sludge annually over a period of ten years. However, the small concentration factors for Cu and Pb (1.2 and 0.4, respectively) measured by Kruse and Barrett (1985) suggested the possible biological regulation of these metals by earthworms. Furthermore, Kruse and Barrett (1985) detected no significant accumulation of Zn in earthworms collected from sludge-treated soil containing 137 mg Zn kg^{-1} compared with untreated and fertilizer controls. Roberts and Johnson (1978) also concluded that the Zn status of animal tissue appears to be effectively regulated and a homoeostatic situation maintained irrespective of dietary intake even in severely polluted ecosystems (≥11 000 mg Zn kg^{-1}). This would also appear to be the case with Zn for higher animals grazing sludge-treated pasture (Stark *et al.*, 1995).

The bioconcentration of heavy metals may impact the environment because earthworms are an important food resource for several bird and small mammal species. Contrary to expectations, however, Levine *et al.* (1989) did not detect any increase in the metal concentrations in voles inhabiting a field community regularly treated with sewage sludge. On the other hand, Ma (1987) showed that the accumulation of Cd, Pb and Zn in moles reflected the bioavailability of these metals to earthworms in soils contaminated by aerial deposition of metals from smelters. The concentrations of metals present in tissues of earthworms and moles did not necessarily reflect the extent of soil contamination with metals, presumably because bioavailability is also a function of other components in the diet which modulate absorption of toxic metals by animals (Roberts and Johnson, 1978). In addition, other soil chemical properties such as pH also exert a strong influence on bioavailability. Thus, in acidic sandy soils (pH 5.2; KCl) Cd may accumulate in earthworms to a considerable extent and critical levels of Cd toxicity in moles can be exceeded (110-260 mg Cd kg^{-1} is reported as a critical toxic range in kidneys of vertebrates (Nicholson *et al.*, 1983)) even when the Cd content in soil is relatively low (1.8 mg Cd kg^{-1}) (Ma, 1987). Increasing soil pH value and organic matter content may reduce metal accumulation by earthworms (Ma *et al.*, 1983). However, Beyer *et al.* (1982) found that liming soil only decreased Cd concentrations in earthworms slightly and had no effect on the absorption of Cu, Zn or Pb. Furthermore, the concentrations of Cd in earthworms from sludge-treated soils measured by Beyer *et al.* (1982) and Kruse and Barrett (1985) were larger than those reported by Ma (1987) in worms sampled from a smelter contaminated site at similar low levels of Cd in soil. This suggests that Cd availability to earthworms is higher in sludged soil compared with soils contaminated through aerial deposition.

In contrast to Cd, there is little evidence for the transfer of Pb, which is also potentially a zootoxic element, through terrestrial ecosystems. The absence of food chain accumulation of Pb is attributed mainly to deposition and immobilization in skeletal components where it is deposited in place of Ca (Roberts and Johnson, 1978). The enhanced mobility of Cd may be associated with its characteristic preferential accumulation in soft body tissues. Roberts and Johnson (1978) established that the most marked changes in Cd concentration occur between herbivorous and carnivorous invertebrates within the vegetation-invertebrate-small mammal food web. Carnivorous arthropods in particular contained markedly elevated concentrations of Cd compared with herbivorous invertebrates (3-4:1 Cd concentration ratio) indicating stepwise accumulation between trophic levels within the invertebrate biomass. Thus, the potential accumulation of Cd may be associated with changes in chemical behaviour and bioavailability of the element between trophic levels of terrestrial communities. This may explain why numerous workers have reported little or no apparent impact of heavy metals on small herbivorous mammals, such as voles, mice and rabbits (Williams *et al.*, 1978; Anderson and Barrett, 1982; Anderson *et al.*, 1982; Anthony and Kozlowski, 1982; Dressler *et al.*, 1986; Alberici *et al.*, 1989; Levine *et al.*, 1989) or on birds (Hinesly *et al.*, 1976; Gaffney and Ellertson, 1979) which live in sludge-treated agricultural habitats, since the food webs of these higher animals are shorter than for carnivorous mammals and largely or completely omit trophic levels which potentially accumulate toxic metals. Despite the potential for Cd to transfer through terrestrial foodchains, and particularly those including carnivore and detritivore trophic levels, human dietary intake of Cd in plant foods, or *via* soil ingestion, is considered the most sensitive environmental route of exposure to Cd applied to agricultural soil in sewage sludge (Davis, 1984b; US EPA, 1992a; Wagener, 1993).

Chapter nine:

Nutrients

Nutrient Content in Sewage Sludges

Sewage sludges contain appreciable amounts of N and P (Table 9.1) and have significant inorganic fertilizer replacement value for these major plant nutrients (Hall, 1984, 1985, 1986a; Hall and Williams, 1984; Coker and Carlton-Smith, 1986; Coker *et al.*, 1987a, 1987b). Assuming average N and P contents in sludge dry solids of 3.8% and 2.2% respectively, approximately 20 000 t of N and 11 000 t of P are recycled to farmland each year in the UK through the agricultural use of sludge. For the EU as a whole, the quantities are about 91 000 t of N and 53 000 t of P. In relation to equivalent amounts of inorganic fertilizers, in the UK this represents a potential saving of approximately £15 million, which benefits both the farming community and the national economy as a whole, but is probably equivalent to less than 5% of UK fertilizer consumption. The K content of all sludges is low, however, and not sufficient to make any significant difference to the recommended quantities of fertilizer K necessary for most cropping situations (Hall and Williams, 1984). Sewage sludge also has potential as a source of sulphur (S) for crop growth (Elseewi *et al.*, 1978) providing additional potential benefits to agriculture in areas where atmospheric deposition of S has declined and crop deficiencies are possible (MAFF, 1994a). Sludge can supply a number of other nutrients (e.g. Mg, Ca, Zn, Cu, etc.) but their importance will depend on whether these are deficient in the soil. Thus in considering the potential impacts of sludge nutrients on the environment, it is necessary to deal specifically with the problems attributable to N and P in agricultural soils.

Table 9.1. Typical N and P content of sewage sludges (as spread; kg m^{-3} for liquids; kg t^{-1} for cakes).

Sludge type	Dry matter (%)	Total N	Total P	Available[1] N	Available[1] P
Liquid undigested	5	1.8	0.6	0.6	0.3
Liquid digested	4	2.0	0.7	1.2	0.3
Undigested cake	25	7.5	2.8	1.5	1.4
Digested cake	25	7.5	3.9	1.1	2.0

(1) Availability in first cropping year
Source: Hall (1986a); WRc (1985)

Biological, Chemical and Environmental Factors Influencing Soil Nitrogen and Phosphorus

Nitrogen

Nitrogen cycle

Nitrogen applied to soils in mineral fertilizers, sewage sludges and other organic materials, and also indigenous soil organic N, are subject to a complex series of interrelated biochemical and physical processes which collectively form the nitrogen cycle (Fig. 9.1).

Mineral N applied as nitrate may be subject to leaching immediately whereas ammonium-N is retained in the soil. This is because positively charged ammonium ions are bound to exchange complexes on clay minerals and organic matter in the soil whereas nitrate remains as free ions in the soil solution (Brady, 1990). The relationship between the various forms of ammonium ion can be represented as follows:

$$\underset{\text{soil solution}}{NH_4^+} \rightleftharpoons \underset{\text{exchangeable}}{NH_4^+} \rightleftharpoons \underset{\text{fixed}}{NH_4^+}$$

Ammonium fixation may be considered an advantage as it provides a mechanism for conserving soil N which is subsequently released slowly.

In contrast to mineral sources of N, organic forms are not available for plant uptake on application to the soil. Organic N is mineralized into plant-available forms under wide ranging environmental and edaphic conditions by a large diversity of soil micro-organisms (Wild, 1988a). Ammonium ions derived from

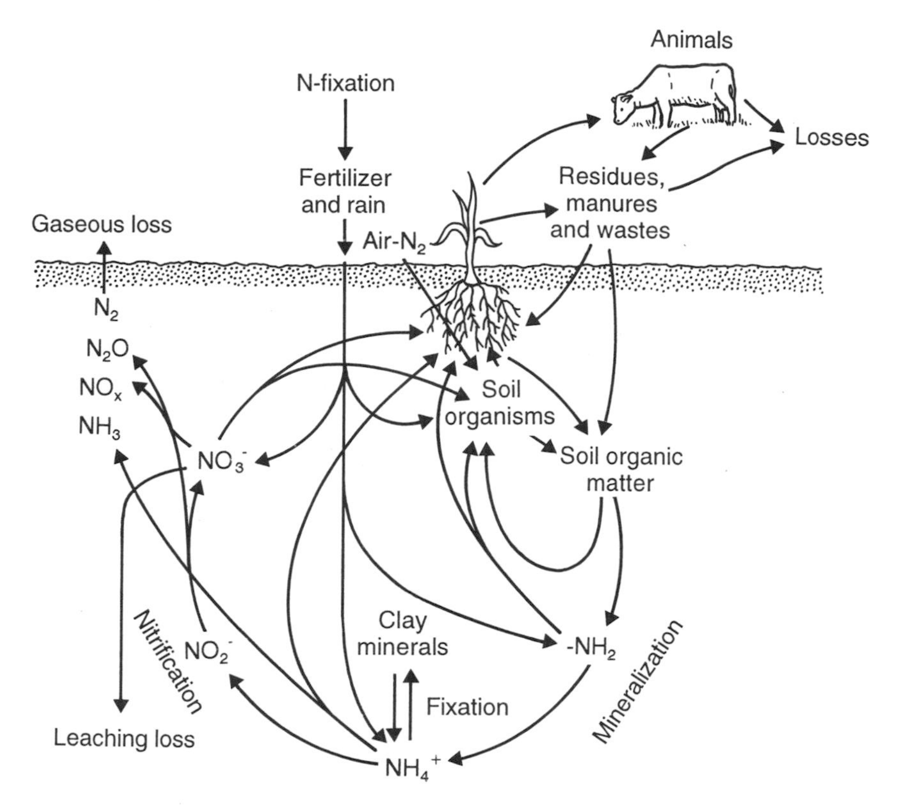

Fig. 9.1. The nitrogen cycle (Brady, 1990).

mineral fertilizers, such as ammonium nitrate, or from the mineralization of residual soil organic matter and organic amendments, are subject to nitrification. This involves a two-stage oxidation of ammonium to nitrate by specialized groups of soil bacteria. The transformations and main nitrifying bacteria are presented below.

Stage	Nitrifying bacteria
1. Ammonium → Nitrite	*Nitrosomonas*
2. Nitrite → Nitrate	*Nitrobacter*

Due to the specialized nature of the nitrifying population the rate of nitrification is sensitive to interacting environmental and edaphic factors including soil temperature, aeration, soil moisture content and pH.

Temperature

Mineralization of organic matter is active all year round, but particularly in spring and autumn when soils are warm and moist (Fig. 9.2). Soil temperature is a principal environmental factor governing the rate of nitrification and increasing nitrate concentrations in soil, produced from applied and mineralized ammonium-N, can be described as a simple function of thermal-time (Honeycutt *et al.*, 1991; Smith and Hadley, 1992; Smith and Woods, 1994). The basal soil temperature for nitrification, below which virtually no nitrate is formed in soil, appears to be close to 0 °C (Smith and Woods, 1994). Nitrification decreases markedly in soil below 5 °C (Addiscott, 1983). Consequently, nitrate accumulates rapidly in soils supplied with a source of ammonium-N (such as liquid digested sludge) until the late autumn to early winter period, under UK conditions, when declining soil temperatures effectively minimize nitrate formation. Nitrate is produced in soil after this period, but generally at a very slow rate that is unlikely to cause significant impact on nitrate leaching.

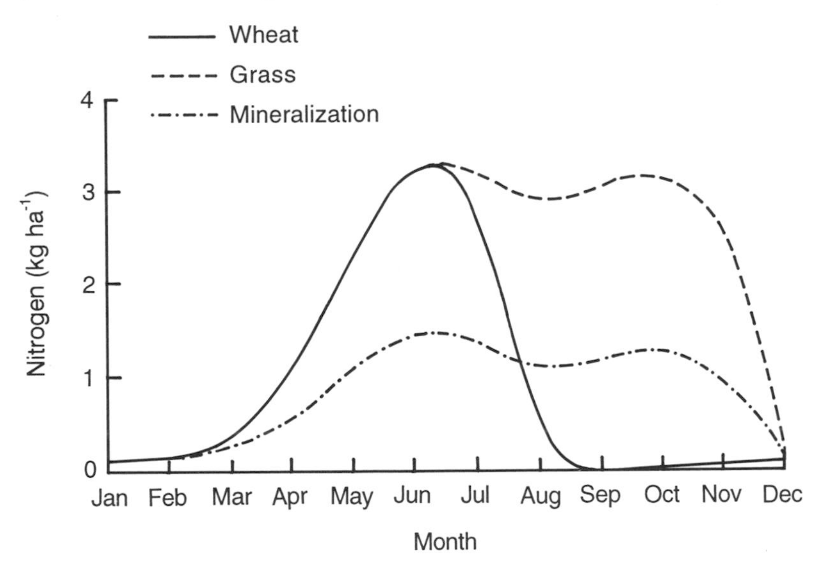

Fig. 9.2. Schematic representation of the nitrogen uptake by winter wheat or grass throughout the year compared with nitrogen produced by mineralization from soil organic matter (Addiscott *et al.*, 1991).

Rainfall

The potential risk of nitrate leaching from soil during autumn increases with increasing rainfall over this period. Nitrate may be leached below the root zone once soil moisture deficits have been satisfied and when soil water holding capacity has been exceeded although this varies with soil type. A schematic representation of the seasonal variation in nitrate loss by leaching is shown in Fig. 9.3.

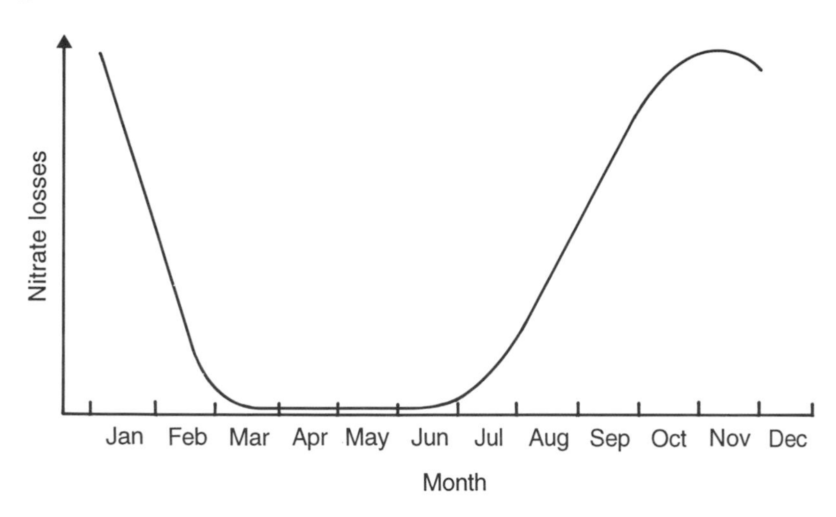

Fig. 9.3. Schematic representation of seasonal variation in nitrate leaching losses (Amberger, 1983).

Rainfall is the vector of transfer of nitrate to groundwater and is therefore an important factor in determining the nitrate concentration in these water systems. During periods of low precipitation the level of effective rainfall which recharges the aquifer is small and nitrate accumulates in the surface strata. Therefore, the nitrate concentration in drainage from soils in areas of relatively low rainfall is greater than in regions which receive more rain on the basis of the same quantity of N applied in all regions. Consequently, the impact of nitrate leaching on groundwater nitrate levels is greater in drier Eastern regions compared with wetter Western areas of the UK. This was demonstrated by Burns and Greenwood (1982) by modelling N leaching losses from arable soils throughout England and Wales during winter and spring. Assuming that 100 kg N ha^{-1} was present in the soil during the previous autumn and nitrate was lost only in drainage water from arable soils, results showed the average nitrate concentration in drainage was in the range 1-5 mg N l^{-1} (4.4-22 mg NO_3^- l^{-1}) in Western regions, 8-9 mg N l^{-1} (35-40 mg NO_3^- l^{-1}) in the Midlands and up to 18 mg N l^{-1} (80 mg NO_3^- l^{-1}) in East Anglia. However, the opposite was true when the total quantities of N leached were considered as the largest total losses occurred in areas of high rainfall in the West, whereas total losses were smaller from soils in the East due to lower rainfall. Burns and Greenwood determined that for the country as a whole 65 kg N ha^{-1} was leached below a depth of 1.0 m in years of high rainfall and 12 kg N ha^{-1} in the driest years during winter. The overall loss from the top metre of soil was reported as 44 kg N ha^{-1}. On the basis that on average 105 kg N ha^{-1} is applied to crops in England and Wales, it was estimated that 42% of this N was lost annually by leaching. This corresponds to 1.9 x 10^5 t of N per year. Burns and Greenwood (1982) concluded that spring and early summer rainfall was unlikely to result in large nitrate leaching losses of spring applied fertilizer.

Immobilization

Nitrogen applied to soils in organic and inorganic forms can be assimilated by soil micro-organisms. This N may be released through mineralization of dead microbial tissues, but much of the N remains in organic forms and is thus unavailable for plant uptake or leaching (Jenkinson and Parry, 1989; Shen *et al.*, 1989). The process of tying up N in organic forms is known as immobilization. Immobilized N becomes an integral component of the soil organic matter and is only slowly mineralized to available forms. The balance between ammonium release and immobilization is a function of the C:N ratio of the organic material undergoing decomposition. This can vary from 5:1 for some animal wastes to as much as 100:1 for cereal straw. Materials with a C:N ratio of less than 25 generally mineralize N when they decompose in soil (Addiscott *et al.*, 1991). Those with a ratio considerably greater than 25 immobilize residual soil mineral N until some of the C in the substrate has been released (as CO_2 through respiration) and the C:N ratio has declined, when a net release of N will occur through mineralization. Smith *et al.* (1992f) reported the temporary immobilization of N for two months by raw dewatered sludge in an incubation study. Chaussod (1981) earlier reported immobilization lasting for a similar time period for undigested and for activated sludges with C:N ratios of >13.5:1 and >10:1, respectively. The mineralization/immobilization characteristics of sewage sludges can be qualitatively described in relation to their C:N ratio as (Chaussod *et al.*, 1978):

1. Sludges with high mineralizable N	Total N>5%	C:N<8
2. Sludges with average mineralizable N	Total N2-5%	C:N10-14
3. Sludges with little mineralizable N	Total N<2%	C:N>15
4. Sludges which temporarily immobilize soil N	C>30%	C:N>15

However, Williams and Hall (1986) argued that immobilization has not been quantifiably demonstrated under field conditions. This is probably explained because soils already contain sufficient inorganic N to compensate for the amount immobilized during the initial stages of the decomposition of sludge in relation to crop requirements. Furthermore, field experiments designed to quantify the fertilizer value of sewage sludges assess N availability in terms of crop responses relative to inorganic sources of N, so the effect of any initial immobilization of N is already compensated for through a lower sludge efficiency value.

Ammonia volatilization and denitrification

Nitrogen in organic soil amendments can also be subject to gaseous loss mechanisms including ammonia volatilization and denitrification (Pain *et al.*, 1990). Significant ammonia volatilization can occur from the soil surface particularly with liquid sludges and animal slurries which may contain appreciable concentrations of ammonium-N. Edaphic and environmental factors are important and volatilization increases under dry and windy conditions and losses are also greater from alkaline soils (Fenn and Hossner, 1985). However, incorporation into soil significantly reduces ammonia volatilization losses.

Denitrification occurs under anaerobic conditions and leads to gaseous losses of nitrogen largely as N_2 gas and as nitrous oxide (N_2O) (Fig. 9.1). These conditions can develop in poorly drained soils and in saturated zones resulting from liquid waste application, particularly when the material is injected into the soil.

Clearly, the potential for nitrate leaching from sewage sludges applied to soil is reduced by ammonia volatilization and denitrification losses of N. However, these mechanisms also reduce the fertilizer value of sludge and gaseous losses of N from sludge-treated soil may be unacceptable in relation to potential atmospheric pollution.

Phosphorus

The P requirement of agricultural crops is typically only about 10-25% of the quantity of N removed by crop plants from soil (Cooke, 1982) whereas sewage sludges generally contain approximately half as much P as they do N (Table 9.1). Consequently, the current practice of applying sewage sludges according to the N requirements of crops will supply P in excess of crop needs. Advice on the agricultural use of sludge (WRc, 1985) indicates that liquid sludges can supply all the phosphate requirements for grass or cereal crops. A single application of sludge cake can provide sufficient P for most 3-4 year crop rotations. However, there is concern about the potential environmental effects of possible soil enrichment with P following the repeated and long-term application of sewage sludge to agricultural land.

A principal source of P in wastewater and sewage sludge is derived from detergents although in global terms this is thought not to be a major input of P into receiving surface waters (Wilson and Jones, 1994). A best estimate of P inputs to surface waters in the EU concluded that 50% comes from agriculture, 41% from wastewater (24% from human sources, 10% from detergents, and 7% from industry), and 9% from background sources (Morse *et al.*, 1993). Consequently, the use of alternative detergent builders to phosphate is considered unlikely to reduce the extent of freshwater pollution with P in the UK (Wilson and Jones, 1994). In any case, the available alternatives all have potentially adverse impacts on the environment. It appears that there will be little opportunity to reduce P concentrations in sewage sludges in the short to medium term. To the contrary, P levels in certain sludges may well increase in the future at those treatment works where P removal from wastewater is implemented (Furrer and Bolliger, 1981; Morse *et al.*, 1993).

Phosphorus in soil is subject to a variety of fixation and precipitation processes which render the element relatively insoluble and immobile. Soil pH value is critical to the availability of P according to the following equation (Brady, 1990):

$$\underset{(\text{pH } 2\text{-}6.5)}{H_2PO_4^-} \underset{H^+}{\overset{OH^-}{\rightleftharpoons}} \underset{(\text{pH } 6.5 - 12.5)}{H_2O + HPO_4^{2-}} \underset{H^+}{\overset{OH^-}{\rightleftharpoons}} \underset{(\text{pH} > 12.5)}{H_2O + PO_4^{3-}}$$

At pH 6.0 both $H_2PO_4^-$ and HPO_4^{2-} ions exist simultaneously, but the former is considered the more available. However, the presence of soluble Fe and Al under acid conditions, or Ca at high pH values, markedly influence the availability of P. In acid soils dominated by $H_2PO_4^-$, precipitation with soluble Fe, Al or Mn usually present under these conditions renders the P insoluble. Adsorption of $H_2PO_4^-$ by clay and insoluble Fe and Al oxides and hydroxides also results in fixation in unavailable forms. In calcareous soils, however, phosphate precipitation is caused principally by calcium carbonate. The optimum pH for P availability to crops is in the range pH 6.0-7.0.

The classification of phosphate compounds can be represented in three major groups presented in Fig. 9.4. At any one time about 80-90% of the soil P is non-labile. Most of the remainder is in the slowly available labile form since only <1% would be expected to be readily available. Applied phosphates are quickly converted from readily available to slowly available forms by chemical reaction in the soil. These are rendered less available with ageing and eventually enter the non-labile pool.

An important consequence of the reaction of P with soil is that P movement is very restricted. Thus, there is very little phosphate leached out of soils in water drainage (Sawhney, 1978; Sharpley and Manzel, 1987; Wild, 1988b). Concern that P saturation of sludge-treated agricultural land may result in possible leaching losses of P has resulted in legislation to control applications to soil in certain European countries (Hall and Dalimier, 1994). No experimental information could be found to justify such restrictions even from long-term applications of sewage sludge to sandy soils potentially vulnerable to leaching (de Haan, 1981; Furrer, 1981; Johnston, 1981; Furrer and Gupta, 1985). The wastes of farm animals are also rich in phosphate and may contain broadly similar amounts of P generally compared with sewage sludge (MAFF, 1994a). However, P mobility in soils treated with animal wastes is reported to be much higher than it is in sludge-amended agricultural land (Johnston, 1981; Furrer and Gupta, 1985). In contrast to the minimal leaching losses of P from sludged soil, surface run-off and soil erosional losses of P applied in fertilizers, animal manures and sewage sludge, potentially cause the greatest risk of environmental pollution due to the contamination of surface water with phosphate (Chaussod *et al.*, 1985).

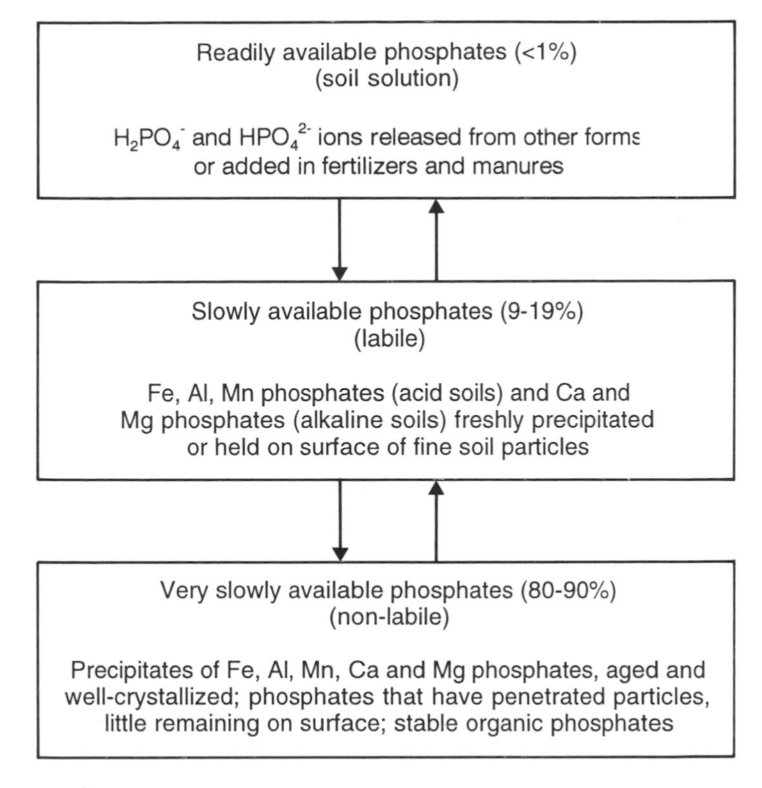

Fig. 9.4. Classification of phosphorus compounds in soils (Brady, 1990).

Environmental Problems Arising from Nitrogen and Phosphorus in Agricultural Soils

General Comments

The application of nutrients to soil in the form of mineral fertilizers or as organic manures such as sewage sludges, FYM or farm slurries is necessary to provide optimum crop yields. However, the loss from soil of N and P by various biological processes and physical mechanisms can result in significant pollution of the aerial environment by ammonia volatilization and denitrification, and contamination of ground and surface water supplies through leaching of nitrate and surface run-off of both nitrate and phosphate.

Nitrate Leaching

A trend of increasing concentrations of nitrate in ground water and surface waters during the last 30 years has raised concerns over possible health and environmental risks associated with nitrate pollution from diffuse agricultural sources (DoE, 1986). Increasing nitrate concentrations in water sources have arisen principally due to intensification of agricultural crop production methods and increased use of inorganic N fertilizers (Smith and Powlesland, 1990). To avoid potential health problems, however, the CEC has ratified a Drinking Water Quality Directive (CEC, 1980b) which sets a maximum permissible level of 50 mg NO_3^- l^{-1} (11.3 mg NO_3^- N l^{-1}) in potable water supplies and a guide value is also included at 25 mg NO_3^- l^{-1}. If the limit is exceeded, the Water Undertakings are required to treat affected sources or to blend them with lower nitrate water, so that the concentration is reduced below the maximum. In some areas in the UK groundwater supplies have already exceeded the 50 mg l^{-1} limit for nitrate and blending or source replacement is being carried out. In many other areas, nitrate concentrations in groundwater are less than 50 mg l^{-1}, but long-term predictions (Oakes, 1989) indicate nitrate levels may approach the limit within 20 to 30 years.

Recent legislation (CEC, 1991b) has established a series of measures designed to control the pollution of water from diffuse agricultural sources of nitrate in areas potentially susceptible to groundwater contamination with nitrate. These will become obligatory on farmers and sludge producers at the latest by December 1999. In particular, the maximum annual application rate of N (as total N) in organic manures, including sewage sludge, will be restricted to 210 kg N ha^{-1} initially, possibly reducing to 170 kg N ha^{-1}, and there will be periods when the application of organic manures is not permitted. However, the difficulty in attempting to regulate the nitrate content of groundwater below 50 mg l^{-1} by controlling agricultural sources of N is emphasized not least because of the extent of atmospheric contributions of N to potential leaching losses. For example, Goulding (1990) estimated that N deposition from the atmosphere could be as high as 40 kg N ha^{-1} y^{-1}, and if this were balanced exactly by N leached, then the drainage water would contain 20 mg N l^{-1} as nitrate, which is almost twice the CEC limit.

The UK Government set up the Pilot Nitrate Sensitive Areas (NSAs) Scheme in 1990 to examine the efficacy of different management practices in reducing diffuse nitrate pollution from agricultural land (MAFF, 1990b). An NSA is an area where nitrate concentrations in drinking water sources exceed or are at risk of exceeding the CEC limit and where voluntary agronomic measures have been introduced as a means of reducing those levels. The report on the first three years of the Pilot Scheme (MAFF, 1993b) indicated the most effective remedial measures were:

1. Avoiding the application of liquid sewage sludge, poultry manure or animal

slurries between 30 June-1 November to arable land and between 31 August-1 November to pasture.

2. Restricting the rate or organic manure application to 175 kg N ha^{-1} (total N).

3. Conversion of arable land to low-intensity grassland which does not receive fertilizer or manure (although grazing may be allowed).

4. The use of green cover crops to avoid bare land in the autumn.

The Pilot Scheme originally covered 10 NSAs although this number has been extended and a further 22 NSAs have been introduced (MAFF, 1994b).

The CEC Nitrate Directive (CEC, 1991b) requires an action programme of measures to be introduced in areas designated as Nitrate Vulnerable Zones (NVZs), where water sources are high in nitrate. An Outline Action Programme has been published for England and Wales (MAFF, 1994c) which includes the following measures:

1. A maximum application of 210 kg N ha^{-1} y^{-1} as total N (farm average).

2. An exclusion period between 1 August-1 November for arable land and 1 September-1 November for grass when the application of liquid digested sewage, poultry manure and animal slurries will not be permitted on sandy or shallow soils.

The Nitrate Directive also requires that a code of good agricultural practice is prepared for areas outside of the NVZs. In England and Wales, the MAFF *Code of Good Agricultural Practice for the Protection of Water* (MAFF, 1991a) already includes recommendations designed to reduce nitrate pollution of water supplies which apply to all areas irrespective of whether there is a requirement to protect water resources or not. The sister document for Scotland is the *Prevention of Environmental Pollution from Agricultural Activity: Code of Practice* (SOAFD, 1991). The MAFF Code stipulates a maximum annual limit of 250 kg N ha^{-1} (total N) for application of organic manures, including sewage sludge. In Scotland, the allowable rate of N application is defined according to crop needs and soil nutrient status. Both Codes of Practice advise that application of organic manures with high available N content, including liquid sewage sludges, poultry manure and animal slurries, to arable land in the autumn or early winter should be avoided whenever practicable. Organic manures low in available N, including FYM and dewatered sewage sludges, can be applied at any time. Whilst the restrictions on timing generally distinguish between different types of sewage sludge with low and high nitrate leaching potential, restrictions on the total amount of N which can be applied do not make this important distinction. Fixing a maximum rate of N which can be applied does not consider the availability of different forms of N contained in sludges or effects which different sludge treatment processes have on N release, which may greatly influence the actual risk of nitrate leaching from sludge-treated agricultural land (Smith *et al.*, 1992f).

The nitrate ion itself is non-toxic, but bacteria found mainly in the mouth can reduce nitrate to nitrite, which is the major source of exogenous nitrite for most humans (Forman *et al.*, 1985). Apart from nitrate-induced methaemoglobinaemia (the so-called 'Blue Baby Syndrome'), which has been virtually eliminated from the UK with the last case being recorded in 1972 (White, 1983), nitrite represents a potential hazard through its involvement in nitrosation and the formation of N-nitroso compounds which are carcinogenic in animals. The stomach is commonly regarded as being at risk from endogenous N-nitroso compound synthesis and suggestive, but often contradictory evidence, has linked high levels of nitrate ingestion with gastric cancer mortality. Ironically, recent studies in the UK and Denmark (Forman *et al.*, 1985, NFU, 1987) have shown that populations exposed to increasing nitrate levels apparently have a lower incidence of deaths from gastric cancer (Fig. 9.5).

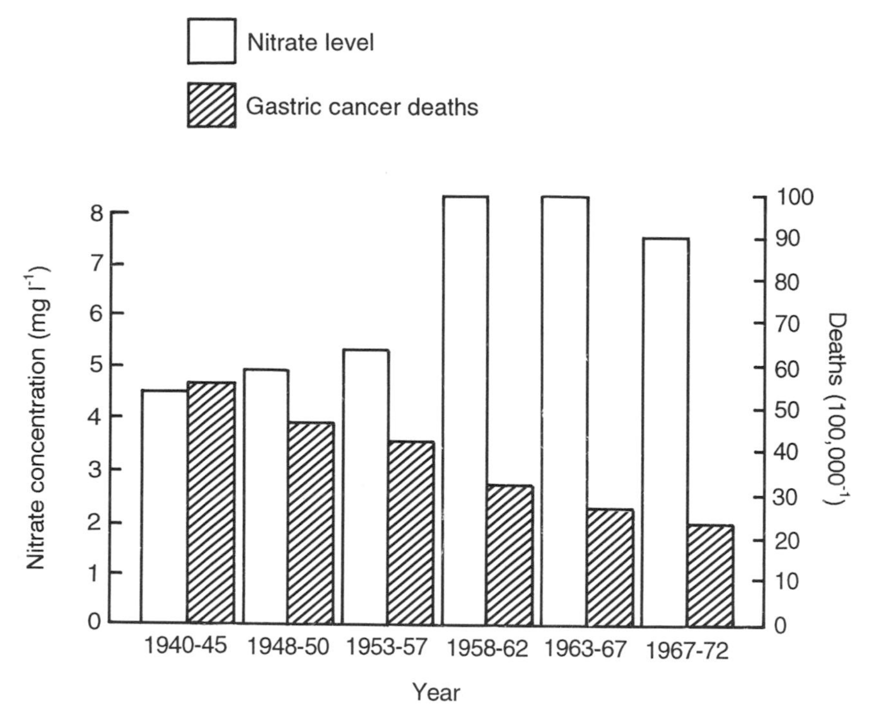

Fig. 9.5. Nitrate concentration of drinking water and gastric cancer mortality in Denmark (NFU, 1987).

Effects of Nitrogen and Phosphorus on Surface Water

Nitrate and phosphorus occur normally in fresh and marine waters and are essential nutrients for plant growth in aquatic environments. Increasing levels of these nutrients improve the productivity of fresh and marine waters by enhancing the growth of algae and other flora thus indirectly improving populations of fish and other animals. Neither nitrate nor phosphate is directly toxic to marine or fresh water animals. However, raising nutrient levels, a process known as eutrophication, may ultimately reduce biota diversity through oxygen depletion (Cartwright and Painter, 1991). Other detrimental impacts of eutrophication include (Cartwright and Painter, 1991):

- production of toxic substances irritating to swimmers and potentially resulting in the death of animals;
- loss of important aquatic plant food sources for water fowl;
- development of anaerobic zones liberating methane, ammonia and hydrogen sulphide;
- impaired operation of drinking water treatment processes;
- unpleasant tastes and odours in drinking water;
- increased population of nuisance organisms including mosquitoes and parasites;
- diminished recreational value.

Nitrate is implicated in eutrophication of freshwater although P is the major contributor (Cartwright and Painter, 1991; Archer, 1992). Archer (1992) indicated that nitrate-limited eutrophication was unlikely to be a major factor in establishing vulnerable zones in England and Wales. Only very limited eutrophication is apparent in estuary waters and is not necessarily related to agricultural nitrate (Archer, 1992).

The Drinking Water Directive (CEC, 1980b) includes limits for P (guide level: 400 μg l^{-1}; maximum admissible concentration 5000 μg l^{-1}, both as P_2O_5) and the P concentration in drinking water is also controlled under UK Regulations (maximum admissible concentration: 2200 μg P l^{-1} (SI, 1989b)). There is no European legislation restricting the application of P to agricultural soils at the present time. However, the European Commission is currently examining the need for a Directive on phosphate reduction (CEC, 1992) which may form the subject of future environmental legislation within the EU. Furthermore, a number of individual Member States have introduced standards on sewage sludge application to agricultural land which incorporate limits on P addition (Hall and Dalimier, 1994).

It should be noted that the CEC limits for nitrate and phosphate protect the quality of water for human consumption, but do not consider potential impacts on eutrophication of surface waters. Water quality standards preventing eutrophication in fresh waters are necessarily much more stringent than for drinking water, especially in the case of phosphate, because nuisance growth of algae can occur in rivers with very low P status. Cartwright and Painter (1991) proposed that an annual mean Environmental Quality Standard (EQS) for N of 100 $\mu g\ l^{-1}$ (total N) in fresh waters would be necessary to maintain oligotrophic conditions which restrict eutrophication. On a similar basis, an EQS for P of 10-30 $\mu g\ l^{-1}$ (total P) was considered appropriate. In the UK, the proposal is to protect riverine ecosystems by Statutory Water Quality Objectives (SWQOs) for P according to conservation value (C. Mainstone, WRc Medmenham, personal communication). The SWQOs for P in river water are likely to be in the range <20 - <200 μg P l^{-1} (as orthophosphate). Stringent controls on P in fresh waters may be impracticable in predominantly agricultural catchments, even with best available land management practices, given that P concentrations in run-off are often higher than those limiting eutrophication, even from unfertilized grassland soils (180 $\mu g\ l^{-1}$ soluble P) (Withers and Sharpley, in press).

Gaseous Losses

Ammonia volatilization

A number of reports and reviews have been recently published indicating the role of ammonia as an atmospheric pollutant and its potential impact on the environment (Hall, 1988; Istas *et al.*, 1988; Apsimon and Kruse-Plass, 1991; Roelofs and Houdijk, 1991; van Breemen, 1991; Fangmeier *et al.*, 1994). About 50% of the ammonia emitted from agricultural activities is deposited near to the source and prolonged exposure can potentially reduce crop yields due to excessive soil acidification from nitrification of the added ammonia (Speirs and Frost, 1987). Direct toxicity of gaseous ammonia to plants has also been reported within short distances (100-200 m) of point sources such as piggeries. The effects may be sublethal or chronic depending on the concentration-exposure time relationship and sensitivity of the plant species.

The average residence time of ammoniacal-N in the atmosphere is relatively short and has been estimated as about thirteen hours (Istas *et al.*, 1988). Ammonia reacts quickly with hydroxyl radicals in the atmosphere to produce NO_x contributing to the acidification of rain, and also combines with sulphur compounds, usually in the form of aqueous sulphuric acid aerosols, liberating ammonium sulphate (Galbally and Roy, 1983; Istas *et al.*, 1988) which will also ultimately contribute to soil acidification. Furthermore, the uptake of gaseous ammonia water into droplets accelerates many of the oxidation processes that produce acidity in cloud water. For example, dissolved SO_2 in cloud water can be oxidized to dissolved H_2SO_4 by ozone (O_3). The rate of reaction decreases with decreasing pH, but ammonia absorption by cloud droplets raises the pH such that more SO_2 is oxidized. This phenomenon can give rise to very acidic clouds producing acid mists over forested hill-tops causing damage to foliage (Apsimon and Kruse-Plass, 1991). In addition, the deposition of ammoniacal-N on surface waters impacts oligotrophic environments and contributes to the problem of eutrophication, although Istas *et al.* (1988) indicated this probably occurred only to a limited extent.

Forest ecosystems tend to be intrinsically more sensitive compared with agriculture to environmental damage caused by atmospheric pollution with ammonia although the links between forest decline and air pollution are complicated and causal relationships have not yet been proven. Nevertheless, van Breemen (1991) considered that high levels of N deposition, and intensive soil acidification associated with inputs of ammonium sulphate, aggravated by nitrification in soil, were probably major factors influencing forest decline.

Roelofs and Houdijk (1991) are more explicit about the effects of ammonia on forest and oligotrophic environments. They state that direct effects of ammonia toxicity occur only in cold climates such as northern Scandinavia or in cold winters when ammonia damage results in red colouring of coniferous trees because of the slow ammonia detoxification capacity of plants at low temperatures. In a mild climate, however, trees normally recover during warmer summer conditions. More widespread and serious are the indirect effects of ammonium deposition on ecosystems. In weakly buffered ecosystems a high deposition of ammonium leads to acidification and N enrichment of the soil. As a consequence many plant species characteristic of poorly buffered environments disappear. Among the acid-tolerant species there will be competition between slow growing plants and faster growing nitrophilous grasses or grass-like species. This process contributes to the often observed change from heath and peatlands into grassland. In forest ecosystems a high input of ammonium leads to leaching of K^+, Mg^{2+} and Ca^{2+} from the soil, often resulting in increased ratios of NH_4^+ to K^+ and Mg^{2+} and/or Al^{3+} to Ca^{2+} in the soil solution. Field investigations show clear correlations between these increased ratios and the poor condition of tree species. Increased ratios of NH_4^+ to K^+ inhibit the growth of symbiotic fungi and the uptake of K and Mg by the root system. At high NH_4^+/K^+ and Al^{2+}/Ca^{2+} ratios, there is a net flux of Mg^{2+} and Ca^{2+} from the root system to the soil solution. Furthermore, coniferous trees can absorb ammonium through the needles and maintain the ionic balance by excreting K^+ and Mg^{2+}. These combined effects may result in K and/or Mg deficiencies, severe N stress, and as a consequence premature shedding of leaves or needles. The trees also become more susceptible to other stress factors such as ozone, drought, frost and fungal diseases, as well as insect infestations.

Dieback of trees has occurred in areas with the highest animal waste N production which include The Netherlands and Belgium (Istas *et al.*, 1988). Damage has also been reported in other regions not closely located to a high ammonia production, but it is difficult to predict which of these remote areas are high risk ones.

Denitrification

The dominant gases liberated during denitrification in soils are gaseous N_2 and nitrous oxide, although nitrification is also an important source of nitrous oxide (Bremner and Blackmer, 1978). Both N_2 and nitrous oxide are chemically inert. However, nitrous oxide absorbs radiation in the infra-red band and in this way accounts for 5-10% of the total greenhouse effect (Jenkinson, 1990b; Kroeze, 1994). In contrast to ammonia, the residence time of nitrous oxide in the atmosphere is 100-200 years. Small amounts of nitric oxide and nitrogen dioxide (NO_x) are also evolved by the soil-plant system largely due to chemodenitrification, but biological processes can also be involved (Chalk and Smith, 1983). Like ammonia, these gases have a short residence time of only a few days in the atmosphere. The NO_x gases are chemically active in the atmosphere, catalysing various reactions involving ozone, carbon monoxide and methane and they contribute directly to the acidity of rainfall. They may also erode the efficiency of the ozone layer as a protective shield of the earth's surface from harmful short wavebands of solar radiation (Johnson, 1971). Jenkinson (1990b) stated that the global problems associated with NO_x were therefore of concern to atmospheric chemists, as well as to agronomists and environmentalists. However, agricultural utilization of sludge is unlikely to have any significant impact on ozone depletion. On the other hand, ammonia volatilization and nitrous oxide emissions arising predominantly from livestock production, may be of more concern.

Sewage Sludge and Farm Wastes

To place the environmental risk from nutrients, particularly N, contained in sewage sludge into context, it is an interesting exercise to compare the amount of N applied to agricultural land in sludge with that supplied in the wastes of farm animals. Approximately 1.5×10^6 t of N are produced each year in the UK in the wastes of cattle, sheep, pigs and poultry (Addiscott *et al.*, 1991). Thus, the contribution made by sludge represents only about 1.0% of the N potentially spread on farmland in animal wastes and is insignificant against this enormous background level. On a national basis, the atmospheric and water pollution resulting from such large quantities of farm waste is likely to be considerably greater than from the small additional inputs made from sewage sludge. This is demonstrated by Jenkinson (1990b) in a global perspective of the N cycle. Animal excreta provide the largest input of ammonia to the atmosphere with global production on average being estimated at 29×10^6 t N y^{-1}, which represents about 54% of the total ammonia emissions. Emissions of ammonia from animal excreta represent about 80% of the total 2.68×10^6 t of ammonia released into the atmosphere each year in Europe (Istas *et al.*, 1988). From a mass balance of estimated ammonia emissions and deposition, Whitehead

(1990) calculated that the UK was a net exporter of atmospheric ammonia. Since atmospheric pollution with ammonia represents a transboundary problem, in that emissions from one country may influence the air quality of another, forthcoming legislation across the EU could place restrictions on ammonia release to the atmosphere (CEC, 1992). Current policy in The Netherlands, where the problems of ammonia pollution are particularly acute, is to reduce emissions in 2000 by 70% relative to 1980 levels (MHPPE, no date) and constraints on methods and rates of application of organic manures to land are already in place.

Other important factors also demonstrate the minimal risk of environmental pollution caused by sludge nutrients compared with farm wastes. For example, the uncontrolled distribution of urine and dung from grazing livestock in highly concentrated patches in the field can provide a potent source of N resulting in large losses of ammonia by volatilization to the atmosphere and of nitrate by leaching to water supplies (Simpson and Steele, 1983; Whitehead, 1990; Jarvis, 1992, 1994). The N deposited by a grazing cow in a urine patch can be equivalent to an application rate as high a 1200 kg N ha^{-1} which compares with a maximum application of 400 kg N ha^{-1} which can be safely exploited by grass (Addiscott *et al.*, 1991).

Approximately 75% of the sludge applied to agricultural land in the UK has received some form of stabilization treatment by anaerobic digestion or other methods (CES, 1993). In particular, the *Code of Practice for Agricultural Use of Sewage Sludge* (DoE, 1989a) emphasizes the benefits of treating sludges before their application to farmland for reasons of pathogen and odour control. The scale of sludge treatment by engineered processes, which are monitored and controlled to prescribed standards, generally improves the uniformity of sludge products for agricultural utilization. This also has the benefit of converting most of the sludge organic matter into forms which decompose very slowly in soil compared with other types of untreated manure thus reducing the potential risk of N pollution (Smith *et al.*, 1992f). Consequently, the N fertilizer replacement value and potential environmental fate of N applied to soil in stabilized sludge is relatively predictable (Smith and Hadley, 1992; Smith and Woods, 1994) and is associated principally with its ammonium and easily degradable organic N content. However, some sludge products may be stabilized so effectively and contain only traces of mineral N such that they have minimal N fertilizer replacement value, but are useful as soil conditioners. The N mineralization characteristics of different sewage sludges are discussed in more detail in the following section.

In contrast to sewage sludges, farm wastes are rarely treated to the same extent, level of control or scale (Berglund and Hall, 1988). Livestock wastes are produced on individual farms, rather than centrally as is the case with sludge. Therefore, the properties of animal wastes requiring disposal will inevitably vary according to local management practices. The wastes of farm animals are inherently heterogeneous principally due to differences in the type, composition and quantities of bedding materials used in housing livestock, and the lengths of time the waste may be stored before disposal. Furthermore, there may be large seasonal differences in the composition and behaviour of livestock manures applied on agricultural land. As a consequence, their N release characteristics may be intrinsically more variable compared with stabilized sewage sludges (Chaussod *et al.*, 1986), posing potentially a greater risk of environmental pollution. Highly concentrated animal excreta, such as poultry manure for example, are a particular concern and leaching of nitrate resulting in the pollution of groundwater and large volatilization losses of ammonia have been frequently reported for poultry waste spread on farmland (Giddens and Rao, 1975; Jackson *et al.*, 1977; Liebhardt *et al.*, 1979; Ritter and Chirnside, 1984; Bitzer and Sims, 1988; Unwin *et al.*, 1991; Nathan and Malzer, 1994; Sims and Wolf, 1994).

The application of sewage sludge to farmland is carried out either directly by the sludge producer or by appointed contractors with the necessary expertise and specialist application equipment. This option is preferred to the alternative, whereby the farmer is supplied with sludge and given the responsibility of applying it, because the operation can be more carefully monitored by the sludge disposal operators ensuring that all the necessary regulations and codes of practice are adhered to (DoE, 1989a; SI, 1989a; WRc, 1989; MAFF, 1991a) and potential environmental problems avoided. Consequently, sludge recycling is a highly organized and structured operation with a central management base within each of the responsible organizations in most countries. In contrast, animal excreta are disposed of in many countries in an uncoordinated and so far unregulated way on individual farms all of which are subject to a range of socio-economic pressures influencing the method and effectiveness of disposal route. For example, approximately 48 000 individual holdings in the UK support a dairy herd of cattle, over 70 000 have a beef herd, 20 000 have pigs, 91 000 have sheep and 2000 have broilers (MAFF, 1990a). Farmers are becoming increasingly conscious of the important role they play in protecting the environment (Hemington, 1992). Given the extent and diversity of the animal waste problem, however, it is not surprising that the National Rivers Authority (NRA, 1992a) has reported a total of 2400-3500 pollution incidents involving organic farm wastes each year between 1985 and 1989. In contrast, there is little evidence of water pollution incidents arising from the application of sewage sludge to agricultural land.

Nitrate Leaching

General

Environmental effects of nitrate leaching from agricultural land are concerned with the contamination of potable water supplies. Where public drinking water supply is not at risk, then N application to soil in inorganic fertilizers and organic manures is probably not a significant issue, but should be in accordance with crop requirements for N for the purposes of good agricultural practice. The comments in the following sections refer to those areas of land within the catchment of direct public water supply abstraction, or an exposed aquifer, which may be impacted by nitrate leaching losses from agricultural land.

Inorganic Fertilizers

Inorganic N fertilizers contribute to the problem of nitrate leaching through indirect and direct effects. Directly, and in the short-term, residues of inorganic fertilizer may be left in the soil which have neither been taken up by the crop nor immobilized in soil organic matter (Glendining *et al.*, 1992). Indirectly, and in the long-term, immobilization and the return of increased amounts of crop residues due to higher yields resulting from fertilizer use increase the soil organic N content and the level of mineralizable N, which also raises the potential risk of nitrate leaching (Glendining *et al.*, 1992). MacDonald *et al.* (1989) and Shen *et al.* (1989) found that 14-17% of spring applied fertilizer N was immobilized in the soil at harvest. However, Scaife (1975) measured a 60% potential lock-up of applied mineral N in immobilized forms.

Early work by MacDonald *et al.* (1989) on the Broadbalk Wheat Experiment at Rothamsted was very supportive of spring applications of fertilizer N not directly contributing to nitrate leaching over the winter period. However, recent studies (Glendining *et al.*, 1992) clearly showed that apparently economic optimum rates of addition of N recommended for Broadbalk would result in unused fertilizer N remaining in the soil for three years out of six. The presence of inorganic residues of N arising from a given recommended rate of application is dependent on:

1. Weather conditions after fertilizer N application - dry soil conditions restrict yield and uptake;
2. Supply of N from the soil - potentially highly variable.

Glendining *et al.* (1992) stated that the risk of leaving large N fertilizer residues in the soil at harvest could be reduced by:

1. Providing realistic predictions of yield;
2. Developing predictive models to estimate the N supply from soils on a site specific basis;
3. Splitting applications of fertilizer N;
4. Applying less than the recommended amounts of N.

Furthermore, Roberts (1987) showed that nitrate from fertilizers could also be leached from soil during spring. These spring losses of fertilizer N are intermittent, but can potentially be very large, depending on the length of time soil was at field capacity in the preceding winter. Thus, maintenance of soil near to field capacity during the winter increases the risk of leaching by spring rainfall. For the catchment studied by Roberts (1987), which comprised about 23% of arable land, spring leaching of fertilizer N occurred in two years over the seven year monitoring period (1978-1984).

Sewage Sludge

The problems of nitrate leaching from agricultural soils treated with sewage sludge can be separated into eight principal areas: (1) land availability; (2) cropping; (3) rate of application; (4) time of application; (5) application method; (6) type of sludge; (7) soil texture; and (8) long-term and residual effects.

Land availability

The availability of agricultural land for sewage sludge application varies through the year and depends on land use and type of sludge. Cake sludges are applied and incorporated in soil prior to drilling spring and autumn sown arable and grass crops. Cake sludges are rarely applied to established grassland although application is permitted under the UK Code of Practice provided certain physical criteria are met (DoE, 1989a). In general, 30% of cake sludge is typically applied in the spring to crops such as maize, spring cereals, linseed and fodder beet prior to sowing (R. Griggs, Thames Water, personal communication). The remaining 70% is applied almost exclusively to stubble during the period late July to October prior to drilling winter crops (R. Griggs, Thames Water, personal communication).

In areas where there is a mixture of arable cropping and grassland, liquid sludges can be applied throughout the year utilizing pasture when there is no arable land available. As with cake sludges, however, the principal period for applying liquid sludges still remains the late summer and autumn (Smith and Powlesland, 1990). The importance of this window increases in Eastern areas, in the UK such as Lincolnshire, which are predominantly dedicated to arable cropping. The large variation in seasonal availability of land for sludge application which occurs in these arable areas is shown in Fig. 6.1. Virtually no sludge can be applied between January and June. Land becomes available from July onwards reaching a maximum in August and September following the harvest of cereal crops providing accessible stubble land for spreading sludge. However, the peak in land availability corresponds to the period of rapid nitrate accumulation in soil potentially at risk from leaching by winter rainfall (Fig. 9.2). Furthermore, the Eastern areas and Midlands of the UK are located over potentially sensitive groundwater sources (DoE, 1986). Thus in these areas, where rainfall is relatively low and cropping predominantly arable, the potential impact of sludge recycling to agricultural land on nitrate leaching and groundwater quality will be significantly larger than in Western areas, which receive higher rainfall and are largely grassland (Smith and Powlesland, 1992).

Organic manures have been traditionally applied to agricultural soil during the autumn (MAF, 1937) when stubble land is accessible to spreading equipment. Clearly, the autumn period is also important to the Water Undertakings for applying sewage sludges for the same reasons. However, this approach is now at odds with current thinking on nitrate pollution. Indeed, it has been argued that autumn applications of N to agricultural land should be refrained from altogether (Prins *et al.*, 1988). Consequently, it is of critical importance that the measures adopted to reduce nitrate leaching losses from agricultural land are appropriate and that they allow flexibility for recycling sewage sludge, according to good agricultural practice, where the risks of impacting water resources are minimal. Some of the considerations and possible management options on nitrate for the agricultural use of sludge are discussed in the following sections.

Cropping

Arable crops remove most N from the soil in spring and summer (Fig. 9.2). Therefore, nitrate accumulated during the late summer and autumn period in arable soils is potentially susceptible to leaching loss by winter rainfall because there is insufficient crop cover of the soil to retain the available N. In contrast, grass provides a permanent cover and generally absorbs N whenever mineralization is occurring (Fig. 9.2) such that grassland is potentially a less 'leaky' system than arable farming. For example, Smith *et al.* (1994a) compared the leaching losses of N from arable and grassland soils treated with sewage sludge in field experiments on the same loamy sand soil. Nitrogen leaching from grassland was 88-96% lower than from arable soil maintained in a bareground condition until the spring.

Rate of application

Increased losses of nitrate by leaching following the application of sewage sludge and animal wastes to soil have been frequently reported (Melanen *et al.*, 1985; Berg *et al.*, 1987; Bertilsson, 1988; Chang *et al.*, 1988; Duynisveld *et al.*, 1988; Powlson *et al.*, 1989; Froment *et al.*, 1992; Wadman and Neeteson, 1992). The loss of nitrate by leaching may be expected to increase generally as a linear function of increasing sludge application rate in the absence of any crop cover, where crop cover is insufficient to absorb appreciable amounts of N, or where crop N requirements have been exceeded, as demonstrated by the root zone simulation model developed by O'Brien and Mitsch (1980). The arable leaching model of Jansson *et al.* (1989) supported this view for sandy soils, but indicated that nitrate leaching from clay soils only occurred above a threshold level of N application in organic manures (120 kg N ha^{-1}). This may be explained because coarse-textured sandy soils are much more sensitive to variation in factors such as climate, form and timing of applied N, and composition of manure than are fine-textured clay soils. In particular, sandy soils generally have a shallower rooting depth and smaller water storage capacity compared with clays and this has important effects on soil water movement and nitrate loss (see later). Steenvoorden (1986) provided experimental evidence of the linear increase in nitrate leaching losses with increasing rate of pig slurry application for a sandy soil under arable crop production. Nitrate leaching also increased approximately linearly with the amount of N fertilization for sand and loam soils under a corn-root arable rotation (Walther, 1989). Chang *et al.* (1988) showed that large concentrations of nitrate accumulated in profiles of sludge-treated soils when rates of sludge application exceeded crop requirements for N. Supra-optimal rates of sludge addition increased nitrate leaching from coarse and fine loamy soils as linear functions of increasing total N inputs.

O'Brien and Mitsch (1980) estimated that leaching rate increased by 21 kg N ha^{-1} y^{-1} t^{-1} of digested sludge (dry solids) incorporated into agricultural soils. Assuming 200 mm of effective rainfall, this value equates to an average groundwater recharge concentration of 10.5 mg NO_3-N l^{-1}. Powlesland and Frost (1990) calculated a standard application rate of raw liquid sludge (4 t ha^{-1} dry solids at 3% total N) injected into soil might be expected to increase the groundwater concentration by approximately 0.84 mg NO_3-N l^{-1}. This value would probably be more than doubled for digested sludges due to the higher N content and N availability. For digested cake sludge an estimated increase of only 0.12 mg NO_3-N l^{-1} was anticipated, although experimental data suggest this may underestimate the potential impact of nitrate from this sludge source on groundwater quality at rates normally applied (12.5 t ha^{-1} dry solids) in accordance with crop requirements for N (WRc, 1985).

Large losses of nitrate may be anticipated from soil when rates of application of organic manures exceed crop requirements for N due to the accumulation of unused mineral N in the soil at the end of the growing season (Soon *et al.*, 1978b; Sims and Boswell, 1980). However, adjusting sludge applications to meet crop requirements for N, ensuring that the additional available N supplied is taken into account in the fertilization programme minimizes the potential risk of nitrate loss (Steenvoorden, 1986; Chang *et al.*, 1988; Steenvoorden, 1989; Lorenz and Steffens, 1992; Wadman and Neeteson, 1992). As indicated earlier, grass can efficiently utilize up to 400 kg N ha^{-1} before the 'break-point' is significantly exceeded above which N will accumulate in the soil and be potentially at risk of leaching by winter rainfall (Prins *et al.*, 1988). Soon *et al.* (1978b) previously showed that three years of sludge applications to grass at 400 kg N ha^{-1} did not result in the accumulation of nitrate in soil. This 'break-point' value also corresponds closely with the economic optimum N application recommended for grassland (MAFF, 1994a).

Prins *et al.* (1988) noted that apparent N recoveries of 70-90% are commonly achieved by grass at annual rates of up to 400 kg N ha^{-1}. Furrer and Stauffer (1986) showed that increasing N application up to 700 kg N ha^{-1} in sewage sludge (which equated to approximately 400 kg ha^{-1} of available N) did not influence the nitrate concentration in the drainage water from lysimeters with grass cover. However, other studies (Walther, 1989) suggest a lower 'break-point' value of 200 kg N ha^{-1} is applicable for grassland before significant leaching of nitrate occurs. Recent work at WRc (Smith *et al.*, 1994a) has measured nitrate leaching losses from raw and digested liquid sewage sludges injected into grassland soils. The results show that concentrations in drainage water generally remain below the CEC limit value of 50 mg l^{-1} of nitrate in drinking water (CEC, 1980b) at operational rates of sludge application of 100 m^3 ha^{-1} when the N fertilizer regime is adjusted to account for the available N in the sludge. These levels of sludge addition fall well within the limit of 250 kg N ha^{-1} set by MAFF (MAFF, 1991a), but correspond with recommended application rates based on crop N requirements for typical sludges (WRc, 1985).

Leaching losses of nitrate from grazed grassland can be large and are unavoidable under current management techniques even where strict regulation of fertilizer and sludge application is practised (Prins *et al.*, 1988; Jarvis, 1992; Jarvis, 1994). This is because up to 80% of the N in herbage consumed by cattle, which can amount to about 330 kg N ha^{-1} y^{-1} under typical stocking densities, is returned unevenly to the soil. For example, nitrate leaching was over five times higher, at 162 kg N ha^{-1}, for a grazed system receiving 420 kg N ha^{-1} y^{-1} compared with a cut sward supplied with the same rate of fertilizer N on a sandy loam soil over chalk (Garwood and Ryden, 1986). However, improving the N efficiency of grazed systems could reduce the N requirement from 400 kg N ha^{-1} to 200-250 kg N ha^{-1} without a loss in production (Steenvoorden, 1989). One way forward here may be to make tactical adjustments to N additions to provide a uniform profile of N supply to the crop (Jarvis, 1992). Spring applications of slurry and sludge by injection in preference to surface spreading would also improve the N budget of grazed systems (Steenvoorden, 1989).

Managing the nitrate concentration in drainage from arable soils below the CEC limit concentration for drinking water is particularly difficult because of the relatively short duration of crop uptake of N early in the growing season and the accumulation of nitrate which occurs in soil during late summer and autumn when crop uptake is minimal (Fig. 9.2). Various reports estimate the 'break-point' for cereals is in the range 100-160 kg N ha^{-1} (Bergström and Brink, 1986; Jansson *et al.*, 1989) which, again, is close to recommended rates of N fertilizer application (MAFF, 1994a). The nitrate concentration in the drainage water from arable soil remained below the CEC drinking water limit (CEC, 1980b) at the optimum N application of 100 kg N ha^{-1} estimated by Bergström and Brink (1989), whereas at 200 kg N ha^{-1} the level rose to almost three times the limit at 30 mg NO_3-N l^{-1}. However, the minimum concentration of nitrate in drainage from the arable crop rotation studied by Walther (1989) exceeded the CEC limit even if no fertilizer was applied. Bertilsson (1988) similarly measured 25 mg NO_3-N l^{-1} in drainage water without any N application for 10 years. Nitrate concentrations in autumn and winter drainage measured by Smith *et al.* (1994a) were 3-4 times the permitted limit for drinking water following a cereal crop fertilized in the previous spring according to recommended practice (MAFF, 1994a). Nitrate levels in drainage reported by Shepherd (1993) also exceeded the drinking water standard for a loamy sand soil under conventional arable management. In one example, the N concentration in drainage was as high as 83 mg l^{-1}. Clearly fundamental changes in arable cropping practices and rotations may be necessary to achieve the level of water quality required by the regulations. Indeed, at certain arable sites (mainly sandy) Duynisveld *et al.* (1988) suggested that it may not be possible to reduce the nitrate concentration in the groundwater recharge to values below the CEC standard for drinking water, in spite of all measures taken to reduce nitrate leaching. Consequently, sludge application to soil when land is accessible in the autumn will almost certainly exacerbate the problems of nitrate leaching, unless management practices are found which are effective in retaining in the soil the potentially large amounts of residual mineral N as well as the applied sludge N, until it can be utilized by the crop in the following spring.

The concept of a 'break-point' rate of N application to crops, which prevents nitrate leaching, provides a practical method for regulating sludge additions to maximize recycling opportunities without placing the environment at risk. However, the N content of sewage sludges can vary, reflecting a range of operational conditions at different sewage treatment works and also at any single works. Consequently, accurate targeting of rates of N application in the field is relatively difficult. Most disposal operations have therefore spread a fixed application rate of sludge generally in accordance with the typical sludge N contents given in the WRc Farmers' Guide (WRc, 1985) and farmers supplement the application with mineral N. However, there is considerable scope to improve the advice to farmers on the N fertilizer replacement value of sludge, which would benefit the environment by ensuring crop requirements for N are not exceeded through over-fertilization with chemical N. Tactical adjustment of sludge application rates according to site specific requirements and crop needs would also be beneficial, although perhaps difficult to achieve in practice. To establish a greater level of accountability with respect to the amount of N applied to agricultural soils in sludge and potential leaching losses of nitrate, closer monitoring of sludge N content would be desirable. This could be achieved by using rapid N analysis techniques in the field (Titchen and Scholefield, 1992) but these require detailed evaluation and testing. Such analyses could form the basis of quantitative, tactically adjusted N fertilizer recommendations for specific cropping, sludge and site conditions. At a more fundamental level, however, closer regulation of operating conditions at sewage treatment works could provide a more uniform product with respect to its N content and application rate for recycling to agricultural land as a fertilizer. Nevertheless, there is a better understanding of sludge N content and availability compared to soil N in relation to fertilizer requirements of crops and leaching losses of nitrate. The uniformity of sludge products as fertilizers will increase with respect to their N content, with improving operational practice at STWs.

Time of application

The efficiency of utilization by crops of N supplied in sewage sludges and animal slurries is highly dependent on the time of manure application. Thus, from a review of data from field fertilizer experiments, Williams and Hall (1986) established that the N efficiency value (total N) of sludges and slurries was 50-70% of equivalent rates of inorganic fertilizer N when applied in the spring and worked into the soil. Autumn or early winter applications can reduce the efficiency to 20-30% on arable land. For grassland, the average N efficiency is about 50% when surface applied in spring, but can be as low as 30% from autumn application. The lower efficiencies obtained from the late season applications probably reflect the losses of N which may occur due to nitrate leaching.

Comparing recommendations on the fertilizer value of sewage sludges (MAFF, 1994a) with experimental evidence on N availability from autumn and winter applied sludges, suggests that the guidance levels may underestimate the amount of available N in soil for the purposes of calculating additional N fertilizer needs (Table 9.2). This may be partly explained because it appears that the values have been estimated on the basis of available N remaining in the soil for spring growth. However, early sown winter crops can be very effective in N uptake (see later) and even October drillings are able to utilize a significant proportion of the applied available N. In operational practice it is unlikely that sewage sludge will be applied in the autumn to arable soil which will not be cropped until the spring. Winter crops will be treated first with sludge since spring crops are available for application later on when winter crops have been installed. Consequently, it may be prudent to derive higher, and possibly more realistic, N availabilities than those given (MAFF, 1994a), thereby adjusting supplementary fertilizer applications downwards. This should avoid exceedence of crop requirements for N and minimize the risk of leaving unused fertilizer residues in the soil in the following autumn which are susceptible to leaching by winter rainfall.

Direct measurements of nitrate concentrations in drainage generally confirm that excessive leaching can occur from autumn applied sludges and slurries depending on management practices. For example, Bertilsson (1988) recorded peak concentrations of more than 100 mg NO_3-N l^{-1} in drainage from lysimeters receiving 200 kg of ammonium-N as slurry in the autumn. However,

Table 9.2. Proportion of total N available to the next crop following applications of sewage sludges (% of total N).

Timing:		Autumn[1] (Aug-Oct)		Winter (Nov-Jan)		Spring[2] (Feb-April)	Summer use on grassland
Soil type:	Dry solids (%)	Sandy/ shallow	Other mineral	Sandy shallow	Other mineral	All soils	All soils[2]
Sludge type							
Liquid digested	4	5	10	15	35	60	40
Dewatered digested	25	5	5	10	15	15	n/a

(1) Where average or actual excess winter rainfall is significantly below 250 mm (annual rainfall 750 mm), the values for autumn application should be increased to those given for winter application which assume 150 mm excess winter rainfall after application.
(2) Following injection or rapid incorporation, ammonia volatilization losses will be reduced and nitrogen utilization may be improved. Grass yield response will depend on the extent of injector tine damage to the sward.
Source: MAFF (1994a).

the extent of these losses is not surprising since the ground was left bare over the winter prior to drilling spring wheat. There was a trend for less nitrate to be leached from later applications due to lower soil temperatures and the smaller amount of drainage. Similar patterns of nitrate leaching have been found in field experiments on coarse-textured soils receiving digested sludges (liquid and dewatered) from October onwards (Shepherd, 1993; Smith *et al.*, 1994a). Winter cereals drilled immediately after application absorbed some of the applied available N, but were unable to prevent excessive nitrate concentrations accumulating in drainage water. Therefore, Shepherd (1993) recommended that a very stringent regime of controls on the timing of sludge applications would be necessary to minimize the risk of nitrate leaching from sludge-treated agricultural land. Proposed limits on application times for liquid digested sludge were:

1. Arable - 1 July-30 December
2. Grassland - 1 September-30 November

If this period were considered to be too long for practical purposes, Shepherd (1993) suggested a compromise might be to allow spreading to recommence one month earlier. Adopting a more pragmatic approach, Smith *et al.* (1994a,b) argued that the declining soil temperatures from November onwards should restrict nitrification to a level which reduces the potential risk and extent of nitrate leaching from sludge-treated soil. For example, a mass balance showed that all of the available N applied in a mid-November treatment of liquid digested sludge was recovered in the soil profile (1 m) in the following spring even with above average drainage at the experimental site of 258 mm. Consequently, Smith and co-workers suggested that the application of liquid digested sludge should only be restricted between 1 September to 31 October provided that certain requirements on cropping could be met for the later start to the exclusion period, which will be discussed shortly. However, this restriction need only apply to coarse-textured soils considered potentially sensitive to nitrate leaching. These recommendations are in-line with those proposed in the Outline Action Programme for NVZs in England and Wales (MAFF, 1994c).

In contrast, Hann *et al.* (1992) reported a maximum nitrate concentration in drainage water about half of the CEC limit value for a sandy loam soil in lysimeters treated consecutively for three years in autumn with liquid digested sludge at a rate of 100 m^3 ha^{-1} y^{-1}. A winter wheat crop reduced the nitrate level in drainage by up to 50% although variability was high between the different years. The relatively small concentrations of nitrate in drainage from sludge-amended soil were partly due to the low background level, which was generally ≤2 mg N l^{-1} except for the third year when the background concentration was ≤4 mg N l^{-1} for the untreated control. Furthermore, the quantity of N applied to the lysimeters in sludge was relatively modest (70 kg N ha^{-1}) although the volume supplied was typical of field practice (WRc, 1985).

Field estimates of the extent of nitrate leaching from sludge-treated agricultural land are usually highly variable. For example, Shepherd (1993) estimated that 59% of the N supplied in liquid digested sludge, injected into a loamy sand soil in September, was leached from the soil profile with 180 mm of drainage. In the following year, only 22% of the sludge N was leached at the same site even though the drainage was nearly 30% higher. Smith *et al.* (1994a) measured a N leaching loss of 19% from injected liquid digested sludge in October from a similar sensitive soil type in a relatively dry year with only 97 mm of drainage. In a later trial (Smith *et al.*, 1994b) with a markedly higher drainage (258 mm) there was only a trace of nitrate detected in leachates which could be attributed to sludge injected in August, September or October. In both cases the low recoveries of N in winter drainage were probably associated with denitrification processes, which may be prevalent in wet conditions occurring at the time of sludge application, although few direct measurements of this route have been made. However, application method also influenced the differences observed in N leaching and is considered in the next section.

Whilst there is a recognized potential risk of nitrate leaching from soil treated with liquid digested sludge during the autumn period, management techniques are available which may provide opportunities for minimizing leaching losses possibly allowing the continuation of this practice, or at least the adoption of less stringent control measures. Whilst the former is almost certainly over optimistic, the latter may be feasible in practice. The options available with recognized merit in reducing leaching losses include:

1. The early sowing of winter crops (Widdowson *et al.*, 1987).
2. Use of cover crops to minimize the periods of time that arable land is in the bareground condition (Christian *et al.*, 1992).
3. Chemical inhibition of nitrification (Slangen and Kerkhoff, 1984; Smith *et al.*, 1989).

The suitability of a particular technique will depend on the timing of sludge application which will govern the efficacy of the method in minimising potential leaching losses of N.

Sowing date is a critical parameter if winter crops or cover crops are to prevent nitrate leaching. Bertilsson (1988) showed that nitrate loss from August applied slurry was almost prevented by a cover crop of rape which was ploughed down in November. Early sowing is essential for the efficient utilization of the available N supplied in sewage sludge and avoidance of large nitrate leaching losses by winter rainfall. For example, new varieties of winter wheat are becoming available which can be sown early in autumn and which readily assimilate available N present in the soil (NSDO, undated). However, delayed drilling from early September until mid-October can significantly reduce the level of N absorbed by more than 50% (Table 9.3). Other winter crops also assimilate N present in the soil during the autumn, but species vary in their effectiveness at N uptake (Shepherd, 1990). For example, winter rye is more effective than rape (Fig. 9.6) although, as with wheat, sowing date is critical (Fig. 9.7).

Table 9.3. Effect of sowing date on the nitrogen uptake and yield of winter wheat (variety Bounty).

Sowing date	Total dry weight[1] ($t\ ha^{-1}$)	Nitrogen content[1] ($kg\ N\ ha^{-1}$)
3 September	4.4	116
20 September	3.0	78
11 October	1.8	57
30 October	1.0	34
3 December	0.2	9

[1] Determined from samples harvested on 28 March, prior to N fertilizer application
Source: National Seed Development Organization

Smith *et al.* (1994b) found an early sown crop of winter wheat, drilled on 2 September, was highly effective in preventing N leaching giving a seasonal average concentration of nitrate in drainage water of only 3.0 mg N l^{-1}. Delayed sowing markedly increased the nitrate content in leachates. Indeed, postponing

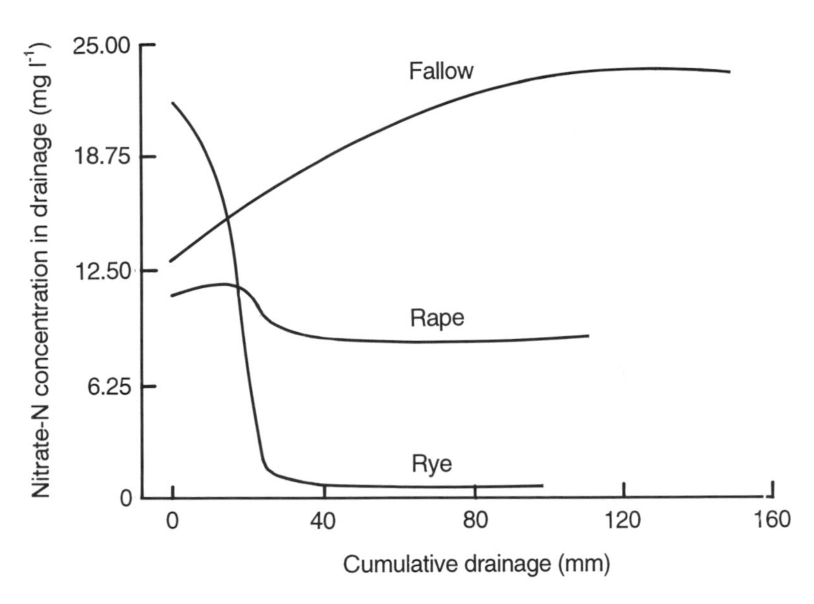

Fig. 9.6. Comparative nitrate leaching losses in drainage from cropped and fallowed soil (Shepherd, 1990).

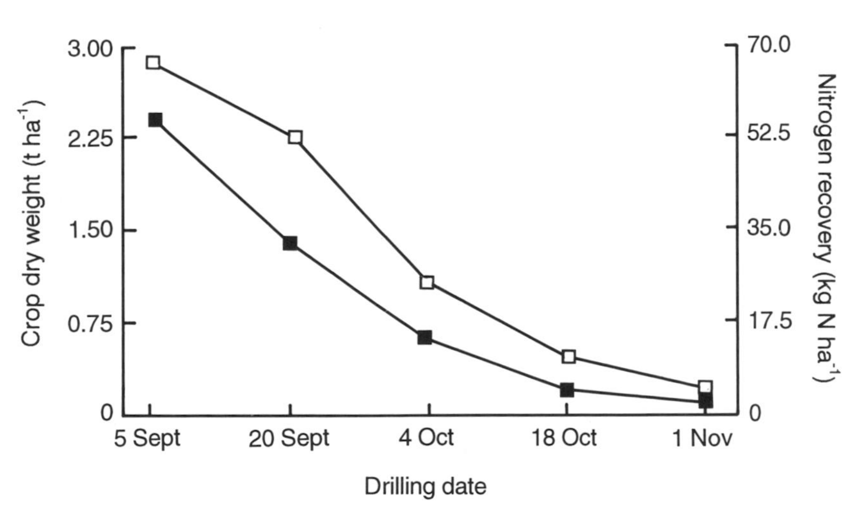

Fig. 9.7. Effect of drilling date on total crop yield [■] and nitrogen uptake [□] of winter rye, sampled on 31 Jan. 1989 (Shepherd, 1990).

the sowing date for a further two weeks caused exceedence of the drinking water standard at 21 mg N l^{-1} on average in the drainage. Winter wheat drilled at the conventional time in mid-October gave an average leachate concentration of 38 mg N l^{-1}. Intermediate cover cropping with winter rye after harvesting the previous crop and prior to drilling wheat in October was evaluated as a potential option for late summer applications of sludge. The cover crop was highly effective in drawing down nitrate levels initially in leachates. However, concentrations soon rose to an unacceptable level when the cover crop was removed and winter wheat was drilled, due to the lower effectiveness in N uptake of conventional autumn sowings of winter cereals. On this basis, Smith *et al.* (1994b) recommended that liquid digested sludges should not be applied to potentially vulnerable soils after 1 September. Earlier applications to these soils could only be justified if a winter crop (e.g. oil seed rape) was drilled immediately afterwards. Other work (Hann *et al.*, 1992) has shown that the closer the time of application and sowing, the more effectively N is retained by the crop, and the smaller the potential leaching loss, such that this approach is commended for late summer applications of sludge.

Whilst early sowings have considerable potential in limiting N loss from soil, this approach does not necessarily avoid the possible risk of nitrate leaching altogether. For example, dry soil conditions can restrict establishment and early growth of crops (Catt *et al.*, 1992; Fielder and Peel, 1992), thus possibly reducing the extent of N uptake. Despite this apparent problem, however, early sown crops have been highly effective in N uptake during dry seasons (Christian *et al.*, 1992). Another important consideration is that early sown winter cereals may require greater pesticide inputs to minimize the incidence of disease problems (Green and Ivins, 1985).

Nitrification inhibitors have shown some potential in reducing losses of nitrate from soils by leaching, as well as denitrification. They act upon nitrification processes in soil by blocking the conversion of ammonium into nitrite. In particular, the inhibitor dicyandiamide (DCD) has been effective in reducing N losses when supplied in combination with animal slurries (Amberger, 1983; Wadman *et al.*, 1989). Smith *et al.* (1994b) assessed the efficacy of DCD in reducing N loss from liquid digested sludge applied to agricultural land in the critical period during autumn to early winter. The DCD was mixed with the sludge and injected into arable soil at different times and the residual mineral N in the soil profile the following spring was measured by sectional (20 cm) sampling to 1 m (Table 9.4). There was no effect of sludge treatment or inhibitor addition on soil mineral N content for the September application compared with an untreated control due to crop uptake and N loss. On the other hand, sludge injection in November significantly increased the mineral N content of bare soil in the spring compared to the control irrespective of DCD incorporation, which had no significant effect. However, DCD was effective in retaining available N in the soil from October applied sludge.

Table 9.4. Average soil mineral N (5x20 cm sections of the profile to I m) in the spring (kg N ha^{-1}) after injecting liquid digested sludge at different times supplied with and without DCD.

Time of application	Untreated control	Sludge-treated no DCD	Sludge-treated with DCD	Least significant difference (P=0.05)
September	5.6	4.9	3.8	ns
October	6.1	6.4	10.8	1.9
November	7.7	17.4	17.3	4.4

No. of replicates for each treatment mean = 15
ns, not significant
Source: Smith *et al.* (1994b)

The effectiveness of chemical inhibition can be highly variable and difficult to predict in the field due to the influence of environmental conditions on inhibitor activity and nitrification rate. Degradation of DCD in soil, for example, is highly temperature dependent (Slangen and Kerkhoff, 1984). Activity of the inhibitor appears to be relatively short-lived in warm soils which may explain its apparent ineffectiveness when supplied with liquid digested sludge in September (Smith *et al.*, 1994b). Unless the inhibitor can remain active beyond the period that soil temperatures promote nitrification then nitrate will ultimately accumulate in the soil and be potentially vulnerable to leaching by winter rainfall. However, there may be a relatively narrow window in October when declining soil temperatures are high enough for nitrification, but allow the persistence of DCD activity until later in the season when nitrification rate is reduced by low temperature conditions. By November, decreasing soil temperatures slow the rate of nitrification providing a natural mechanism for protecting the available ammonium-N in sludge from leaching and denitrification, obviating the need for chemical means of inhibition.

A further disadvantage in the use of nitrification inhibitors is the cost of available products. For example, at recommended rates of application the cost of treatment is currently £40 ha^{-1}. However, to improve the effectiveness of inhibition it is probably advisable to supply larger rates than this (Shepherd, 1993; Smith *et al.*, 1994a). The high cost of using these materials with sewage sludge may be difficult to justify given the uncertain benefits. More importantly, however, nitrification inhibitors are not featured within the restrictions on manure applications to farmland intended to reduce nitrate leaching from diffuse agricultural sources.

In contrast to the potential impacts on drainage water quality of autumn applications of sludge and slurries, significantly improved efficiency in N utilization by crops, with a concomitant reduction in nitrate leaching, is achieved by spring applications to actively growing crops. This is because the N is being supplied when the crop is best able to utilize it (see Fig. 9.2). For example, Lorenz and Steffens (1992) combined N applications in animal slurry (100-160 kg N ha^{-1}) and mineral fertilizers (0-60 kg N ha^{-1}) and measured only minimal accumulation of N in soil at harvest following treatment of winter cereals during the period end February-early March or immediately before sowing of sugar beet, potatoes and maize. Kemppainen (1990) recovered almost all the soluble N in slurry (68 kg N ha^{-1}) applied to a spring barley crop 1-9 days before sowing and 11-13 days after sowing. Later treatment (17-24 days after sowing) resulted in lower recoveries, presumably due to crop damage sustained through trafficking with the application equipment. The method of application was also important with lower recoveries being achieved by surface spreading compared with injection (see later). Smith and Chambers (1992) did not detect any difference in the efficiency of slurry N surface-applied to winter cereals between mid-February (early tillering) and late April (early stem extension). They recommended that later applications towards flag leaf emergence should be avoided due to the risk of yield reductions caused by scorch and smothering. The improved N balance obtained by sludge application shortly before or after drilling spring wheat, compared with sludge treatment in autumn and winter, was demonstrated by Hann *et al.* (1992). However, there are few quantitative data available on the effects of spring treatment of winter cereals with sludge on crop N utilization, soil mineral N status or potential leaching losses of nitrate. In a preliminary study, Smith *et al.* (1994a) showed that liquid digested sludge spread over a growing crop of winter wheat in the spring had more or less comparable fertilizer value to inorganic N on an available N content basis. Treating the crop with sludge as late as the end of May produced a similar yield response to that obtained with mineral N, but without the adverse effects caused by leaf scorch apparently observed with animal wastes.

Opening the period when crops are established in spring and summer to sludge application could make the single largest contribution to managing nitrate loss from sludge-treated arable land whilst still allowing flexibility in agricultural outlets for sludge recycling (Smith and Powlesland, 1990). There is an urgent need for research in this area, and into new application technologies for sludge, to maximize the opportunities for sludge recycling on growing crops to gain the full benefit from the nutrient content of sludge and minimize risk to the environment.

Application method

The importance of application method was briefly mentioned in the previous section. The largest difference in nitrate leaching is obtained when surface application is compared with soil incorporation techniques. Surface spreading can result in significant losses of N by ammonia volatilization (see later) compared with incorporation, where substantially more of the applied N is retained in the soil which improves fertilizer efficiency, but is also potentially at risk from leaching. Thus, O'Brien and Mitsch (1980) estimated that surface application of sludge resulted in leaching rates 50% lower than following sludge incorporation. Anticipated losses of N by leaching from surface applications will depend principally on the extent of ammonia volatilization and on crop uptake of the remaining N. However, encouraging gaseous losses of ammonia in this way, in order to reduce nitrate contamination of water supplies, is considered unacceptable because volatilized ammonia itself has a significant environmental impact.

Soil injection of sludge is widely practised and effectively prevents ammonia loss by volatilization (Williams and Hall, 1986). However, deep placement of sludge (>20 cm) in grassland on light sandy soils may potentially lead to increased N leaching because most grass roots are in the upper 5 to 10 cm (Garwood and Sinclair, 1979; Steenvoorden, 1989), although the effective rooting depth of grass may extend to 80 cm down the soil profile (Garwood and Sinclair, 1979). Shallow placement of sludge in grass and arable crops is likely to benefit the uptake and retention of N by putting the sludge N in close proximity to crop roots. Consequently, the leaching losses of N measured by Shepherd (1993) from liquid sludges injected into arable soil in September were probably aggravated to some extent by the excessively deep placement of sludge to 40 cm in the soil profile. The recommended optimum working conditions for winged tine injection equipment used in the trials work is less than half this depth at 15-20 cm (WRc, 1989). In the following year, N leaching was reduced by up to 80% when a shallower injection depth of 20 cm was used despite an increase in the quantity of drainage compared with the previous winter. In contrast, Smith *et al.* (1994b) observed only small losses of N by leaching from liquid digested sludge applied in August, September and October by injection to an arable soil at a depth of 15 cm and cropped with winter wheat. Indeed crop uptake of N measured in the spring accounted for more than 50% of the available N supplied in the sludge. Shallow slot injection and incorporation techniques for applying sludge more efficiently to grassland and growing arable crops (Sneath and Phillips, 1992) will also lead to better utilization of applied N and smaller leaching losses (Smith and Powlesland, 1990).

Type of sludge

Sewage sludges vary in their inorganic, readily mineralizable and recalcitrant N contents, depending on the method of sewage and sludge treatment (Smith *et al.*, 1992f). Therefore, an assessment of the potential impacts of sewage sludge utilization in agriculture on nitrate pollution of the environment requires an understanding of the N release characteristics of sludges commonly applied to agricultural soils.

Hall (1986a) described the N availability characteristics of sewage sludges based on several years' data from extensive field fertilizer trials with sludge and these are presented in Table 9.1. The information has been published in the form of a practical guide to farmers in *The Agricultural Value of Sewage Sludge: A Farmers' Guide* (WRc, 1985), offering advice on the N fertilizer replacement value of the different types of sludge which are available for application to agricultural land. Over 70 000 copies of the Farmers' Guide have been circulated within the UK water industry. Similar guidance on sludge from the same source is given in the *Fertiliser Recommendations for Agricultural and Horticultural Crops (RB209)* (MAFF, 1994a).

Nitrogen is present in sludge in ammoniacal and organic forms. The ammonium-N is readily available for crop uptake and nitrification. The rate of mineralization of the organic component varies with the sludge treatment process and environmental and edaphic conditions when the sludge is applied to soil. However, the organic forms are generally regarded as providing a slow-release source of N effective over several years (Fig. 9.8).

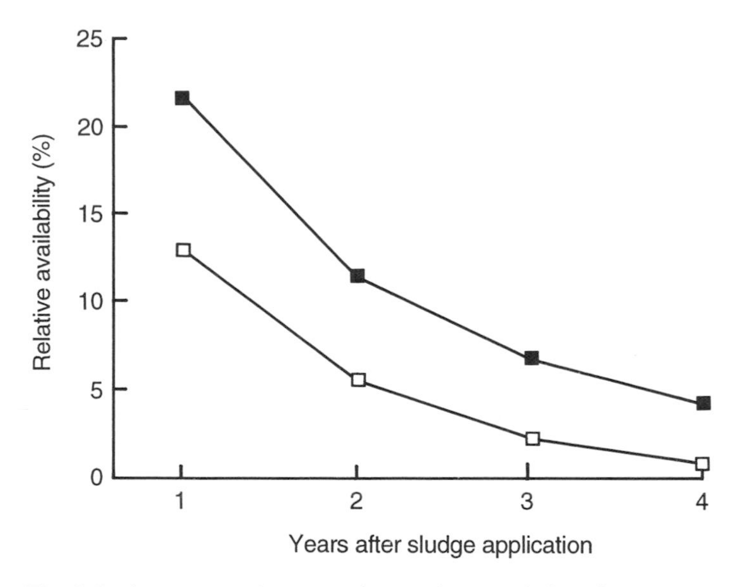

Fig. 9.8. Long-term nitrogen release characteristics of dewatered undigested [■] and digested [□] sewage sludges relative to inorganic nitrogen fertilizer (Hall, 1985).

Nitrogen contained in liquid raw sludge is largely in organic forms with only 5-10% as ammoniacal nitrogen. The organic matter is unstabilized and so mineralizes more rapidly than that in digested sludge. In general, approximately 35% of the organic N will be mineralized in the first year. Liquid digested sludge contains a high proportion of ammoniacal N since anaerobic digestion mineralizes much of the readily degradable organic matter. On average, 60% of the total N will be ammonia but this varies from 30 to 70% reflecting the range of operating conditions on different sewage works. Organic nitrogen remaining in the sludge after digestion is highly resistant to mineralization in the soil and Coker (1983) estimated only about 15% becomes available in the first year following application.

In contrast to liquids, cake sludges lose a large proportion of their soluble nutrients during dewatering so they contain relatively little inorganic N on a dry solids basis. The rate of mineralization in soil of the remaining organic N depends on whether or not the sludge has been previously digested. Hall (1986a) estimated the N availabilities of raw and digested cake sludges were 20% and 15% in the first year, respectively. Thus, at operational rates of sludge application in the field (100 m^3 ha^{-1} for liquids and 50 t ha^{-1} for cakes; WRc, 1985), the quantity of available N supplied to soil in digested dewatered sludge may be comparable to that applied in liquid digested forms due to the larger dry solids addition to soil. This reflects the variation in digestion conditions and effectiveness in dewatering achieved during sludge treatment.

Laboratory incubation studies of the N release characteristics of sewage sludges mixed with soil have also proven useful in developing recommendations for farmers on the fertilizer replacement value of sludges and are remarkably consistent with field observations. For example, the mineralizable N content of anaerobically digested sludge has been estimated as 15% of the organic N fraction by this approach (Sommers and Nelson, 1981; Parker and Sommers, 1983). Laboratory determinations offer some major advantages in predicting N mineralization from organic manures because they are relatively quick to perform, allow careful control of soil and environmental conditions and are relatively inexpensive compared with field experimentation.

Composting processes similarly produce residues which are highly resistant to N release. Indeed, Epstein *et al.* (1978) suggested that the mineralization patterns of digested sludge organic matter resembled those of native soil organic N, whereas N in composted sludge was even less available. Incubation estimates of compost mineralization in soil suggest that <10% of the organic N content is released in available form (Tester *et al.*, 1977; Sommers and Nelson, 1981; Parker and Sommers, 1983; Smith and Woods, 1994). Thus the risks of N leaching are minimal. For example, Chang *et al.* (1988) detected no significant build-up of nitrate in the profile of soil receiving 2500 kg N ha^{-1} of composted sludge. Field trials investigating the manurial value of sludge composts similarly show the limited extent of N release from these stabilized products. Smith *et al.* (1992a) supplied a total application of up to 800 kg N ha^{-1} as sludge compost and detected no increase in uptake of N by cabbage or onions.

Consequently, stabilized sludge organic matter is unlikely to cause excessive nitrate leaching unless very large quantities, well above normally applied operational rates, are regularly spread on farmland. The principal benefit to agriculture of organic manures with low N availabilities is from the improvement of soil physical properties by supplying organic matter. Indeed, organic matter destruction in soil is a continuing concern of modern intensive crop production practices (MAFF, 1993a). The problem may become particularly acute on certain arable soils (e.g. sandy loams) such that yields and fertility may progressively decline due to loss of structure which cannot be rectified by applications of high rates of N fertilizer (Tyson *et al.*, 1990). Indeed, it is those soil types most at risk from nitrate leaching which are also highly susceptible to structural damage. These soils would benefit from the application of organic materials to correct the decline in soil quality and agricultural productivity. However, it is necessary to apply bulky soil conditioners at large rates exceeding the recommended maximum level of N application in manures of 250 kg N ha^{-1} (MAFF, 1991a) to gain a significant benefit. For example, sewage sludge has a measurable effect on soil physical conditions following the application of 20-30 t ds ha^{-1} (Hall and Coker, 1983), which is equivalent to 600-900 kg N ha^{-1} in sludge compost at a typical total N content of 3% (dry solids). Nevertheless, the amount of available N which is applied to the soil is less than 40% of the maximum recommended limit.

The *Code of Good Agricultural Practice for the Protection of Soil* (MAFF, 1993a) recommends that measures are taken to maintain soil organic matter status to achieve high yielding crops although there is no advice given on how much manure should be supplied. However, the required amounts of organic matter cannot be applied without also applying seemingly large rates of N. Consequently, it would appear that there could be a conflict emerging between maintaining soil quality by spreading bulky organic manures and the limits on total N application. Furthermore, farmer acceptance of these types of material may decline if no discernible benefit is observed at the permitted rate of N application thereby perhaps unnecessarily restricting the opportunities and benefits of recycling sludge products (and other organic wastes) to agricultural land in the future.

On the basis of reported N availabilities for sludge in the first year after application and the patterns of N mineralization determined by laboratory incubation, the decreasing potential risk of nitrate leaching from the various types of sewage sludge supplied at operational rates in the field is likely to be:

Liquid digested $\geq$ Dewatered digested $>$ Liquid undigested $>$ Dewatered undigested

At equivalent rates of N application the decreasing level of risk is likely to follow:

Liquid digested $\gg$ Liquid undigested $>$ Dewatered digested $>$ Dewatered undigested

Smith *et al.* (1994a) and Shepherd (1993) have provided experimental evidence confirming these relative differences in leaching risk between the principal types of sewage sludge currently recycled on agricultural land.

In the longer-term, however, the decreasing potential risk of nitrate leaching due to mineralization of the residual N content of the sludges at operational rates and frequency of N application is likely to be reversed and follow the sequence:

Dewatered undigested $\geq$ Liquid undigested » Dewatered digested > Liquid digested

A similar pattern is probably also appropriate when long-term residual effects are considered on the basis of equivalent rates of N application in sludge. However, C:N ratio may play a significant role in the decomposition of raw cake sludges (see earlier), and Jansson *et al.* (1989) showed that increasing the C:N ratio of solid manures resulted in decreased leaching of nitrate particularly for sandy soils. Further aspects of the residual properties of sewage sludges are discussed in more detail later.

Research with organic farm wastes has also made the distinction between manures containing a high proportion of available N, such as animal slurries and particularly poultry manures, from those materials with only limited N mineralization potential such as FYM. For example, Unwin *et al.* (1991) reported that only limited mineralization of N occurred from autumn applied FYM (193-432 kg N ha^{-1}) and there was no evidence of nitrate movement to depth in the soil profile. Losses of nitrate from FYM averaged about 5% of the total N applied from October-December treatments. In contrast, nitrate leaching losses from poultry manure (215-360 kg N ha^{-1}), spread in October and November, were 40-50% of the total N applied. However, applications made in January and February gave no increased loss compared with the control. No crop was established on the plots so these data represent potentially a 'worst-case' leaching loss of nitrate. Interestingly, assuming a typical level of drainage from the experimental site of 200 mm, even the control plot (which had received no N since the previous crop of spring barley) would have exceeded the CEC drinking water standard for nitrate in the drainage water.

Chambers and Smith (1992) showed that autumn applied FYM (300-500 kg N ha^{-1}) resulted in little increase in soil nitrate content, whereas animal slurries (pig and cattle) (100-300 kg N ha^{-1}) and poultry manure, which are rich in ammonium-N, raised soil nitrate content appreciably. Chambers and Smith (1992) concluded that autumn/early winter applications of 'high' available N manures (poultry manure and animal slurries) should therefore be avoided to reduce leaching losses of nitrate. Froment *et al.* (1992) also measured large leaching losses of N from animal slurries applied before December. Up to 40% of the slurry N was lost equating approximately with the mineral-N fraction applied whereas losses from October applied FYM were small in comparison.

All available data indicate that the principal sludge factor which determines the potential extent of nitrate leaching from stabilized digested sludges, irrespective of dewatering, is the rate of ammonium-N application to soil in the sludge (Smith *et al.*, 1992f). Chambers and Smith (1992) demonstrated a similar relationship for farm wastes. Consequently, sludges with low N availability are likely to pose a smaller risk of nitrate pollution at comparable rates of N application to high availability materials, and Froment *et al.* (1992) argued that this provided justification for separating high- and low-availability manures when considering restrictions on the application of organic amendments to the soil which attempt to reduce nitrate leaching.

Soil texture

Soil texture and structure affect the amount of water retained at field capacity and consequently influence the extent of nitrate leaching (Wild, 1988a). If this water content is expressed as a fraction, θ, of the soil volume and it is assumed that all incoming water displaces the resident solution, and that Q represents the amount of drainage, the downward movement can be represented as Q/θ. Thus nitrate is more readily leached from sandy soil which has a value of θ of approximately 0.1 than from a loam which has a value of 0.35. In other words, approximately three times as much drainage might be needed to leach nitrate out of a loam soil compared with a sandy soil (Wild, 1988a). However, from a comprehensive survey of 112 reports dealing with N leaching, Walther (1989) found that the nitrate concentration below a loamy soil was not much smaller than below a sandy soil.

Nitrate losses from clay soils are generally expected to be small due to high soil water retention. Using a simulation model of nitrate leaching from arable soils, Jansson *et al.* (1989) predicted that clay soils were relatively insensitive to manure treatment in the autumn. Sandy soil was much more sensitive to nitrate leaching from manure applications in comparison due to shallower rooting depth and small water-holding capacity. However, macropores (large natural cracks and channels) in clay soils, can provide pathways for the rapid movement of water and substantial leaching of nitrate (Haigh and White, 1986). Thus, Hann *et al.* (1992) measured marginally higher levels of nitrate leaching from a clay soil than from a sandy soil for autumn applied sewage sludge. Haigh and White (1986) reported that nitrate concentrations in drainage water from a clay soil under grazed grassland could exceed 20 mg N l^{-1} throughout the winter period. Nitrate content in drainage measured by Roberts (1987) from a predominantly clay catchment was always greater than the CEC drinking water standard (CEC, 1980b) on resumption of flow each autumn, but subsequently declined. A similar pattern of drainage water nitrate was observed by Catt *et al.* (1992) for a clay soil at the Brimstone Farm Experiment. The initial winter drain flows at Brimstone contained 24-75 mg N l^{-1}.

Nitrate leaching from sewage sludges can be influenced by soil textural properties for other reasons. Mills and Zwarich (1982) monitored the accumulation, movement and losses of nitrate from fine-textured and predominantly poorly drained soils treated with heavy applications of digested sewage sludge. Sludge was applied during the winter at rates of 5500-12 000 kg N ha^{-1}. The fields were left fallow for the following summer and were then cropped with cereals. Despite the massive additions of N made in sludge, downward movement of nitrate was very slow and did not occur below 1.5 m. Consequently the nitrate concentration in groundwater collected adjacent to the sites remained at a low level. However, there was a continuous loss of nitrate over the study period which was attributed largely to denitrification. Therefore, in contrast to high risk permeable soils with good internal drainage, poorly drained soils favour denitrification and may reduce the likelihood of groundwater pollution by nitrate. However, nitrous oxide gas generated during denitrification has a significant impact on the environment through its potential contribution to global warming and may be considered unacceptable in the long-term. Denitrification is discussed in more detail in relation to sewage sludge application to agricultural soil in a subsequent section.

The vulnerability to nitrate leaching of the different soils in England and Wales has been classified according to textural properties and soil depth (NRA, 1992b). Controls on certain types of manure and sludge are justified to protect vulnerable groundwater resources from contamination with nitrate in catchments where soils have been designated as having potentially a high risk of nitrate loss. However, outside of these catchment areas, and where soils have lower risk of leaching due to increasing clay and silt contents, less stringent measures on sludge applications may be acceptable without placing potable water supplies at risk.

Long-term and residual effects

The application of ammonium-rich and readily mineralizable sewage sludges to soil during the autumn is an obvious source of potentially leachable nitrate. Organic-N supplied in treated sludges, on the other hand, is only mineralized slowly and until relatively recently this has been seen as a positive benefit because the slow release of N reduces reliance on chemical fertilizers (Hall, 1985). However, there is concern that repeated applications of organic matter leading to greater mineralization of N will result in large leaching losses of nitrate (Powlson *et al.*, 1989). This is because the production of nitrate from the mineralization of soil organic matter is not necessarily synchronized with crop uptake so accumulated nitrate in the autumn is potentially at risk of leaching by winter rainfall (Fig. 9.2). Indeed, Harding *et al.* (1985) showed that the amount and efficiency of N uptake by crops depended more on the residual mineral N content of the soil at the time of planting than on the N mineralized during the growing season. Mineralization of soil organic N has therefore been considered the principal cause of nitrate loss from agricultural soils, although evidence indicates chemical fertilizer N is also a major contributor (Glendining *et al.*, 1992). However, the agronomic importance of soil organic-N should not be overlooked because the productivity of agricultural soils depends heavily on the mineralization of organically bound residual N. For example, Christian *et al.* (1985) showed that more than 70% of the total N uptake of winter cereals could be derived from non-fertilizer sources.

Powlson *et al.* (1989) reported large leaching losses of nitrate from the Broadbalk winter wheat and Hoosefield spring barley experiments at Rothamsted, which have received large applications of FYM at 35 t ha^{-1} (238 kg N ha^{-1}) annually for more than 130 years. The FYM plots contained about 6.9 t N ha^{-1} whereas the unmanured control was less than half this value at 2.9 t N ha^{-1} (Wild, 1988a). Model predictions showed that 124 kg N ha^{-1} y^{-1} could be leached from the FYM treatment compared to 25 kg N ha^{-1} y^{-1} from the inorganic fertilizer plots. Other long-term studies (Johnston *et al.*, 1989) at Woburn similarly showed that more than 100 kg N ha^{-1} y^{-1} can be lost from soils when large applications of FYM, sewage sludge and sludge compost are made. Manure-amended soils initially contained 8.0 t N ha^{-1} and the average annual loss of N was 219, 166, 125 and 95 kg N ha^{-1} during the first four 5-year periods after soil amendment had ceased.

The long-term experiments reported by Powlson *et al.* (1989) and Johnston *et al.* (1989) represent extreme cases of organic matter application to soil and nitrate loss. In practice, the rate of organic N addition to soil in sludge is probably much smaller by comparison. For example, a typical application of liquid digested sludge (100 m^3 ha^{-1} at 0.08% resistant organic N as spread (WRc, 1985)) would supply approximately 80 kg N ha^{-1} of organic N whereas digested cake (50 t ha^{-1} at 0.64% resistant organic N as spread (WRc, 1985)) may add up to 320 kg N ha^{-1}. Liquid sludges are often applied annually, but sludge cakes are usually spread only once in a crop rotation cycle. Consequently, only frequent, or large, applications of sludge are likely to raise the organic N status of the soil significantly (Stark and Clapp, 1980; Griffin and Laine, 1983; Hall, 1985; Harding *et al.*, 1985; O'Keefe *et al.*, 1986; Lindemann *et al.*, 1988). Nevertheless, one-off high rates of application (e.g. 20-30 t ha^{-1} dry solids) are desirable from the point of view of improving the organic matter status and productivity of agricultural soil (Hall and Coker, 1983; MAFF, 1993a). It is inevitable that this may also raise the background level of nitrate leaching from amended soil due to mineralization of organic matter.

The long-term implications of repeated sludge applications were considered by O'Brien and Mitsch (1980) using a 50-year simulation model. The model compared long-term impacts on soil N status and nitrate leaching of rapidly decomposing sludge (50% in the first year) with more stable sludge (20% decomposition in the first year) applied at a rate of 10 t ha^{-1} y^{-1}. Stable sludges built up organic N in the crop root zone to 10 t ha^{-1} over the 50-year simulation from an initial value set of 4 t N ha^{-1}. In contrast, unstable sludges increased soil organic N content to 7 t N ha^{-1}. Application of unstable sludges resulted in the highest crop yields because the rapid initial decomposition resulted in more mineral N being available for crop uptake, but they also gave the highest leaching rate. O'Brien and Mitsch (1980) concluded that stabilized sludges were more amenable to higher application rates since they liberate less inorganic N compared with unstable sludges. This supports the view that slowly mineralized sludges pose a relatively low risk of contaminating water supplies with nitrate (Chang *et al.*, 1988), and contrasts with the long-term predictions of N release made at Rothamsted and Woburn for manures and sludges (Johnston *et al.*, 1989; Powlson *et al.*, 1989).

Clearly the impacts of repeated applications of sludge and mineralization of residual N on potential leaching losses of nitrate are complex and involve a multiplicity of crop, soil, sludge and environmental factors. If crop requirements for N are exceeded, then more nitrate is lost from readily mineralized and ammonium-rich sludges than from slowly mineralized types. On the other hand where crop requirements are accurately met, slowly mineralized sludges (digested and raw cakes) could potentially cause greater leaching of nitrate due to the dichotomy in N release and crop uptake of N during the autumn, although large and frequent additions of organic N would be necessary to raise losses of nitrate significantly above that released from the native soil organic matter (Chang *et al.*, 1988; Johnston *et al.*, 1989; Powlson *et al.*, 1989; Unwin *et al.*, 1991; Froment *et al.*, 1992; Chambers and Smith, 1992). However, increasing the pool of mineralizable N in the soil through repeated applications of sludge can be particularly beneficial to grassland. For example, Hall (1985) determined that annual applications produce progressive increases in mid- and late-season grass yields relative to early growth since the longer growing season of grass permits uptake of a much larger proportion of the mineralized N compared with arable crops. Coker *et al.* (1984) showed that the sludge N value to grass was about twice that to cereals.

Smith *et al.* (1994a) assessed the leaching losses of N from bare-ground soil (loamy sand) in the second year after treatment with liquid undigested and dewatered digested sludges supplying 67 kg N ha^{-1} (at 100 m^3 ha^{-1} as spread) and 338 kg N ha^{-1} (at 50 t ha^{-1} as spread) respectively. These types of sludge exhibit slow N release properties (Hall, 1985; WRc, 1985) and could therefore constitute a potential long-term risk of nitrate leaching. A winter wheat crop was grown after the initial application of sludge although there was no additional mineral N provided. Average nitrate concentrations in drainage from sludge-amended soil did not differ significantly from the untreated control in the second year. The concentration of nitrate in drainage from the control was 17.5 mg N l^{-1} and was 14.9 and 20.1 mg N l^{-1} for the liquid undigested and dewatered digested sludges, respectively. The quantity of drainage from the field site was relatively high in the second year and was estimated to be 276 mm. Consequently, these results provided confirmation of the minimal risk of N loss by leaching from sludge-treated soil in subsequent years after application. This is emphasized since the field plots were maintained in a bare-ground state in the second winter after application of the sludge which represents potentially a 'worst-case' condition for nitrate leaching. They also imply that, provided supplementary mineral N applications are properly adjusted to account for the available N supplied in sludge, there is a low probability of leaving unused residues of the applied fertilizer N in the soil in the following autumn at risk from leaching by winter rainfall.

Ammonia Volatilization

Since Beauchamp reported in 1983 that relatively few studies had been done on the volatilization losses of ammonia from organic manures, a large body of data has been published in recognition of the extent of the problem, particularly in relation to the disposal of animal wastes (Nielsen *et al.*, 1986, 1988, 1991; Hansen and Henriksen, 1989). Digested sewage sludges contain appreciable amounts of ammonium-N, often in excess of 50% of the total N content in the case of liquid sludges and, when spread on land, appreciable losses of ammonia by volatilization can occur. The extent of these gaseous losses depend greatly on cropping, environmental and soil conditions, sludge properties and also on the method of application.

Various estimates of ammonia volatilization from surface-spread sludge obtained from field and laboratory studies indicate that between 10-60% of the applied ammonium-N may be potentially lost by this route (Ryan and Keeney, 1975; Stewart *et al.*, 1975; Beauchamp *et al.*, 1978). The simulation model of O'Brien and Mitsch (1980) also predicted a loss of 50% from surface-applied sludge. Amberger (1991) estimated that ammonia losses from animal slurries can increase up to 80% of the applied ammonium-N; similar losses from sewage sludge may also be anticipated under certain conditions. Indeed, there are examples where all the ammonium supplied in animal waste and sewage sludge was lost through volatilization (Kolenbrander, 1981).

Clearly, ammonia volatilization losses in the field are highly variable. To understand the importance of specific factors controlling ammonia volatilization from sludge, Donovan and Logan (1983) conducted a series of laboratory studies varying only a single experimental parameter. As expected, ammonia loss increased with increasing soil pH value and temperature. However, losses from air-dried soil were much lower than have been reported for inorganic fertilizers. This may be because high soil moisture tensions resulted in the sludge being absorbed into the soil reducing volatilization losses. Donovan and Logan (1983) found that volatilization losses were greatest from lime-stabilized sludge (pH 12). Acidification of animal slurries with sulphuric acid prior to spreading has reduced ammonia loss (Pain and Thompson, 1989) although it is difficult to envisage this as an acceptable pretreatment of sludges or slurries in practice.

In general, gaseous losses of ammonia from sludge decrease with time after incorporation, and they also decline by shortening the period of time between application and incorporation (Donovan and Logan, 1983). This indicates that timely incorporation is the best management practice available to reduce volatilization losses of sludge N. The principal losses of ammonia occurred within the first 24 h in the experiments of Donovan and Logan (1983). Therefore, for effective reduction in ammonia volatilization it appears likely that sludge should be incorporated within hours of application in the field.

These data have largely been confirmed in studies on farm wastes using wind tunnels and micrometeorological methods. For animal slurries, Pain and Thompson (1989) concluded that 70% or more of the total ammonia loss occurs within 24 h of application although losses may continue over a period of 15 days. Losses during spreading with a conventional vacuum tanker and splash plate arrangement represented <1% of those which occurred from land after spreading. However, machines with a low trajectory spreading pattern gave smaller losses during application than those from conventional spreaders.

Slurry and sludge composition has a major influence on the extent of ammonia loss. Volatilization losses increase with increasing total solids content of slurry since infiltration into the soil is reduced. Smith and Chambers (1992) found an inverse correlation between dry solids content and N efficiency of surface applied slurries, implying increased losses of ammonia were responsible for the lower fertilizer value of thick slurries. There is also a time-dependent increasing importance of dry solids content on ammonia loss from slurry left on the soil surface (Sommer and Christensen, 1991). Donovan and Logan (1983) reported increased ammonia losses when sludge containing large particles (>1 mm) was applied to straw covering the soil surface. Apparently, the surface cover filtered out the solids and prevented the rapid movement of the sludge liquid into the soil. Amberger (1991) similarly observed that the application of highly viscous slurry onto stubble and straw, or onto compacted soil or grassland, impaired infiltration into the soil thereby promoting ammonia volatilization. Pain and Thompson (1989) considered that the dry solids content of slurries was more important than any other environmental or management factor in determining the extent of volatilization losses from soil.

The dry solids content of animal slurries can vary widely between 1 and 15% whereas liquid sewage sludges are generally more uniform and fall in the region 3-5%. Therefore, dry solids may not be as important for sludges except for lagoon-matured sludges which may be thickened to >10% dry solids. Extrapolating published data for animal slurries (Sommer and Christensen, 1991; Smith and Chambers, 1992) to sewage sludge would suggest ammonia losses from thickened sludges may approach 70-80% of the applied ammonium-N, twice that from thinner liquids.

Soil injection of sludge is an effective method of minimizing ammonia volatilization losses from soil in addition to improving pasture hygiene for grazing animals and reducing odorous emissions (Hall, 1986b; Hall and Ryden, 1986; Istas *et al.*, 1988). Hoff *et al.* (1981) measured the proportion of applied ammonium-N lost as ammonia from pig manure over a 3-5 day sampling period using microplot volatilization chambers in the field. Losses were 14.0, 12.2 and

11.2% of the ammonium-N applied in 90, 135 and 180 $m^3 ha^{-1}$ surface spread, respectively. However, only 2.5% was lost when 90 and 180 $m^3 ha^{-1}$ were injected. From micrometeorological measurements, Pain and Thompson (1989) showed that injection of pig slurry into grassland gave a 94% reduction in ammonia loss during a 73-hour monitoring period compared with surface spreading with a conventional vacuum/pressure tanker at a rate of 100 $m^3 ha^{-1}$ (Table 9.5). Soil injection (or rapid incorporation) is recommended as an effective measure for reducing ammonia losses to the atmosphere from liquid manures applied to agricultural land to protect air quality (MAFF, 1992).

Hann *et al.* (1992) calculated a N mass balance for liquid digested sewage sludge surface-applied or injected into arable soils. Results indicated that, in overall terms, losses of N to the environment were significantly larger from surface applications - i.e. by ammonia volatilization, than those arising from injection - i.e. by nitrate leaching and possibly denitrification. Ironically, it is nitrate leaching which is receiving most attention, due to concerns over the potential effects on human health of nitrate in groundwater, which have not been proven. By contrast, the environmental effects of ammonia emissions into the atmosphere are widespread and can be serious. In the near future, the profile of ammonia volatilization could well increase on the environmental agenda for legislative controls (Istas *et al.*, 1988; CEC, 1992).

Table 9.5. Rates of ammonia loss following injection or surface application of pig slurry to grassland.

Period (hours)	Ammonia loss ($g NH_3$-N $h^{-1} m^{-3}$ slurry applied) Injection	Surface
0-1	0.80	24.0
1-24	0.16	1.8
25-46	0.01	0.7
47-73	0.02	0.3

Source: Pain and Thompson (1989)

Denitrification

There has been little research on measuring the gaseous losses of N by denitrification from sludge-treated soils. The available data for sludge provide only a fragmentary indication of what the potential losses of nitrogen and nitrous oxide to the atmosphere may be through denitrification. Environmental concern is focused on emissions of nitrous oxide. Mosier *et al.* (1982) reported that nitrous oxide emissions from a calcareous sandy loam treated with 290 and 1440 kg N ha^{-1} in sludge accounted for only 0.8 and 1.0% of the applied mineralizable N respectively under field conditions. Thus, it was concluded that nitrous oxide emissions from soils amended with organic or inorganic sources of N were relatively short-lived and that only a very small fraction of the N amendment contributed to increasing the concentration of nitrous oxide in the atmosphere. Incubation studies on the other hand, have suggested that gaseous losses of N arising through denitrification could potentially be large and in the range 15-20% of sludge-applied N (King, 1973; Sommers *et al.*, 1979).

Considerably more work has been done on denitrification losses from soils treated with animal slurries. Research has shown that cropping, time and method of application, prevailing weather conditions and soil type are all important factors in determining the extent of denitrification losses from animal wastes. In grassland soils for example, denitrification may account for a larger proportion of the ammonium-N applied in slurry during autumn and winter than in the spring. Summarizing recent research on denitrification losses from grassland treated with cattle slurry, Thompson and Pain (1989) indicated that up to 30% of the ammonium-N in slurry surface-applied during the autumn and winter period could be lost by this route from freely drained soils. In the spring, however, this was reduced to a value of about 5% because sward depletion of soil nitrate was more rapid in the spring than during the autumn-winter period and the soil was also drier in spring thus promoting adequate oxygen diffusion. Van den Abeel *et al.* (1989) reported similar denitrification losses from pig slurry surface-applied to a well-drained fallow soil in the spring which accounted for 7% of the ammonium-N content in the slurry. In contrast, Maag (1989) considered spring as the period of highest denitrification activity in slurry-treated soil cropped with spring barley. The maximum rate of denitrification measured by Maag (1989) for a spring application of 80 kg ha^{-1} of ammonium-N in pig slurry was 0.98 kg N ha^{-1} day^{-1}, which is broadly comparable with the maximum value of 0.6 kg N ha^{-1} day^{-1} given by Thompson and Pain (1989) for 100 kg ha^{-1} of ammonium-N applied in cattle slurry in winter. Maag (1989) drilled spring barley and applied slurry in April and large concentrations of nitrate, susceptible to loss by denitrification, accumulated in the soil before the crop was sufficiently established to absorb the available N. Furthermore, application of slurry to the soil surface, rather than onto grass, facilitated the movement of a larger proportion of ammonium-N into the soil. In years with average weather conditions Maag (1989) indicated that up to 50% of applied ammonium-N could be lost by denitrification from spring applications of slurry. High rates of denitrification (>0.5 kg N ha^{-1} day^{-1}) were only observed in periods with frequent rain and when soil nitrate contents were >10 mg N kg^{-1}, or in periods with alternating frost/thaw. However, moderate rates of denitrification (0.1-0.5 kg N ha^{-1} day^{-1}) were also measured in relatively dry soils amended with large amounts of slurry. Restricted diffusion of oxygen in wet soils is considered a principal factor which enhances the rate of denitrification (Arah and Smith, 1989). However, these data suggest that high levels of available organic substrate can also promote denitrification activity irrespective of soil moisture status.

Soil type has an important influence on denitrification activity depending on the presence of available nitrate. Thompson and Pain (1989) found that losses were negligible from slurry applications to a poorly drained clay soil in the autumn and spring. This was apparently because saturated soil conditions over the experimental period prevented nitrification of the slurry ammonium. Therefore, denitrification was low because very little nitrate was formed. However, the level of denitrification activity and loss could potentially be enhanced in heavier textured soils under conditions that permit some nitrate accumulation because the aeration capacity of these soils is lower than more freely draining soil types (Arah and Smith, 1989).

The benefits of injecting slurries and sludges in retaining ammonium-N in the soil by reducing volatilization losses were discussed in detail in the previous section. However, injection can considerably enhance denitrification losses of N compared with surface applications. This is because a larger quantity of inorganic N is present in the soil together with added organic material in a zone of more restricted aeration (Thompson and Pain, 1989). Thus, injection reduced ammonia volatilization from cattle slurry from 74 to 2% of the applied ammonium-N in the experiments described by Thompson and Pain (1989). However, the total denitrification loss increased concomitantly from approximately 30% of the ammonium-N, obtained by surface-spreading, to 40-54% for injected treatments. The nitrification inhibitor dicyandiamide (DCD) was very effective in controlling denitrification, reducing losses by up to 90%. Pain *et al.* (1990) reported that DCD reduced denitrification losses from autumn applications of cattle slurry to a free-draining soil by approximately 70%. There was little practical benefit in using inhibitors on a poorly drained soil, or for spring applications of slurry to a free-draining soil.

In quantitative terms ammonia volatilization represents the most important route of gaseous N loss from slurry-treated soils, with comparatively smaller emissions generally being measured for denitrification. However, extrapolating the results from animal slurries to sewage sludges suggests losses of nitrous oxide through denitrification may be appreciable under certain conditions. It is likely that practices aimed at reducing nitrate leaching losses, such as closing the autumn period for spreading and applying sludges to growing crops, would also reduce denitrification. Research is needed in the first place to quantify the extent of denitrification losses from sludge-treated soils and also to assess the impact that modifications in management and application practices may have on denitrification emissions.

Surface Run-off

Movement of water across the soil surface, known as overland flow, can be highly erosive removing not only soil, but also seeds, fertilizers and pesticides (Briggs and Courtney, 1989). Moreover, because it provides a major input of water, sediment and solutes to streams, it plays a vital role in controlling surface water discharge and quality. Thus, under potentially erosive conditions, run-off may be particularly hazardous immediately after the application of liquid sludge, but it may also occur after the application of dewatered sludge or following incorporation because of erosional transport of sludge solids and run-off of soluble sludge constituents, particularly nutrients (Kladivko and Nelson, 1979).

Briggs and Courtney (1989) give an account of the processes which affect overland flow under conventional agricultural practices. Overland flow does not always occur spontaneously, but is often preceded by a period of surface storage in depressions on the soil surface. With further rainfall, the discrete ponds of water eventually link resulting in a downslope flow. The relative rates of rainfall and infiltration are important factors in controlling the onset of overland flow. However, storage capacity is related to surface roughness and under arable cultivation is greatest after soil is ploughed, but declines as the soil surface is smoothed by weathering and soil settlement. Thus, marked seasonal variations in surface storage capacity and incidence of overland flow are seen at a single site.

Pollution of watercourses through surface run-off of applied organic manures generally results from one or more of the following factors: excessive rates of application, heavy rainfall following spreading, impermeable soil and sloping ground (Sherwood and Fanning, 1981; Hall, 1986b,c; Steenvoorden, 1989). Research assessing the problems of run-off have focused principally on the disposal of farm wastes with comparatively little being carried out with sewage sludges.

The concentrations of N, P and other compounds in surface run-off from soils treated with animal manures generally increase with increasing rate of application (Steenvoorden, 1981). However, Sherwood and Fanning (1981) noted that the maximum concentration of phosphate-P in run-off from slurry-treated soil was unlikely to exceed 30 mg l^{-1}. This value is orders of magnitude higher than the suggested EQS for P in surface water (Cartwright and Painter, 1991). Nutrient levels in surface water flow decline as the time elapsed between application and rainfall increases. For example, Steenvoorden (1981) detected little increase in the N and P concentration of surface water 15 days after application of cattle slurry at rates up to 40 m^3 ha^{-1}. Sherwood and Fanning (1981) showed that the biochemical oxygen demand (BOD) can decline to <100 mg l^{-1} if the first run-off event does not occur for 2-3 weeks after application, reflecting the ability of soil micro-organisms to quickly utilize the slurry carbon. Ammonium concentrations in run-off also decreased with time (but not to the same extent as BOD), depending on the level of ammonia volatilization and nitrification. In contrast, concentrations of phosphate-P were found to remain at a high level in run-off water for at least six weeks after application. Consequently, Sherwood and Fanning (1981) concluded that the rate of slurry application has an important adverse effect only if run-off occurs within six weeks of application. A maximum application rate of P in slurry of only 26 kg ha^{-1} was apparently a safe limit for most soil types when spread in late spring. In comparison, standard applications of sewage sludge to farmland supply significantly more than this amount (Table 9.1). For example, a typical application of liquid digested sludge of 100 m^3 ha^{-1} provides about 70 kg P ha^{-1} whereas digested cake sludge spread at 50 t ha^{-1} supplies approximately 200 kg P ha^{-1}. However, Misselbrook *et al.* (1995) found that P concentrations in run-off declined fairly rapidly following the application of different livestock wastes to grassland. They concluded that losses of nutrients *via* run-off from treated impermeable grassland soils (7° slope) in autumn and winter may be relatively small, although peak concentrations of nutrients may have implications in pollution of surface waters. The NRA has identified pig slurry as a particularly important source of phosphate contamination in surface waters (NRA, 1992a).

Compared with applications of phosphate fertilizers, and the enormous quantities of P-rich farm wastes spread on agricultural land (Fleming, 1993), the contribution to surface water contamination with P in run-off from sludge-treated soil is probably of relatively minor significance.

Surface run-off from sludge-treated soils in the main has been demonstrated only for sites receiving sludge at loading rates which exceed those normally spread on agricultural soils. For example, Matthews *et al.* (1981) applied 63 t ha^{-1} of sludge (dry solids) to a catchment area, which provided on average 1130 kg N ha^{-1} (of which 440 kg was ammonium-N) and 1180 kg P ha^{-1}, and monitored effects on surface water quality. Mean values for conductivity, dissolved solids, total N, ammonia, total oxidized N, total P and total organic carbon in surface water samples all significantly increased with time from the sludged soil. However, due to changes in the run-off water quality which occurred in the control treatment, only the increases in ammonia, oxidized N and total organic carbon could be attributed to sludge application.

Melanen *et al.* (1985) reported on the quality of surface and subdrainage waters from field experiments with sludge conducted in Finland. Sewage sludge was applied at 16-20 t ha^{-1} (dry solids) supplying 750-950 kg N ha^{-1} and 400-500 kg P ha^{-1}. Application of dewatered sludge to non-frozen clay soil increased the total P and phosphate concentrations, but concentrations of nitrate (5.3 mg NO_3-N l^{-1}) and total N and the chemical oxygen demand (COD) were lower following sludge treatment compared with the control. Only 0-10% of precipitation percolated to groundwater in this soil so nutrient losses to surface waters in subdrainage were also important. Analyses of subdrainage water showed significant increases in total-N, nitrate-N, total-P and phosphate concentrations for sludge-treated plots, but no effect of sludge on COD compared with the control. In particular, nitrate-N concentration increased from the background level of 9.4 mg l^{-1} in the control to 32 mg l^{-1} with applied sludge. Melanen *et al.* (1985) concluded that most of the total leaching, generally over 80-90%, was through subdrainage in this soil. However, the level of precipitation during the monitoring period was 40-50% higher than the average levels normally received which produced excessive subdrainage and run-off from the plots. These factors together largely explained the increases in nutrient concentrations recorded. However, sludge application to frozen ground significantly increased the concentrations of nutrients in surface run-off. Thus, the problem of run-off can be exacerbated where infiltration of rainfall is impeded by, for example, frozen or saturated ground conditions or where the soil is severely compacted.

Soil structural stability plays an important role in controlling erosion of soil by rainfall (Smith and Wischmeier, 1962). Reduced stability makes the soil highly susceptible to erosion by rainfall (Johnson *et al.*, 1979). Soil stability is related directly to organic matter content and instability of structure can develop in soils when organic matter falls below 3% (Strutt, 1970). In simulated rainfall experiments, Guerra (1994) showed that soils with less than 3.5% organic matter content have unstable aggregates although soil texture, and high clay content in particular, may protect the aggregate stability of soils low in organic matter. Levels of 2-3% organic matter are typically found in soils under arable cultivation in which the only returns of fresh organic material are from crop residues (roots and tops).

The beneficial effects of organic matter in reducing erosion and sediment losses from soils have been reported. Run-off and sediment loss are both reduced with increasing soil organic matter content and aggregation (Wischmeier and Mannering, 1965, 1969; Guerra, 1994). Surface residues of crops and organic mulches are also important in reducing erosion because they protect the soil surface from raindrop impact (Jacks *et al.*, 1955; Meyer and Mannering, 1963; Briggs and Courtney, 1989). Although improving crop cover reduced total P loss in run-off, Sharpley *et al.* (1992) reported that the proportion of total P loss which was in bioavailable form had increased. This was attributed to the greater contribution of bioavailable P leached from plant material and the preferential transport of clay sized particles in run-off increased the contribution of bioavailable particulate P, both of which were accentuated by vegetative cover. Few data are currently available at this level of understanding on the implications for surface water quality of P losses in run-off from sludge-treated agricultural land. Nevertheless, the importance of sludge in improving soil aggregate stability and thereby protecting soil against erosion losses is widely reported (Kelling *et al.*, 1977b; Metzger and Yaron, 1987).

Kladivko and Nelson (1979) used a simulated rainfall technique to measure surface run-off from silt loam and sandy loam soils treated with up to 89.6 t ha^{-1}

of sewage sludge (5.9% dry solids). The soils studied contained 2-3% organic matter and the rainfall regime represented two consecutive 30-minute storms at a rainfall intensity of 5.72 cm h^{-1}. Results showed that sludge can help stabilize potentially sensitive soils against erosion and promote greater water infiltration by protecting the soil surface against raindrop impact and increasing the stability of soil aggregates. The degree of erosion achieved was related to the amount of surface cover provided by dry sludge itself. Run-off from sludge-treated plots contained larger concentrations of nutrients compared with run-off from control plots. However, the total loss of nutrients was lower than that originating from untreated soil because of the decreased gross erosion from sludge-treated soil. This study demonstrated the effectiveness of sludge in stabilizing the soil surface thereby minimizing run-off erosion which is emphasized by the severity of the rainfall intensity applied. To put this into perspective, it is unlikely that daily rainfall in excess of 2.5 cm occurs more than twice a year in the main agricultural regions of the UK (Ward, 1975).

Pagliai and Sequi (1981) similarly showed that the application of pig slurry to soil increased water infiltration because slurry treatment maintained the porosity of the soil surface layers thereby reducing potential run-off. They concluded therefore, that run-off was likely to occur only at the time of spreading if excessive rates of slurry were applied.

The data of Kladivko and Nelson (1979), Pagliai and Sequi (1981) and Melanen *et al.* (1985) are particularly interesting because they reflect the importance of soil permeability on surface run-off. The extent of nutrient loss in run-off from soil increases as a function of increasing volume of run-off water (Sherwood and Fanning, 1981). Thus, freely drained soils generate a smaller volume of run-off water compared with soils of low permeability, and therefore have a lower risk of causing surface run-off problems. In addition, sludges and slurries of low dry matter content readily infiltrate the soil, thereby reducing the risk of direct surface run-off provided the rate of application does not exceed the infiltration capacity of the soil, which again is a function of the soils' permeability. However, where the absorption capacity is exceeded (characterized by ponding on the soil surface), liquid sludges and slurries with low dry solids contents are much more susceptible to run-off than thicker forms (Vetter and Steffens, 1981).

The analysis of run-off waters collected from sludge-treated soil of the Rosemount wastewater sludge watershed in Minnesota showed the quality of run-off water was only marginally impacted in relation to the quantities of nutrients being supplied (Duncomb *et al.*, 1982). However, application method was particularly important in determining the extent of run-off contamination with nutrients. Over the five year monitoring period the total quantities of nutrients supplied in digested liquid sludge were 4-5 t N ha^{-1} and 2 t P ha^{-1} to both corn and grass crops. The highest nutrient loss was from the sludge-treated grass areas, with average concentrations of 19.7 mg N l^{-1} and 5.0 mg P l^{-1}, compared with 6.8 mg N l^{-1} and 1.1 mg P l^{-1} for the untreated control. Nutrient concentrations in run-off from the corn areas were lower at 9.7 mg N l^{-1} and 0.8 mg P l^{-1} with sludge, and 3.9 mg N l^{-1} and 0.4 mg P l^{-1} without sludge. The higher losses from grass areas were explained because sludge was more frequently applied and by surface spreading compared with the arable crop where sludge was introduced by injection. Cultivation of the arable soil also reduced surface run-off losses compared with the grassland by aiding permeability. The slope of the watershed varied from 2 to 10%, but this did not greatly influence the quality of run-off from sludge-treated soil. In general, however, the extent of potential overland flow from grass crops is considered to be lower than from arable land (Briggs and Courtney, 1989). This is because a grass sward provides a continuous vegetation cover which interrupts rainfall and impedes any overland flow which does occur. The improved rooting and organic matter accumulation in grassland additionally means that infiltration capacities tend to be higher than in arable soils.

Kladivko and Nelson (1979) also demonstrated the importance of method of application on nutrient losses in run-off. Sludge remaining on the soil surface after drying was apparently much more effective in decreasing losses by erosion than sludge incorporated by rotary cultivation or by discing. In contrast, Mostaghimi *et al.* (1989) determined that the incorporation of sludge reduced both the concentration and total loss of N in run-off from arable soils compared with surface applications, and Gavaskar *et al.* (1990) recommended the incorporation of sludge into soil to avoid direct run-off problems. Incorporation of animal wastes in soil is also recognized as the principal method of reducing surface run-off (Uhlen, 1981). Incorporation of sludge where possible immediately after application is also recommended in the UK Code of Practice and other advisory literature (SAC, 1986; DoE, 1989a) but mainly for reasons of odour control and to minimize potential risks from pathogens.

Soil injection of animal slurries can significantly reduce the risk of surface run-off occurring provided there is sufficient soil cover to contain the slurry. Ross *et al.* (1979) monitored run-off quality for a range of pollution parameters from grass and cultivated contour plots which had received cow slurry to the surface or by injection and found that injection eliminated any pollutant yield in the run-off. Injected sludges and slurries with low dry solids contents are more at risk to run-off than thicker forms, particularly as the slope increases, and a particular concern is that the injection slots may act as drains. Injection along the contour can prevent this, but operationally this is not always possible and on gradients of over 10% it is normal practice to inject down the slope. Hall (1986c) showed that application rates should be decreased for sludges with less than 5% dry solids as the slope increases. Above 5% dry solids slope had no effect on sludge movement when injected with winged tines (Fig. 9.9). Injection into dry, cracked soils, or into drained land with shallow gravel backfill is inadvisable due to the possible risks of sludge running directly into land drains and watercourses (Hall, 1986c).

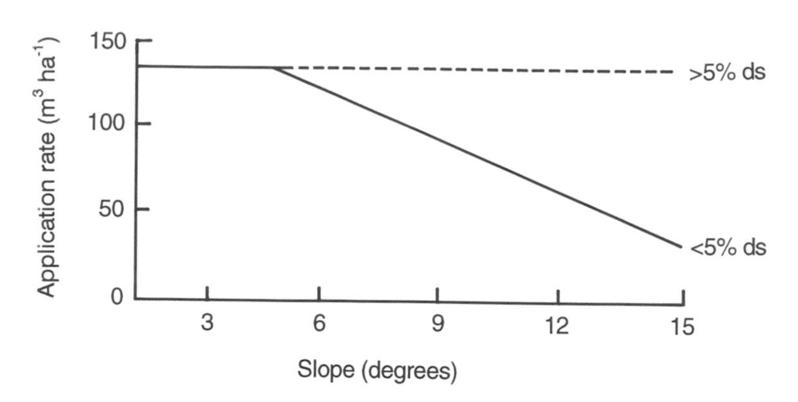

Fig. 9.9. Effect of slope upon the maximum application rate of sludge injection in relation to sludge dry solids (ds) content (Hall, 1986c).

Recommendations given in national codes of practice (MAFF, 1991a; SOAFD, 1991) should minimize the risk of polluting surface waters from organic manures spread on farmland (Webb and Archer, 1994), but these can only provide a general guide to the measures which can be followed. In consequence such codes of practice tend to be highly precautionary. However, the potential for pollution to occur is dependent on many local factors and the National Rivers Authority (NRA) advocates the development of practical waste management plans at the farm catchment level to ensure sound disposal regimes of wastes to land and hence minimize pollution risk (NRA, 1992a). Factors which need to be considered include topography, geology and soil type, land use, location of drains, watercourses and boreholes, season and rainfall (NRA, 1992a). These factors can be considered together using catchment farm waste pollution models such as the FARMS (Farm Activity and River Management Simulation) model developed by WRc (Oakes and Slade, 1992), to provide a quantitative basis for developing sludge application strategies which minimize the risk of surface water pollution where sludge is used in agriculture.

Chapter ten:

Organic Pollutants

General Comments

The production and use of organic chemicals has risen rapidly over the last four decades and the widespread application of these compounds, for example as agricultural pesticides, industrial solvents, dyes, plasticizers, detergents and heat exchangers, has caused concern about their potential impacts on the environment and particularly on human health (Rogers, 1987). In general, any compound which is released into the environment potentially may be found in sewage sludge either through direct discharge into the sewer system, or indirectly by aerial deposition (Halsall *et al.*, 1993) through run-off from roads and other surfaces. Consequently compounds such as polynuclear aromatic hydrocarbons (PAHs), which are not synthesized for industrial purposes but are liberated into the environment from the combustion of fossil fuels are also present in sewage sludge.

The range of organic compounds known to exist in sewage sludge is extensive and diverse and is potentially transferred to sludge-amended agricultural soils. From a review of literature, Drescher-Kaden *et al.* (1992) reported that 332 organic pollutants, with the potential to exert a health or environmental hazard, had been identified in German sewage sludges. Forty-two of these were regularly detected in sludge by researchers. Concern about the environmental effects of recycling sewage sludge on agricultural land would appear to be legitimate given that many of the organic chemicals designated as 'priority pollutants' (CEC, 1976; CEC, 1986b; CEC, 1988; CEC, 1990b) due to their potentially toxic effects, are known to occur in sludge. Anxiety may be fuelled because less is known about the environmental consequences of trace organic pollutants in sludge-treated soil, which have been investigated only to a limited extent, compared with the voluminous amount of information which is available for PTEs. This is partly explained by the fact that analytical techniques which can extract, fractionate, identify and quantify the often very small concentrations of complex mixtures present have only been developed and applied to sewage sludge in recent years, and also because these analyses are laborious and very expensive. However, the likely fate and behaviour of organic pollutants in sludge-treated soils has been examined in detail on the basis of their known physico-chemical properties (Jones and Wild, 1991; Wild and Jones, 1992a). The occurrence, fate and potential impact of trace organic contaminants in sewage sludges and sludge-treated agricultural land have also been extensively reviewed. (Lester, 1983; Overcash, 1983; Davis *et al.*, 1984; Jacobs *et al.*, 1987; Rogers, 1987; O'Connor *et al.*, 1991; Sweetman, 1991; Drescher-Kaden *et al.*, 1992). This information has been collated and is presented in summary form in Table 10.1.

In contrast to the perceived concerns, however, the general consensus based on the limited information currently available is that organic contaminants applied to agricultural soils in sewage sludge are unlikely to cause significant environmental problems (Wild and Jones, 1991; Sweetman *et al.*, 1994). Indeed, the WHO Working Group on the Risk to Health of Chemicals in Sewage Sludge Applied to Land (Dean and Suess, 1985) recognized that organic chemicals may be ingested with soil by grazing animals, but concluded that: "the total human intake of identified organic pollutants from sludge application to land is minor and is unlikely to cause adverse health effects".

Furthermore, the increasing level of investigation in recent years has still not identified the ecotoxicological significance of organic contaminants in the soil-plant-water system and in the food chain (Sauerbeck and Leschber, 1992). In spite of this, organic pollutants have achieved a level of notoriety because of the high toxicity and carcinogenic properties of some organics (Demirijian *et al.*, 1987) and their apparent widespread and uncontrolled abundance and persistence in the environment. They are perceived as being particularly harmful because they are entirely artificial materials. However, it may be anticipated that the level of trace organic contamination of sludges should decline as actions taken to reduce levels generally in the environment are implemented (DoE, 1989b). For example, restrictions on the use of polychlorinated biphenyls (PCBs) have markedly reduced the environmental burden of these contaminants, although the rate of decrease has diminished in recent years due to the recirculation of PCBs in the environment from contemporary releases to the atmosphere (Harrad *et al.*, 1994). Harrad *et al.* (1994) estimated that the bulk (93.1%) of the UK environmental burden of total PCBs (ΣPCBs) was associated with soils. Volatilization from soil is a principal source of PCBs in the UK environment (88.1%) whereas volatilization from sewage sludge-amended land represents a comparatively small proportion of the annual flux (0.2%), the significance of which is debatable within the error of the assumptions in the global model for the UK.

Table 10.1. Summary of the properties, occurrence, fate and transfer of the principal organic contaminant groups found in sewage sludge and sludge-treated soils.

Compound group	Physico-chemical properties	Concentration in sludge (dry solids)	Degradation	Leaching potential	Plant uptake	Transfer to animals
Polynuclear aromatic hydrocarbons (PAHs)	Limited water solubility volatile to lipophilic	1-10 mg kg^{-1}	Weeks to 10 years Strongly adsorbed by soil O.M.	Low	Very poor Foliar absorption Root retention	Possible but rapidly metabolized Not accumulated
Phthalate acid esters	Generally lipophilic, hydrophobic and non-volatile	High 1-100 mg kg^{-1}	Rapid Half-life <50 days	Low	Root retention Not translocated	Generally very limited
Linear alkylbenzene-sulphonates (LAS)	Amphiphilic[(1)]	Very high 50-15000 mg kg^{-1}	Very rapid	Low	Minimal	Minimal
Alkylphenols	Lipophilic	100-3000 mg kg^{-1}	Rapid <10 days	Low	Minimal Root retention	Minimal
Polychlorinated biphenyls (PCBs)	Complex, 207 congeners, low water solubility, highly lipophilic and semi-volatile	1-20 mg kg^{-1}	Very persistent Half-life several years Strongly adsorbed by soil O.M.	Low	Root retention Foliar absorption Minimal root uptake and translocation	Possible into milk/ tissues *via* soil ingestion Long half-life

Table 10.1 continued

Compound group	Physico-chemical properties	Concentration in sludge (dry solids)	Degradation	Leaching potential	Plant uptake	Transfer to animals
Polychlorinated dibenzo-*p*-dioxins and furans (PCDD/Fs)	Complex, 75 PCDD congeners, 135 PCDF congeners, low water solubility, highly lipophilic and semi-volatile	Very low <few μg kg^{-1}	Very persistent Half-life several years Strongly adsorbed by soil O.M.	Low	Root retention Foliar absorption Minimal root uptake and translocation	Possible into milk/ tissues *via* soil ingestion Long half-life
Organochlorine pesticides	Varied, lipophilic to hydrophilic (e.g. 2,4-D), some volatile	<Few mg kg^{-1}	Slow, half-life >1 year Loss by volatilization	Low to moderate	Root retention Translocation not important Foliar absorption	*Via* soil ingestion Persistent in tissues
Monocyclic aromatics	Water soluble and volatile	<1-10 mg kg^{-1}	Rapid	Moderate	Limited due to low persistence Rapidly metabolized	Rapidly metabolized
Chlorobenzenes	Water soluble/volatile to lipophilic	<0.1-50 mg kg^{-1}	Lower mol wt types lost by volatilization Higher mol wt types persistent	Low to high	Possible *via* roots and foliage May be metabolized	Important for persistent compounds
Short-chained halogenated aliphatics	Water soluble and volatile	0-5 mg kg^{-1}	Generally rapid loss by volatilization	Moderate	Foliar absorption Possible translocation	Low

Table 10.1 continued

Compound group	Physico-chemical properties	Concentration in sludge (dry solids)	Degradation	Leaching potential	Plant uptake	Transfer to animals
Aromatic and alkyl amines	Water soluble and low volatility	0-1 mg kg^{-1}	Slow	Moderate to high	Possible	Low
Phenols	Varied, lipophilic, high water solubility and volatile	0-5 mg kg^{-1}	Rapid	Low to moderate	Possible *via* roots and foliage	Generally very limited

O.M. - organic matter
(1) i.e. partly hydrophilic and partly hydrophobic

Gilbert (1994) examined the general implications for food quality of organic pollutants in the environment and concluded that the contamination of foodstuffs, which is the principal route of human exposure to these chemicals (Duarte-Davidson and Jones, 1994; Wild *et al.*, 1994), was a legacy of past ignorance in the use and disposal of certain substances, and a misjudgement of the likely long-term effects of such environmental pollution. The situation has improved markedly because the manufacture of substances such as PCBs is no longer permitted and there is more careful attention given to the disposal of waste materials to ensure minimal emission. There are also improvements generally in waste management through incineration ensuring the lowest achievable production of toxic gases. Indeed, the concentrations of PCBs in UK sewage sludges have probably decreased by more than ten-fold (C.D. Watts, WRc Medmenham, personal communication). Gilbert (1994) confirmed that analytical methods for measuring trace levels of organic pollutants in foodstuffs were well refined and that surveillance programmes in the UK are being used to monitor anticipated declines in some contaminants in the food supply and also to provide an early warning of any future problems.

Perceived public concern that sludge recycling to agricultural soil can seriously affect the quality of the human diet, due to contamination with trace organic compounds, has restricted the agricultural use of sludge in Sweden. In the Federal Republic of Germany, public and political concerns have recently resulted in highly precautionary regulations for organic contaminants and there is some doubt as to whether the agricultural utilization of sludge in Germany will be permitted at all in the future (Sauerbeck and Leschber, 1992). The recent discovery in Sweden that dioxins are apparently formed during sewage treatment may further increase public concern about potential food chain impacts of sludge utilization in agriculture (ENDS, 1992).

Sewage Sludge Regulations for Organic Compounds

In the UK currently there are no limit values set for trace organic contaminants in sludges or soils. The organic contaminants of principal concern are polyhalogenated compounds for which limits in sludge have been set in German regulations (Table A4). These regulations have set highly precautionary values which have no toxicological basis (Sauerbeck and Leschber, 1992). Current knowledge of background levels and the behaviour of trace organics in soils was considered insufficient for limit values for soils to be established. It is very likely, however, that proposed permissible levels in soils will be equally precautionary and will not be based on toxicological information. Sauerbeck and Leschber (1992) noted that permitted amounts of adsorbable organo-halogens (AOX) and PCBs were unlikely to restrict agricultural utilization of sewage sludges given the level of contamination with these compounds normally found in German sludges. However, Sauerbeck and

Leschber (1992) considered the limit for polychlorinated dibenzo-*p*-dioxins (PCDDs) and furans (PCDFs) could significantly curtail the application of sludge to agricultural land due to the cost and complexity of the chemical analysis. Ironically, the most toxic congener 2,3,7,8-T_4CDD had hardly been detected at all in sludge (Sauerbeck and Leschber, 1992). Concentrations of 2,3,7,8-T_4CDD generally fall in the range 0-1.0 ng kg^{-1} (Wild *et al.*, 1994). The total TEQ (the Toxicity Equivalent Quotient is a weighted toxicity factor for PCDD/Fs relative to the most toxic compound, 2,3,7,8-T_4CDD) for the PCDD/F content of sewage sludge is typically in the range 20-40 ng kg^{-1}. Information on the dioxin content of UK sludges is very limited. Data produced in a recent report (DoE, 1993) is somewhat suspect due to analytical problems and in particular, the low recoveries of internal standards obtained, in spite of apparently adding these *after* solvent extraction of the sludge which is against normal analytical convention. Sewart *et al.* (1995) recently reported the ΣTEQ values for eight digested sewage sludges from north-west England ranged from 19 to 206 ng kg^{-1} (mean 72 ng kg^{-1}) with the higher values being detected in the samples from urban/industrial areas. Comparison with archived samples of sludge applied to the Woburn experiment between 1942 and 1960 provided some evidence that there had been a decline in ΣTEQs since the 1950s. In Germany, limits for other synthetic organics including organochlorine pesticides, PAHs, alkylphenols (surfactants), phthalates, or the polychlorinated terphenyls have not been set either because environmental risk is considered to be small or sufficient ecotoxicological information is not yet available. A recent report to the Department of the Environment in the UK (Sweetman *et al.*, 1994) on the occurrence, fate and behaviour of some of the principal 'priority' organic pollutants in sewage sludge concluded that there is no evidence of any significant problems arising from organic contaminants in sludges applied to agricultural land. The report recommended that there was no need for change in the current regulations on the agricultural use of sewage sludge with respect to organic contaminants.

In a review of 20 years of land application research with sludge Chaney (1990b) stated that, except for the peel of carrots, there is little evidence for plant uptake of PCBs from sludge-treated soils. Transfer to the food chain is principally through direct ingestion of sludge by cattle, but Chaney considered that the transfer was so low that sludges containing several mg kg^{-1} of PCBs could still be surface applied. He pointed out that in all cases organics did not occur in US sludges at concentrations which would fail the pathway analysis (US EPA, 1989b), since the median concentration was 0.21 mg kg^{-1} (PCB-1248, dry solids) and the 98th percentile concentration in sludge was 1.5 mg kg^{-1} (dry solids) compared to the derived NOAEL (see Appendix) sludge limit for PCBs of 2.13 mg kg^{-1}. Consequently, the proposal to regulate PCBs in sludge (US EPA, 1989a) was considered unnecessary in the Final Part 503 (US EPA, 1993) due to the very low risk to the human food chain (R. L. Chaney, USDA-Agricultural Research Service, personal communication). The PCB content of UK sludges compares closely with the US values. For example, McIntyre and Lester (1984) analysed 444 sludge samples from sewage treatment works in the UK and obtained median and 98th percentile concentrations of PCBs in sludge of approximately 0.14 and 1.5 mg kg^{-1} (dry solids), respectively. More recently, Alcock and Jones (1993) reported the total PCB content of 12 UK sludges from rural, urban and industrial STWs ranged between 0.106 and 0.712 mg kg^{-1}, with a mean value of 0.292 mg kg^{-1}. This implies that the overall concentrations of PCBs in UK sludges have dropped markedly in the last ten years. On this basis apparently there is minimal risk to the environment from PCBs applied to agricultural soils in sewage sludge. In a recent study of the occurrence of PCBs in sludge-amended soils, Sweetman *et al.* (1994) confirmed that it is unlikely that sludge recycling would be limited in practice due to PCB content.

In developing the recent *Standards for the Use or Disposal of Sewage Sludge* (US EPA, 1993), the US EPA screened 200 pollutants and selected 18 organic pollutants of principal concern for further evaluation by the pathway risk analysis of environmental exposure (Table A6) (US EPA, 1992a).

The selection criteria considered: frequency of occurrence, aquatic toxicity, phytotoxicity, human health effects, domestic and wildlife effects, and plant uptake. With the exception of certain chemicals, such as trichloroethylene, emphasis was placed on organic contaminants exhibiting a moderate to high level of persistence in soil. Those compounds with a high propensity to degrade or volatilize from soil are generally considered of less concern in sludge-amended agricultural land. During the public comment period for *Technical Standards for Use and Disposal of Sewage Sludge; Proposed Rule* (US EPA, 1989a) deletion of some of the organic pollutants was recommended because of restrictions on manufacture or use in the US. The Agency re-evaluated the pollutants by the revised pathway risk analysis and deleted *all* of the organic pollutants from the final rule for land application (US EPA, 1992a,b). For an organic pollutant to be deleted from the regulation, one of the following criteria had to be satisfied (Ryan, 1993):

1. The pollutant was banned from use in the US; has restricted use or is not manufactured for use in the US.
2. Based on the results of the National Sewage Sludge Survey (NSSS), the pollutant has a low level of detection in sewage sludge.
3. Based on data from the NSSS, the limit for an organic pollutant in the Part 503 exposure assessment (US EPA, 1992a) is not expected to exceed the concentration in sewage sludge.

The US EPA is currently reassessing the health risks of exposure to 2,3,7,8-T_4CDD and related dioxins in response to emerging knowledge of the biological, human health and environmental effects of dioxin (US EPA, 1994a,b). Round 2 of the Part 503 regulation is scheduled for completion by the end of 1999. Included in the potential list of pollutants for the Round 2 rule are dioxins. However, based on the findings of the final Dioxins Study, proposal of dioxin standards in sewage sludge could be accelerated by up to two years. Consequently, US EPA could propose dioxin standards for sludge as early as 1997 (A.B. Rubin, Water Environment Federation, personal communication).

Maximum concentrations of organic pollutants in sludge-treated soil have been calculated by Chang *et al.* (1993) following a modification of the methodology used by the US EPA for the development of its regulations on sewage sludge reuse on the basis of environmental exposure pathways. The soil limit values were estimated for the sludge→soil→plant→human route (Table A6) and considered only the pollutant intake from plant foods since food chain transfer is the primary route of human exposure to environmental pollutants. Chang *et al.* (1993) argued that the potential for sludge-borne pollutants to be transferred to humans through other food groups including dairy and animal products is relatively small, which is at odds with the view that ingestion of contaminated herbage and soil by grazing animals is the principal route of exposure (Stark and Hall, 1992; Wild et al., 1994). In the risk assessment, it was assumed that all of an individual's plant food intake was from sludge-treated soil. From the earlier discussion (Chapter four) this is accepted as a highly precautionary and somewhat unrealistic assumption in developing soil limit values for sludge-borne contaminants. Maximum pollutant concentrations were calculated on the basis of:

1. Acceptable daily dietary intakes (ADI) of pollutants.
2. A standardized global diet of plant food intake.
3. Pollutant exposure derived from consuming the main food groups in the global diet (grain, vegetables, root/tubers and fruit).
4. A limit on daily intake of pollutants in the plant food groups to 50% of the ADI.

The maximum concentrations of organic pollutants estimated for sludge-treated soil are listed in Table 10.2. Clearly, there are potential limitations in deriving 'global' soil limits which protect human health when sewage sludge is applied to agricultural land under widely contrasting agronomic, soil and environmental conditions. Furthermore, the patterns in food consumption vary considerably from region to region which will also influence the amounts of pollutants consumed.

It could be argued that highly conservative coefficients have been used in the risk assessment for calculating the transfer of pollutants from soil into crops because the bioavailability to plants of sludge-borne organic contaminants is regarded as being virtually negligible (O'Connor *et al.*, 1991).

Table 10.2. Estimated maximum concentrations of organic pollutants in sludge-treated agricultural land.

Compound group	Compound	Maximum concentration in soil (mg kg^{-1} dry soil)
PAH	Pyrene	480
	Benzo[a]pyrene	3.0
PCBs	PCBs	30
Polychlorinated dioxin	2,3,7,8-T_4CDD	30
Organochlorine pesticides	Aldrin	0.2
	Dieldrin	0.03
	Chlordane	0.3
	Lindane	0.6
	Heptachlor	1.0
	2,4-D	10
	Methoxychlor	20.0
	Toxaphene	9.0
Monocyclic aromatics	Benzene	0.03
	Toluene	50
Chlorobenzenes	Hexachlorobenzene	40
Short chained halogenated aromatics	Chloroform	2.0
	Tetrachloroethylene	250
	Tetrachloroethane	4.0
	Hexachloroethane	2.0
Phenols	Pentachlorophenol	320

Source: Chang *et al.* (1993)

Concentrations of Some Persistent Organic Pollutants in Sludge-treated Soil

Increases in the concentrations of certain persistent organic pollutants in sludge-amended agricultural land have been predicted on the basis of a theoretical consideration of sludge concentration, application rates and cultivation depth. For example, Alcock and Jones (1993) estimated that the concentration of ΣPCB would increase by 2 μg kg^{-1} in soil treated with 10 t ha^{-1} of sewage sludge (dry solids) with mean ΣPCB content, and by only 5 μg kg^{-1} at the maximum ΣPCB level measured in sludge when the sludge was incorporated to 15 cm. The UK background soil ΣPCB concentrations are <40 μg kg^{-1}. On the same basis, the application of typical sludge containing 0.72 ng kg^{-1} (dry solids) of 2,3,7,8-T_4CDD may be expected to increase soil 2,3,7,8-T_4CDD concentrations by 0.0048 ng kg^{-1} (0.96%) to 0.5048 ng kg^{-1}, assuming that agricultural soil contains 0.5 ng kg^{-1} of this dioxin (Wild *et al.*, 1994). After twenty annual applications of sludge the soil 2,3,7,8-T_4CDD and TEQ concentrations would be 0.517 ng kg^{-1} and 3.027 ng kg^{-1}, respectively, accounting for the half-life of dioxin in soil.

There are few published results of soil analyses for persistent organics in sludge-treated agricultural land. The available information is mainly from Germany. For example, Kampe and Leschber (1989) reported on the concentrations of organic pollutants in German agricultural soils which had received very large cumulative rates of sludge treatment up to a maximum application of 340 t ha^{-1} (dry solids) commencing as early as 1959. On average, concentrations of PCBs were 5-17 times higher in sludge-treated soil compared with background levels and were in the range 5-49 μg kg^{-1}, increasing to a maximum value of 340 μg kg^{-1} in one extreme case. Sewage sludge treatment increased the concentrations of PAHs in soil by a factor of 5-10. The average concentration of benzo[a]pyrene in sludged soil was 200 μg kg^{-1} and for other PAHs was 30-300 μg kg^{-1}. Chlorinated hydrocarbons (e.g. aldrin, dieldrin, ΣDDT, chlordane, heptachlor and hexachlorobenzene) did not accumulate in soil with the application of sewage sludge. On the basis of these results, which reflected extreme conditions of soil treatment with sludge, Kampe and Leschber (1989) concluded that there would be no critical accumulation of organic contaminants in soil as a consequence of using sewage sludge on agricultural land.

Hembrock-Heger (1992) measured the concentrations of PAHs, PCBs and PCDD/Fs in soils treated with sewage sludge for a period of ten years compared with unsludged control areas. However, there was very little difference in the concentrations of benzo[a]pyrene detected between sludged and unsludged soils and no correlation with application rate of sludge was apparent. A small increase in PCB content of sludge-amended soil was observed that was attributed to land spreading of sludge from mining wastewater. There was no accumulation of PCDD/Fs in the soils due to sludge treatment in practice. The total PCDD/F concentration in soil was raised in various field trials treated with sewage sludge at an extreme rate equivalent to 100 years of application. In spite of the large rate of sludge addition, 2,3,7,8-T_4CDD was not detected (detection limit 0.01 μg kg^{-1} dry solids) in any of the experimental soils.

From a recent survey, Ilic *et al.* (1994) determined that the concentrations of PCDD/Fs and PCBs were within the general background levels for both types of pollutant in arable soils treated with sludge on an operational basis. The results were consistent with the atmospheric deposition of traces of the relevant compounds emitted from the combustion of fossil fuels. There was no evidence of any soil contamination which could be attributed to the application of sewage sludge and Ilic *et al.* (1994) considered that the observed values were far below those likely to create any risk to human health. Sweetman *et al.* (1994) also reported that the concentrations of chlorobenzenes, PCBs and PCDD/Fs in sludge were generally very low and that they are virtually undetectable in sludge-amended soils. Because these compounds are semi-volatile, Sweetman *et al.* (1994) considered that soil inputs from atmospheric deposition were probably far more important than the application of sewage sludge.

The available information suggests that the application of sewage sludge to agricultural land, and especially to arable soils, has a minimal effect on the soil concentrations of key 'priority pollutants' of potential concern including PAHs, PCBs, PCDD/Fs and chlorinated hydrocarbons. Indeed, there are examples where no detectable increases in soil concentrations have been observed in practice. The concentrations of organic pollutants which are normally likely to be achieved in sludge-treated soil will therefore remain significantly below the risk derived limit values of Chang *et al.* (1993). In contrast, Sweetman *et al.* (1992a) reported that soils treated with long-term applications of highly contaminated sludges in the past may contain substantially elevated concentrations of organic pollutants compared with unsludged soil. Nevertheless, even under this potentially 'worst-case' of soil contamination, only dieldrin was identified as potentially exceeding the tentative pollutant levels listed in Table 10.2. The maximum concentration of this compound measured by Sweetman *et al.* (1992a) in historically sludge-treated soils was 204 $\mu g\ kg^{-1}$ (0.204 $mg\ kg^{-1}$). Whilst the maximum soil concentrations devised by Chang *et al.* (1993) are designed to protect the human diet from potentially harmful intakes of organic pollutants in plant foods, other studies strongly suggest that the critical pathway of environmental exposure is the transfer to grazing livestock when sewage sludge is spread on the surface of grassland (Stark and Hall, 1992). Under these conditions, there is greater potential for enrichment at the surface layers of the soil, thereby increasing the concentrations and burden of pollutants in soil ingested by grazing animals, compared with the arable soil scenario where the pollutants may be distributed more evenly (and therefore diluted) to depth due to soil cultivation.

Persistence in Soil

Major transport and fate pathways for organic chemicals applied to the soil-plant system include adsorption by soil, volatilization and degradation (Jacobs *et al.*, 1987). Many organics are strongly adsorbed to soil organic matter and/or undergo degradation, reducing the potential for plant uptake or leaching. These properties are listed for the broad compound groups in Table 10.1. Due to the large number of organic contaminants potentially present in sewage sludge, the development of models which utilize basic physical/chemical properties of an organic compound to predict the environmental fate of sludge organics in soil has been a logical approach and widely used in assessing environmental risk. This has formed the basis of the screening methodology developed by Jones and Wild (1991) to establish the potential transfer to crops and livestock of trace organics applied to agricultural soil in sewage sludge.

Volatilization is an important transport route and many organic compounds of interest may be lost from sludge-treated soil by this pathway (Jin and O'Connor, 1990; Jones and Wild, 1991). However, isomer- and congener-specific differences are important within any one group of compounds. Thus, increasing molecular weight or degree of chlorination within a homologous series generally reduces aqueous solubility and volatilization potential.

Wilson *et al.* (1994) measured the concentrations of 15 important volatile organic compounds (VOCs) (e.g. chloroform, benzene, toluene) in samples of 12 liquid digested sewage sludges obtained from rural, urban and industrial treatment works in north-west England. No apparent relationship was found between sludge VOC concentrations and the volume of industrial input to the sewage treatment works, influent treatment, population served and sludge dry solids content. Wilson *et al.* (1994) concluded that normal rates of sludge application to agricultural land were unlikely to increase the VOC content of the soil to levels which may cause concern for human health and the environment. Webber and Goodin (1992) confirmed from a laboratory incubation study of the persistence of VOCs in sludge-treated soils that sludge VOCs do not present a hazard to agriculture. Indeed, any elevation in soil concentrations resulting from spreading sewage sludge on agricultural land is likely only to be transient due to rapid volatilization losses and degradation of VOCs (Table 10.1).

Microbial degradation has been shown to be the most important loss mechanism for many trace organic contaminants in soil (Gibson and Burns, 1977; O'Connor *et al.*, 1981; Matsumura *et al.*, 1983; Bossert *et al.*, 1984; Fairbanks *et al.*, 1985; Marcomini *et al.*, 1988, 1989; Bellin *et al.*, 1990; O'Connor *et al.*, 1990b; Sweetman *et al.*, 1994). The potential for increased microbial degradation of persistent organic pollutants has been observed in soil possibly due to the activities of earthworms (Reinecke and Nash, 1984). Compounds applied to agricultural soils in sewage sludge which are rapidly lost by volatilization and/or microbial degradation are unlikely to result in any significant environmental impact (Table 10.1). In contrast, there is concern that chlorinated aromatics may pose a significant risk due to their toxicity, bioaccumulation and persistence in the environment. Resistance to microbial degradation is dependent on the type, position and degree of halogenation (Boyle, 1989). This is because most micro-organisms do not possess enzyme systems that are capable of degrading molecules which lack structural regularity (Bumpus and Aust, 1987). The presence of large halogenated groups on aromatic ring systems also tends to interfere with the degradative activity of ring-cleaving enzymes thereby increasing the persistence of these compounds (Boyle, 1989).

Thus, in sludge-treated soils, PCBs can have half-lives of 2.5-6 years (Fairbanks *et al.*, 1987). Other highly persistent compounds include the PAHs. Wild *et al.* (1991) calculated half-lives for PAHs in sludge-amended agricultural soils in the range <2 to >9 years, with multi-ringed species showing the greatest persistence. Over a 15-month period Orazio *et al.* (1992) measured significant degradation of di- and trichloro PCDDs and PCDFs in soil, but the tetra- to octachloro congeners were highly persistent. Nevertheless, even these highly persistent chemicals are susceptible to microbial degradation under certain environmental conditions. For example, Quensen (1988) reported the anaerobic reductive dechlorination of most PCBs by micro-organisms in river sediments. Over 50% of the total Cl was removed in 16 weeks and the dechlorination products were both less toxic and more readily degraded by aerobic bacteria. Quensen (1988) suggested that a sequential anaerobic-aerobic biological treatment system for PCBs could be feasible. In soil, white rot fungi such as *Phanerochaete chrysosporium* have the ability to mineralize a wide variety of structurally diverse and apparently persistent organic compounds including PAHs, PCBs, organochlorine pesticides and PCDDs to carbon dioxide (Bumpus *et al.*, 1985; Eaton, 1985; Bumpus and Aust, 1987). White rot fungi are members of the Basidiomycetes and are equipped with enzymes which are capable of degrading the structurally complex and resistant lignin molecule. It is the ligninolytic enzyme system which apparently provides this type of fungi with unique biodegradative abilities. Bumpus and Aust (1987) suggested that this naturally occurring biodegradative system, which is already in place in the environment, may be responsible, in part, for the decline in some chemicals that has been observed since their introduction to the environment has abated.

Plant Uptake

Table 10.1 shows that the uptake and translocation of trace organic contaminants by plants occurs only to a limited extent or not at all from sludge-treated soil (Dean and Suess, 1985; O'Connor *et al.*, 1991). In particular, uptake is minimal for the polyhalogenated compounds and for the PAHs which are considered the principal sludge contaminants of concern (Sauerbeck and Leschber, 1992; Wild and Jones, 1992b). Kampe and Leschber (1989) showed there was minimal or no transfer of chlorinated hydrocarbons, PCBs or PAHs to crop plants even where very large amounts of sludge were applied to soil at rates up to 340 t ha^{-1} (dry solids). They concluded, therefore, that these principal organic contaminants are not limiting to the use of sewage sludge in agriculture. Wild and Jones (1992b) determined that crop uptake of sludge-applied PAHs does not pose a risk to the human food chain even when sludge is applied to lipid-rich root crops (especially carrots) representing a 'worst case' condition of PAH exposure. Carrots may become slightly contaminated by PCBs due to partitioning from soil to lipophilic root tissues (Offenbacher, 1992), but this may occur only to a limited extent and would be removed by normal culinary practices (O'Connor *et al.,* 1990a). As most root crops are not nearly as good PCB accumulators as carrots, O'Connor *et al.* (1990a) suggested that the potential human exposure to PCBs would be very low under recommended sludge application practice for agricultural utilization.

The minimal risk from agricultural recycling of sewage sludge is further demonstrated by studies in which very large rates of apparently highly industrially contaminated sludge have been applied. For example, Webber *et al.* (1994) generally detected only traces of a very few organic contaminants in different crops grown on reclaimed mine spoil treated with up to 3360 t ha^{-1} (dry solids) of Chicago municipal sewage sludge (Cd content was 76 mg kg^{-1} dry solids - see Table 2.3 for comparison with UK sludges). As expected, carrot peel contained more PCB compared with other plant materials. Except for cabbage 'wrapper' leaves, however, the PCB concentrations in plant tissues were not related to those in the soil. Furthermore, soil PCB concentration accounted for only about 25% of the variance in leaf concentration and the bioconcentration factor was very small (0.0042). Despite the very large rates of sludge application employed in the study, Webber *et al.* (1994) concluded that they did not result in high levels of organic contamination in the treated growing media or represent a significant organic contaminant hazard to the quality of crops for food or feedstuffs.

Hembrock-Heger (1992) determined that plant uptake of PAH, PCB and PCDD/F was very low from sludge-amended agricultural soil. Transfer factors were probably <0.01 and PCBs were barely detectable in the plant materials examined. Offenbacher (1992) detected little or no PCBs in crop plants except for carrot peel and the transfer factor here decreased with increasing degree of chlorination. This is in agreement with earlier work by Moza *et al.* (1979) showing bioconcentration of trichlorobiphenyl by carrots, but not by sugar beet, whereas neither crop accumulated pentachlorobiphenyl. Ye *et al.* (1992) detected no active transport of the tetra- or hexachlorobiphenyl congeners (IUPAC 77 and 169) through the plant xylem system. Schmitzer *et al.* (1988) and Aranda *et al.* (1989) established that plant uptake of phthalate (DEHP) was of minor importance due to rapid biodegradation in soil. Similarly, there is minimal uptake of the more persistent pentachlorophenol (PCP) by crop plants from sludged soil; the reported bioconcentration factors of <0.01 indicate that transfer of PCP to food chain crops should not limit land application of sludge (Bellin and O'Connor, 1990). Accumulation of PCDDs by plants due to uptake from soil is also highly unlikely (Isensee and Jones, 1971). Indeed, the very minor effect of sludge application to arable land on human exposure to PCDD/Fs (Wild *et al.*, 1994) is entirely consistent with the minimal transfer of PCDD/Fs from soil to crop plants.

Plant tissues may contain very low levels of organic pollutants, but there is rarely any relationship between the concentrations of contaminants in sludge-treated soils and the amounts detected in crops (Witte, 1989; Hembrock-Heger, 1992). Therefore, the presence of organic residues in plant tissues is unlikely to be due to sludge-borne organics, but is explained by aerial deposition to foliage, and adsorption of contaminants from the gaseous phase which are derived principally from other sources (Witte, 1989). However, there is also the possibility that foliar contamination may occur due to volatilization and redeposition of trace organics from sludge applied to soil. For PCBs, the magnitude of plant residues is likely to decrease with increasing level of chlorination due to lower saturation vapour pressures limiting volatilization (Fries and Marrow, 1981). Reischl *et al.* (1989) and Sacchi *et al.* (1986) showed that volatilization and absorption by the above ground parts of plants was more important than root uptake processes in shoot tissue contamination with PCDD/Fs. Nevertheless, very little contamination of plants with sludge-borne PCDD/Fs is likely, due to the very small concentrations of these compounds occurring in sludge-treated agricultural land (O'Connor *et al.*, 1991).

O'Connor *et al.* (1991) suggested that the concerns about the possible bioavailability to food chain crops of organic pollutants from sewage sludge applied to agricultural land were largely groundless on the basis that transfer factors for most organic pollutants are small and often <0.01 (dry matter). Even if some compounds are taken up by plants, they are generally metabolized within plants to non-toxic forms or accumulated in plant parts such as carrot peel that would usually be removed by washing and peeling in the course of normal food preparation. However, O'Connor *et al.* (1991) identified that aromatic surfactants had demonstrated some potential to accumulate in plants and that further studies were necessary to evaluate their bioavailability. Unlike most of the xenobiotic compounds which occur in sludge in relatively small (mg kg^{-1}) or very small (μg kg^{-1}) amounts, detergent-derived surfactants are distinguished as a special group because the concentrations in sludge may be several orders of magnitude greater and in the per cent range (Table 10.1). Consequently, the fate and environmental impact of this group of chemicals may be of particular concern in sludge-treated soil. In their assessment of physico-chemical properties of organic contaminants in sludge, Jones and Wild (1991) also highlighted chlorobenzenes as having the potential to transfer to plants by both root uptake and foliar absorption. A recent review by Wang and Jones (1994) of the fate and behaviour of chlorobenzenes in sludge-treated soil similarly suggested that the transfer of chlorobenzenes to crop plants was possible from theoretical perspective, although there was an absence of data specific to sludge-amended soils on actual crop uptake of these organic contaminants.

Sweetman *et al.* (1994) examined the potential accumulation of components present in aromatic surfactants (linear alkylbenzenes (LABs)) and chlorobenzenes by potatoes, cabbage, leeks and carrots grown with increased exposure to these compounds in sludge-treated soil compared with an unsludged control. In most cases the concentrations of the organic contaminants present in plant tissues were below the analytical limits of detection. However, certain chlorobenzenes were detected in unpeeled potatoes although the chlorobenzene content of tubers in the unsludged control was similar, or in some cases greater than for the sludge-treated soil. Very low concentrations (3-4 $\mu g\ kg^{-1}$) of hexachlorobenzene were detected in the peel of carrots grown in sludge-treated soil which may be explained by sorption of the chemical at the root surface. Overall, the results indicated there was minimal risk to food crops from LABs and chlorobenzenes in sludge-treated agricultural land.

No adverse effects of trace organic pollutants on the growth of crops have been observed when sludges are applied to soil at agronomic rates according to crop requirements for nutrients (Jacobs *et al.*, 1987).

Animal Ingestion

The direct ingestion of sludge-treated soil by grazing livestock is considered to be the principal route of trace organic bioaccumulation in the food chain which may result from agricultural utilization of sewage sludge (Stark, 1988; Jones and Wild, 1991; Hembrock-Heger, 1992; Sauerbeck and Leschber, 1992; Stark and Hall, 1992). Organic compounds are potentially ingested by grazing animals following the same routes described for PTEs in Chapter six. Consequently, potential effects on grazing animals are highly dependent on the level of intake of organic contaminants in soil and in sludge adhering to forage. The lipophilic properties of ingested organic compounds may result in significant bioconcentration in tissue fat and also in milk. Once absorbed into the body, organic residues are unlikely to be eliminated rapidly following withdrawal of contaminated feed due to storage in the fatty tissues (Stark, 1988).

Fries (1982) reviewed the available information on PCB residues in animal tissues. He concluded that, in the absence of data to the contrary, PCB ingested from any source (e.g. sludge or soil) produced the same level of contamination in the animal and that tissue residues increased with increasing degree of chlorination of the PCBs. When a constant amount of PCB was fed, the concentration of residue in milk fat increased rapidly, but approached a steady equilibrium state within three weeks. Fries (1982) noted that several studies had shown that the maximum concentrations of PCBs in milk fat were four to five times larger than the dietary content. Residues in the body fat of dairy cows were always less than concentrations in milk fat while PCBs were being fed. Other limited data suggested that concentrations of PCBs in the body fat of non-lactating animals were similar to those in milk fat produced by dairy cows.

The contamination of animal feed and milk by chlorinated pesticides has been monitored and reviewed by several workers. In one set of experiments

(Muir and Baker, 1973), residues in the milk of cattle grazing orchard areas treated with pesticides were twice as high as those in milk from animals grazing uncontaminated pasture. However, Gilbert and Lewis (1982) found that dieldrin and heptachlor residues in milk and body fat showed considerable variation between dairy cows. This was attributed to metabolic difference between animals and variations in grazing behaviour and soil ingestion. Harrison *et al.* (1970) showed a similar pattern of transfer for DDT.

According to Fries (1982) the potential residue problems are more serious for dairy cows than they are for meat animals. This is because lactating animals have a higher feed intake relative to body weight and are thus more likely to consume greater quantities of contaminated soil.

Sewage sludge has been identified as a potential dietary source of organic contaminants in grazing animals by several authors. For example, Lindsay (1983) showed that grazing immediately after sludge treatment resulted in concentrations of organochlorines and PCBs in milk fat five times the level measured in the forage dry matter. Fries (1982) concluded that the greatest potential problems arising from long-term applications of PCB-contaminated sludges were through animal ingestion of contaminated soils when grazing. Alcock and Jones (1993) also suggest that significant elevations in the potential transfer of PCBs to grazing livestock may result from the surface application of sludge to grassland. In practice, however, agronomic and management factors can reduce the effects of PCBs in sludge. For example, feeding concentrates lower the intake of forage crops and a non-grazing period following sludge application reduces the direct contamination of grass with sludge. Sludge application to short swards also minimizes the extent of sludge adherence to herbage (Sweet *et al.*, 1994). Furthermore, injecting sludge into soil also reduces the potential accumulation of PCB residues through grazing.

A number of problems can be identified in attempting to estimate potential tissue concentrations of organic contaminants based on soil levels. This is because the extent of soil ingestion is highly variable (0-30% of total dry matter intake), although it has been estimated as 10% for the purposes of risk assessment (MAFF/DoE, 1993b), and there are insufficient data on the absorption characteristics of different compounds in animals. The presence of organics in the diet of dairy cows is rapidly reflected by changes in concentrations present in milk. However, the effects on meat animals occur over a longer period of time and are complicated by the differential deposition of lean and fat tissue. Indeed, sludge application to permanent grassland is not permitted in Germany because of concerns over the potential bioconcentration of organic contaminants, and PCBs and dioxins in particular, in animal tissues and milk from ingestion of sludge and sludge-treated soil (Sauerbeck and Leschber, 1992). This contrasts with the US view on PCBs based on risk assessment analysis (US EPA, 1992a) which showed there was minimal risk to the human food chain from PCBs in sludge which is surface-applied at agronomic rates to grassland soils (Chaney 1990b). The published information reviewed by Stark (1988) similarly suggested that the environmental effects of organic contaminants in sewage sludge were likely to be minimal since, in most 'normal' situations where animals grazed pasture previously treated with sewage sludge, persistent organics were below detection limits in liver, kidney and muscle tissue, and at normally expected background concentrations in milk. More recently, Stark and Hall (1992) concluded that the risk of adverse effects of sludge-borne organic contaminants on animal or human health is low at agronomic rates of sludge application. However, exceptions may occur when sludges are highly contaminated, or when animals graze pasture immediately after sludge application (which is not permitted by the UK regulations (SI, 1989a)), and thus may ingest relatively large quantities of sludge adhering to plant material.

The possible transfer of PCBs and PCDD/Fs from sludge-treated soil to milk was demonstrated by McLachlan *et al.* (1994) in a comparative study of dairy cattle on different farms which had received sludge in the past, or where there had been no previous sludge treatment. Adherence of sludge-treated soil to foraged crops was identified as a principal mechanism of contaminant transfer in the feed of housed livestock. However, the accumulation of PCB and PCDD/F in sludge-treated soil reported by McLachlan *et al.* (1994) contrasts with the minimal or no accumulation observed by other researchers following repeated applications of sludge (Hembrock-Heger, 1992; Ilic *et al.*, 1994). Unfortunately, there were no reliable records available of the amounts of sludge applied to the different fields and the authors relied upon farmers' recollections of past sludge treatment history which covered a period of at least 30 years. A possible explanation for the observed increases in soil concentrations could be that they reflect the application of highly contaminated sludges in the past. However, a further difficulty encountered in this study was that no archived samples of the applied sludges had been preserved for chemical analysis so it was not possible to provide any information on the composition of the sludges used. Consequently, it is arguable how relevant these data might be in relation to assessing the potential exposure from organic contaminants caused by current and future sludge application practice given that the concentrations of organic pollutants in sludge continue to decline.

Wild *et al.* (1994) identified livestock ingestion of sludge adhering to vegetation as the single most important transfer mechanism of PCDD/Fs from sludge-amended soil to the human food chain. However, their exposure model assumed that an individual consumed produce exclusively from sludge-amended land which is recognized as an unrealistic and highly conservative level of exposure in practice (MAFF/DoE, 1993b). If only 1.3% of consumed produce came from sludge-amended areas, then human exposure would remain below the tolerable daily intake value of 10 pg TEQ kg^{-1} body weight day^{-1} recommended by WHO/EURO. In Pathway 5 (sludge → soil → animal → human) of the risk assessment for land application of sludge (US EPA, 1992a), the US EPA considered the HEI was an individual consuming 3% of their milk and dairy products and 10% of their meat intake from animals ingesting sewage sludge. The US model (US EPA, 1992a) assumed that the diet of cattle grazing sludge-treated pasture contained 1.5% of sewage sludge adhering to herbage and/or ingested directly from the soil surface.

Wild *et al.* (1994) probably overestimated the level of sludge ingestion in herbage for several reasons. Firstly, they assumed a sludge contamination level of 5% on the basis of sludge adherence information produced by Buttigieg *et al.* (1990), with high dry solids (12%) liquid sludge which is atypical of normal practice (normal range 2-5% dry solids). Indeed, Buttigieg *et al.* (1990) showed there was minimal contamination of pasture with sludge containing 2% dry solids within the recommended minimum three week no-grazing period for sludge (DoE, 1989a). Secondly, a constant high rate of sludge consumption was assumed until both milk and body fat are in steady-state with dietary intake which may take very long periods of time, >60 days for body fat. This assumption is necessary where bioconcentration factors are used to relate concentrations of pollutant in the diet to concentrations in the animal or food product (Fries, 1991). However, grass contamination declines rapidly in actively growing crops, particularly when sludge is applied to short swards. This provides a management barrier to the potential transfer of sludge contaminants to animals because grazing is not practicable until grass regrowth occurs (Sweet *et al.*, 1994). Consequently, it is highly unlikely that herbage contamination with sludge will remain at an elevated level during the life of a grazing animal reared for milk or meat production. To the contrary, the periods of exposure to contaminated herbage would be expected to be relatively short and not exceed a few weeks (Fig. 6.2). Thus the implicit assumptions of the steady state condition that the contaminant intake, animal pool size and excretion rate remain constant is highly unlikely in practice under normal animal feeding and management conditions (Fries, 1991).

On balance, there is probably minimal risk of human exposure to PCDD/Fs from sludge-treated agricultural land, even when sewage sludge is surface applied to pasture, and Wild *et al.* (1994) emphasized the conservative nature of their assumptions, especially on the proportion of produce consumed from sludge-amended soil. Furthermore, the exposure assessments used food PCDD/F concentrations in uncooked produce whereas loss of meat fat during cooking may decrease levels. Interestingly, charcoal grilling, on the other hand, may be expected to increase the concentrations of PCDD/Fs in meat. If sludge ingestion by livestock could be eliminated as the source of PCDD/Fs, Wild *et al.* (1994) estimated that TEQ exposure from sludge-treated soil would be <1% higher than the background level.

More data are needed to fully quantify the potential risk to grazing animals of organic pollutants in sludge surface-spread on grassland, although currently available information suggests the risk is likely to be small. It is widely recognized, however, that sludge injection can virtually eliminate problems of animal ingestion of organic contaminants (Chaney, 1990b; Stark and Hall, 1992; Alcock and Jones, 1993; Wild *et al.*, 1994).

Mutagenicity

Researchers in the US have developed techniques to assess whether organic pollutants released into the environment in sewage sludge and municipal waste water have mutagenic properties. Hopke *et al.* (1987) defined an environmental mutagen or genotoxin as an agent that is released into the environment which can alter the genetic material or alter the proper functioning of the genetic material. The presence of such genotoxic agents in the environment has been considered as a potentially serious threat to public health. Depending upon the developmental state of an individual, a genotoxin can cause birth defects, result in coronary disease, produce mutations involving germinal cells or cause mutations in somatic cells that may develop into cancer. The mutagenicity of sewage sludges has been determined using a *Salmonella*/microsome assay by counting the number of revertant colonies produced on exposure to a specially prepared sludge extract (Ames *et al.*, 1975).

Numerous workers (Angle and Baudler, 1984; Brown *et al.*, 1986; Hopke *et al.*, 1987) report decreased mutagenic activity with time of soils treated with potentially mutagenic sewage and other sludges. In contrast, Donnelly *et al.* (1989) recently showed that sewage sludge may contain organic mutagens which persist in the soil for long periods of time. However, this could be explained because very large rates of sludge were applied (150 t ha^{-1} dry solids) to soil in this study. Davis *et al.* (1984) considered that mutagenicity testing appeared relevant to sludge utilization on agricultural land. However, Jacobs *et al.* (1987) concluded that, whilst of interest, mutagenicity tests are difficult to interpret and have not been adequately correlated to mutagenic activity or biological toxicity of soil/sludge mixtures in the field. Indeed, Ramel (1983) reported that there was not always a quantitative correlation between mutagenic and carcinogenic potency such that the definite identification of potential carcinogens had to be verified by a multiplicity of other test procedures. Consequently, mutagenicity tests probably have little relevance for assessing the environmental impact of the agricultural use of sewage sludge due to difficulties in the practical interpretation of the test results in relation to observed effects in the field.

Leaching and Water Source Contamination

Application of sewage sludge to land has been identified as one of several possible ways groundwater may become contaminated by organic substances (Kenrick *et al.*, 1985). Other potential contamination routes include:

- leachate from domestic and industrial landfill sites;
- accidental spillages or leakages;
- agricultural and horticultural usage of organic chemicals;

- farm waste disposal;
- atmospheric washout by rain or snow;
- artificial aquifer recharge or effluent recharge;
- infiltration of contaminated surface water;
- natural processes such as animal or plant decomposition.

In spite of the variety of routes for potential contamination and the ubiquity in the environment of certain trace organic substances, such as PAHs, PCBs and PCDDs, generally there is little evidence suggesting concentrations of organic pollutants in groundwater are of concern in the UK. Kenrick *et al.* (1985) surveyed the quality of groundwater from thirty-two representative source supplies in Britain and identified 30 uncontaminated samples which contained a few compounds at very low levels ($<0.1\ \mu g\ l^{-1}$). However, the numbers of compounds and the presence of individual compounds varied from sample to sample. Kenrick *et al.* (1985) concluded that the presence of individual compounds, even those believed to be anthropogenic, must not be taken as indicative of contamination unless concentrations are very much greater than the reported background levels. Two other sites exhibited some contamination, but the results did not suggest that the quality of the water was unacceptable for drinking. Furthermore, inorganic constituents of groundwater exhibited marked changes along the recharge pathways with the result that inorganic water quality in the unconfined zones of the aquifers was quite different from that in their confined zones. However, Kenrick *et al.* (1985) detected no such trends in the distribution of organic compounds through the aquifers sampled, again suggesting contamination was unlikely to account for the presence of organics in groundwater. Nevertheless, the occurrence in deeply confined groundwater of some halogenated compounds, even at the very low concentrations measured, was difficult to explain and indicated there was an unidentified source of these compounds in the confined zones.

Many of the organic pollutants of principal concern in sludge-treated soils are strongly adsorbed by soil organic matter, thereby retarding leaching (Table 10.1). Those compounds, which are to some extent water soluble, are also generally susceptible to microbial degradation and/or volatilization and are therefore similarly not leached from soil, although there are examples of persistent and soluble organics which are potentially leachable. For example, examination of physico-chemical properties (Jones and Wild, 1991) indicates that certain chlorobenzenes have relatively low degradation and adsorption potentials with a moderate to high risk of leaching.

Laboratory measurements of organic pollutant mobility through soil columns show, not surprisingly, that soil adsorption and water solubility are generally inversely correlated. Seip *et al.* (1986) found that the chlorinated hydrocarbons and some of the light aromatic hydrocarbons (benzene, toluene and xylene) behave in this way. However, there are difficulties in extrapolating results from these studies to establish what the risk may be under field conditions. This is because column measurements can seriously overestimate the potential leaching loss due to the small internal column diameter, producing large mass-flow edge effects. In addition, soil structural integrity is frequently disturbed affecting the water flow characteristics compared with field intact soil and there are difficulties in establishing realistic rates of drainage. For these reasons, soil column techniques are not generally adopted by soil scientists investigating nutrient leaching losses from agricultural soils and should only be used very cautiously by organic chemists in assessing the mobility of organic chemicals through soils.

An alternative approach, which has become increasingly refined in recent years, is to predict the leaching potential of organic contaminants using mathematical models based on known chemical properties of the compounds. For example, Steenhuis and Naylor (1985) developed a simple screening model to determine the relative risk to groundwater of organic contaminants in sewage sludge and pesticides applied to agricultural land in the US. The model used physico-chemical and degradation data inputs for the organic compounds, soil properties (organic matter and hydraulic conductivity) and environmental factors (recharge rate and depth to groundwater) to determine downward movement under a simulated 'worst-case' condition. The following conservative assumptions were also made:

- no plant uptake of the chemical occurs;
- no volatilization takes place;
- no significant adsorption below the root zone occurs due to low organic matter content;
- groundwater directly below the application site is withdrawn for drinking purposes.

Consequently, all of the chemical applied to the soil and/or its degradation products are considered to be potentially leachable to the groundwater in the model. Results of the screening model for a range of organics commonly found in sludge indicated that the small halogenated short-chain hydrocarbons, chloroform, dichloromethane, trichloroethane and trichloroethylene showed a particularly high leaching risk under the assumptions of the screening procedure. Some of the chemicals within this generic group of compounds have a relatively long half-life coupled with a low adsorption coefficient. However, they are also highly volatile and, therefore, are unlikely to be present in sewage sludge such that groundwater supplies are not at risk from these compounds when sludge is applied to agricultural land. This is supported since short-chain halogens are rarely found in British aquifers (Kenrick *et al.*, 1985). Steenhuis and Naylor (1985) concluded that groundwater monitoring of nearby sludge-treated sites should focus on the short-chain halogenated compounds since these are potentially mobile in the soil. On the other hand, higher molecular weight chemicals, such as phthalates and other ring compounds, present little risk to groundwater contamination from the agricultural utilization of sewage sludge. A very extreme example of the immobility of certain trace organics, emphasizing the minimal risk to groundwater, was given by Murphy (1989) who predicted it would require nine million to ten billion years for dioxin from landfilled incinerator ash to migrate significantly through the soil. Orazio *et al.* (1992) confirmed that the movement of dioxins in soil is confined to a very small zone around the contaminated layer and is most likely related to diffusivity of dioxin molecules in the liquid phase.

Relatively few studies have investigated groundwater contamination with trace organic contaminants arising from sludge application under field conditions and none have been published in the UK. Recent studies by Sweetman *et al.* (1992b) and Sweetman *et al.* (1994) have determined the concentrations of trace organics in the soil solution collected with suction samplers from sludge-treated soils to examine the potential movement of organic contaminants by leaching. Levels of LABs, 4-nonylphenol, chlorobenzenes (DCBs, TCBs, HCB), PCBs, PCDD/Fs and lindane in soil water samples were at or below detection limits of analysis indicating that there is minimal risk of these compounds leaching from sludged soil. The limit of analytical detection for PCDD/Fs was typically <2 pg l^{-1} emphasizing the likely absence of any environmental risk due to leaching of these compounds. Further emphasis is placed on the results since the two soils studied were a coarse-textured sandy soil, with a high leaching potential, and a soil from a dedicated sludge disposal site, with elevated soil concentrations of organic contaminants and soluble organic matter, representing 'worst-case' scenarios of potential leaching risk.

The potential mobility of a range of organic contaminants artificially spiked into soil was actually reduced by sludge addition in laboratory leachability tests conducted by Sweetman *et al.* (1994). This was due to the increased sorptive capacity of the soil from the application of organic matter in sludge. However, the field and laboratory studies demonstrated that the contaminants were highly immobile even in soils of low organic matter content. Furthermore, Sweetman *et al.* (1994) showed that the mobility of organic pollutants was unlikely to increase due to the presence of surfactants in sewage sludge at the concentrations of these chemicals occurring in sludge in practice.

Other available information, principally from the US, confirms that there is little risk of transfer of organic pollutants in sludge to groundwater supplies. For example, Demirjian *et al.* (1984) detected no contamination of groundwater with organic compounds resulting from repeated applications of highly industrially contaminated sludge to a dedicated disposal site. This study is of special interest because the soil was particularly susceptible to leaching being a permeable sand. In another survey, Berg *et al.* (1987) similarly showed that groundwater remained free of trace organic compounds when sewage sludge was applied to a sandy agricultural soil.

Agricultural Pesticides

The potential environmental impact from sludge organic contaminants can be put into perspective by comparing the amounts of organic chemicals applied to farmland in sludge with the quantities and toxicities of synthetic organic pesticides used for intensive agricultural crop production. The previously widely used herbicide 2,4,5-T is a particularly interesting example because it contains 2,3,7,8-T_4CDD, regarded as the most toxic dioxin, and other TCDD/Fs, which are formed as trace impurities during the chemical synthesis of 2,4,5-trichlorophenol, the precursor of 2,4,5-T (Hassall, 1982). However, 2,4,5-T was withdrawn from use only recently in 1986 (MAFF, 1991b). In contrast, detectable concentrations of 2,3,7,8-T_4CDD have barely been measured at all in sewage sludge (Sauerbeck and Leschber, 1992), although larger concentrations of other PCDD/Fs have been reported in sludge (Horstmann *et al.*, 1992; Wild *et al.*, 1994).

Assuming that sludge is spread at an agronomic rate of 10 t ha^{-1} (dry solids), the organic chemical loadings for organic concentrations in sludges of 1, 10 and 100 mg kg^{-1} are 0.001, 0.01 and 0.1 kg ha^{-1}, respectively (Jones and Wild, 1991). Thus, in comparison with normal pesticide rates (0.2-4.0 kg ha^{-1} of active ingredient), the anticipated application rates for sludge organics are generally considerably smaller than for agrochemicals. By definition pesticides are potentially highly bioavailable to have a toxic effect on the target organism. In comparison, organic contaminants in sludge have low bioavailability because many are strongly adsorbed by organic matter in sludge and soils, they may be rapidly degraded or are volatilized from soil.

Jacobs *et al.* (1987) considered that for agronomic rates, concentrations of an organic contaminant approaching 100 mg kg^{-1} in sewage sludge should be regarded as having the potential for impact on soil-plant systems, depending on chemical/toxicological properties of that chemical. On this basis the concentration ranges for most organic contaminants in sludge are unlikely to cause concern. According to the median sludge concentration values listed by Jones and Wild (1991), only concentrations of LAS, alkylphenols and phthalates are likely to exceed this tentative threshold value. However, these compounds are very rapidly degraded in soil and are therefore not considered a particular risk to the environment.

In quantitative terms LAS, alkylphenols and phthalates are the principal organic contaminants identified in sludge and represent about 84, 10 and 3% of the total identified organic pollutant load in sludge, respectively. The remaining 3% of identified compounds amount to a total application to agricultural soil nationally that is equivalent to <1.0% of the 20 000 t of pesticide active ingredient applied annually to grassland, forage and arable crops and soils in England and Wales (MAFF, 1990c; MAFF, 1991a). Furthermore, the non-agricultural uses of pesticides are also extensive, and there are other identified sources such as sheep dip chemicals which present another potential hazard (DoE, 1991; NRA, 1992a). In contrast to sludge organics, for example, pesticide applications to agricultural land have been shown to put ground and surface waters at risk due to surface run-off and leaching (Nicholls, 1991; NRA, 1992a; Huang *et al.*, 1994).

Consequently, the environmental effects resulting from the widespread use of pesticides are likely to be considerably more significant than any potential impacts caused by recycling sewage sludge to agricultural land since organic contaminants in sludge present little quantitative risk to the environment in comparison. Nevertheless, further research is necessary to quantify the potential transfer of non-degradable and lipophilic compounds to the food chain *via* animal ingestion of sludge to alleviate anxiety that organic contaminants in agriculturally applied sludge may be detrimental.

Soil Fertility

Relatively little information has been published on the effects of organic contaminants in sewage sludge on the fertility of sludge-treated soils. However, Kirchmann *et al.* (1991) reported that important soil microbial parameters including respiration, N mineralization and nitrification were not influenced overall by common 'priority' sludge contaminants including: toluene, naphthalene, 2-methylnaphthalene, 4-*n*-nonylphenol or di-2-ethylhexyl phthalate. Kirchmann *et al.* (1991) incubated soils spiked with large concentrations of these compounds, up to 1000 times higher than levels expected from normal agricultural use of sewage sludge. Similarly, no significant detrimental effects on soil microbial activity (measured as the reduction of Fe(III)-oxides to Fe^{2+} ions) were detected by Welp and Brümmer (1992) for a range of organic contaminants artificially spiked into soils, except at very large rates of addition to soil up to an application equivalent to 4 t ha^{-1} of chemical. Application rates at the levels used by Kirchmann *et al.* (1991) and Welp and Brümmer (1992) would not conceivably occur from agricultural recycling of sewage sludge (see earlier).

Pentachlorophenol (PCP) was the most toxic compound examined by Welp and Brümmer (1992) followed by the herbicides 2,4,5-T and 2,4-D and the detergent LAS (dodecylbenzenesulfonic acid). The herbicides and LAS resulted in 10% inhibition (effective dose, ED_{10}) at concentrations about 10-20 times higher than for PCP. Picloram, benzene and trichloroethene were 4-10 times less toxic than the herbicides and LAS. Chlorobenzeneamine, phenol and atrazine caused inhibition at relatively low doses in some soil samples, but in others even high concentrations did not cause inhibition, reflecting differences in soil properties. DDT and HCB were not toxic to microbial activity due to their very low solubility in water and strong adsorption potential in soil.

Sewage sludge contains a multiplicity of organic contaminants. However, even the most abundant chemicals found in sludge do not occur at concentrations in treated soil that are likely to significantly affect soil microbial processes when sludge is applied at normal agronomic rates (5-10 t ha^{-1} dry solids). In particular, the detergent-derived surfactant, LAS is the single most abundant contaminant likely to occur in sludge: the maximum concentration of LAS measured by Holt *et al.* (1989) in UK sludges was 14 300 mg kg^{-1} (dry solids). Nevertheless, soil concentrations of LAS are unlikely to exceed the ED_{10} value determined by Welp and Brümmer (1992) for this compound. Moreover, LAS is rapidly degraded with a half-life probably of <10 days (Marcomini *et al.*, 1989) so that any potential effects will only be transient. Similarly, PCP and 2,4-D are only moderately persistent in soil having reported half-lives of 40-50 days and <15 days, respectively (Jones and Wild, 1991).

The study by Welp and Brümmer (1992) is useful because it demonstrates the important influence that soil chemical properties may have on the inhibitory response of organic contaminants. Thus, toxicity is reduced in soils with high organic matter content due to increased adsorption capacity of the soil. Soil pH value also affects microbial toxicity, but the inhibitory action of organic compounds is influenced by soil pH in different ways through effects on speciation and adsorption. The toxicity of weakly adsorbed compounds, such as 2,4-D, decreases with increasing soil pH value as the anion species predominates, which is less toxic compared with the non-ionic form present under acid conditions. However, the opposite is true for strongly adsorbed chemicals, such as PCP, because adsorption increases with decreasing soil pH value even though the more toxic non-ionic form predominates in acid soils.

From a review of literature published up to the late 1970s, Wainwright (1978) concluded that there was little evidence to suggest that pesticide treatment has any prolonged effect on soil micro-organisms. There are other reports in the literature also demonstrating short-lived or no toxic effects of pesticides on soil microbial processes when applied at agronomic rates (Atlas *et al.*, 1978; Kuthubutheen and Pugh, 1979; Tena *et al.*, 1984). For example, Biederbeck *et al.* (1987) found no toxic response to long-term applications of 2,4-D in soil treated annually with the herbicide for 35 years. Ferrer *et al.* (1986) similarly showed no detrimental effects of 2,4-D or 2,4,5-T application to soil. Interestingly, Welp and Brümmer (1992) placed 2,4,5-T and 2,4-D as the second and the fourth most toxic compounds assayed, respectively. These reports help put the minimal risk to soil fertility from sludge organic contaminants into perspective since the concentrations of most of the chemicals in sewage sludges will result in inputs to soil which are considerably lower than recommended pesticide application rates (see earlier).

By contrast there are other published data suggesting that recommended pesticide applications to agricultural land can have detrimental and subtle impacts on soil fertility. Thus, Lewis *et al.* (1978) concluded that, although herbicides apparently did not affect the general decomposition of organic matter in soil, they did affect activities mediated by specific types of microflora as well as certain of the microflora themselves, such as algae. These subtle effects on soil micro-organisms resemble those discussed in Chapter eight for heavy metals in sludge-treated soils. Wainwright and Pugh (1973) reported that the intentional removal of pathogenic fungi by the addition of fungicides also affects the entire soil microflora. Such partial sterilization of the soil brings about changes in the complex microbially mediated soil processes, disturbing the dynamic equilibrium of the soil microflora, and thus affecting soil fertility. For example, long-term applications of atrazine (4 kg ha^{-1}, annually for 14 years) to an orchard soil induced significant changes in the structure of the microbial community, although the total numbers of bacteria and fungi remained unaltered (Voets *et al.*, 1974). In contrast, monochlorophenoxyacetic acid (MCPA) and simazine apparently have caused no detectable effects on the microflora, but repeated applications of paraquat can significantly reduce soil microbial biomass (Duah-Yentumi and Johnson, 1986). Roslycky (1982) showed that glyphosate could also be inhibitory to soil microbial activity with actinomycetes and bacteria being affected to a greater extent than soil fungi. Not surprisingly, fungicidal compounds potentially have a large impact on fungi populations in soil. For example, Duah-Yentumi and Johnson (1986) reported drastic reductions in soil microbial biomass following the application of various fungicides to soil due to a decrease in fungal biomass. Captan and benomyl significantly reduced the species diversity of the rhizosphere fungi of onions, which was associated with large reductions in the yield response of treated plants (de Bertoldi *et al.*, 1978). Nitrogen-fixation may also be particularly sensitive to the toxic effects of pesticides. Wegener *et al.* (1985) reported the total suppression of algal growth and nitrogenase activity following the treatment of soils with a range of herbicides, although fungicide application had no effect. Herbicides may be inhibitory to N_2-fixation by white clover potentially due to toxic effects on *Rhizobium leguminosarum* bv. *trifolii* (Clark and Mahanty, 1991).

More recently, Mårtensson (1992) reported adverse effects of herbicides on *R. leguminosarum* bv. *trifolii* occurring at levels $^1/_{10}$ - $^1/_{10000}$ of recommended applied concentrations. Symbiotic interactions were also adversely affected by all of the agrochemicals screened. Bacterial-induced root hair deformations necessary for nodulation decreased in the presence of fungicides (benomyl, fenpropimorph and mancozeb) and herbicides (bentazone, chlorsulphuron and MCPA). Fenpropimorph and mancozeb did not cause root hair deformations at increasing concentrations, indicating that these chemicals may inhibit nodulation under field conditions. Nodule development was inhibited at increased levels of bentazone, chlorsulphuron, glyphosate and mancozeb. Dry matter production of nodulated plants was adversely affected by bentazone and chlorosulphuron, indicating disturbances in nodule function. Mårtensson (1992) concluded that almost nothing is known about how the common agricultural practices with a variety of agrochemicals repeatedly added to the soil-plant system solely or together, will affect the N_2-fixing symbiosis over time. These results suggest that such practices may be limiting to N_2-fixation and are of concern because of the widespread use and reliance on pesticides in agriculture and also in non-agricultural situations.

Acute and sublethal toxic effects of pesticides on non-target soil animals have been reported (Edwards and Thompson, 1973). Earthworms in particular have been the focus of considerable research due to their importance in maintaining soil fertility and because they form an important link in terrestrial food chains. However, species-specific sensitivity towards pesticides is apparent with *Eisenia foetida* demonstrating greater tolerance compared with other species including *Lumbricus terrestris* and species of the genera *Aporrectodea* and *Allolobophora* (Haque and Ebing, 1983; Heimbach, 1985; Kula and Kokta, 1992). The extent of the toxicity response also varies depending on the chemical and concentration present in the soil. For example, earthworms are insensitive to the acutely toxic effects of DDT and its metabolites at levels higher than those likely to be found in the environment (WHO, 1989). In contrast, other pesticide compounds including dieldrin, azinphos-methyl, parathion, endosulfan, propoxur, paraoxon, carbofuran, lindane, aldicarb, malathion and carbaryl (insecticides); glyphosate, paraquat and atrazine (herbicides); and captan, benomyl and triadimefon (fungicides) all exhibit varying levels of toxicity to earthworms (Haque and Ebing, 1983; Drewes *et al.*, 1984; Kula and Kokta, 1992; Reddy and Reddy, 1992; Senapati *et al.*, 1992; Springett and Gray, 1992; Stenersen *et al.*, 1992). Even biocides with apparently low levels of toxicity can have a substantial effect on the growth rates of earthworms when repeated applications are made (Springett and Gray, 1992). In addition, both the reproductive capacity and the total population of earthworms in the soil could be expected to fall following repeated low doses of agricultural pesticides. However, other studies indicate agronomic rates of treatment are not hazardous to earthworms and that much larger levels of addition are necessary to interfere with the positive role of earthworms towards soil fertility (Senapati *et al.*, 1992).

Comparatively little work has assessed the effects of sludge organic contaminants on earthworms. However, Drewes *et al.* (1984) detected no sublethal neurophysiological effects on earthworms in a solution lethality contact exposure test following treatment with dimethyl phthalate (48 h, LC_{50} was 10 900 mg l^{-1}) which is widely used as a plasticizing agent and is an important trace contaminant of sludge. Sublethal effects of fluorene, a constituent of coal tar and coke-oven tar and also potentially present in sludge, were evident at 100 mg l^{-1}. In another study by Reinecke and Nash (1984), there were no toxic effects of 2,3,7,8-T_4CDD observed on earthworms in spiked soil containing 0.05 to 5.0 mg kg^{-1} of the dioxin. The critical toxic concentration of 2,3,7,8-T_4CDD in soil was estimated to fall between 5.0 and 10.0 mg kg^{-1}. Earthworms were apparently indifferent to the pollutant and did not attempt to move away from it as observed in metal contamination experiments (Spurgeon *et al.*, 1994). Data are scarce for many of the trace contaminants present in sludge and sludge-treated soils, but these results suggest that toxicity to earthworms appears unlikely at the very small concentrations of xenobiotic compounds present in soils receiving sewage sludge at agronomic rates of application. Furthermore, comparative toxicology studies indicate that earthworms are the least sensitive of a range of possible test organisms, including micro-organisms or plants, for assessing potential environmental impacts of hazardous waste contaminated soil (Miller *et al.*, 1985).

On balance, it appears highly unlikely that organic contaminants in sewage sludge impair the fertility of agricultural soils. At the very least, the risk of a potentially detrimental effect on soil fertility due to sludge organic contaminants is considerably lower compared with the likely impacts on soil micro-organisms and invertebrate macro-fauna arising from the extensive use of agricultural pesticides in the environment.

Natural Ecosystems

There is relatively little information available on which to establish the potential environmental impact of sludge organic contaminants on natural ecosystems resulting from the utilization of sewage sludge on agricultural land. As with PTEs (Chapter eight), earthworms have been identified as accumulators of lipophilic trace organics potentially placing at risk higher trophic levels of natural food chains (Diercxsens and Tarradellas, 1983; Reinecke and Nash, 1984; WHO, 1989; Haimi *et al.*, 1992). However, Thiel *et al.* (1989) reported that the application of paper mill sludge containing up to 80 ng kg^{-1} of toxic 2,3,7,8-T_4CDD to a pine forest apparently presented little risk to wildlife. Clutch size, hatching rates and fledging rates of several species of birds, and age distributions of mouse populations, indicated normal reproduction. Earthworm, mouse and insectivorous bird populations were generally higher in sludge-treated areas. Litter invertebrate diversity and density were unaffected by sludge, although soil invertebrate diversity and density were reduced, probably as a result of increased soil moisture and a temporary smothering effect of the sludge. The histopathology, reproduction and population data indicated that effects of exposure to sludge were not discernible and it was concluded that the risk of harm to wildlife appeared to be low.

The concentration of 2,3,7,8-T_4CDD in sewage sludge is significantly smaller than in paper mill sludge. For example, Sewart *et al.* (1995) reported the concentration of 2,3,7,8-T_4CDD in sewage sludges (eight samples) from north-west England was in the range 1.0-3.8 ng kg^{-1} (dry solids) with a mean value of 2.1 ng kg^{-1} (dry solids). Consequently it seems unlikely that sludge application to agricultural soils will have an adverse effect on natural ecosystems due to the presence of trace organic contaminants. Furthermore, sewage sludge is being applied to less than 1% of UK agricultural land and therefore any bioaccumulation effects of sludge-borne organic contaminants on higher animals will be well diluted in comparison with effects of airborne depositions of the same contaminants (Halsall *et al.*, 1993).

Chapter eleven:

Pathogenic Organisms

General Comments

The pertinent data relating to the environmental impact of biological contaminants in sludge, which potentially represent a health risk to animals and man when sludge is spread on farmland, have been comprehensively and frequently reviewed (for example see Burge and Marsh, 1978; Carrington, 1978, 1980; Reddy *et al.*, 1981; Coker, 1983; Block, 1986; Gemmell, 1986; Jakubowski, 1986; Jones, 1986; Pawlowski and Schultzberg, 1986; Pike, 1986a,b; Pike and Carrington, 1986; Wekerle, 1986; Sorber and Moore, 1987; Bruce *et al.*, 1990). Some of the main water-borne pathogens of concern which may be present in sewage sludge are listed in Table 11.1. Disease-causing agents occur in sewage because they are being discharged in the faeces of infected humans or animals. Thus, the species diversity and numbers present will reflect the health of the community and the standards of hygiene which prevail.

Much has been done to minimize the potential transmission of pathogens by reducing infectivity of sludges through effective treatment processes and then matching efficiency of pathogen removal to operational restrictions on application practices and land use. Thus, the CEC Directive on the use of sludge in agriculture (CEC, 1986a) permits only appropriately treated sludges to be surface-spread on farmland. Raw sludge must be injected or immediately worked into the soil providing a barrier to infection. A further constraint specifies a minimum three week no-grazing interval to ensure that levels of any remaining pathogens have been greatly reduced before animals are introduced to sludge-treated grassland. These requirements have been implemented in the UK through the *Sludge (Use in Agriculture) Regulations* (SI, 1989a) which are supported by a Code of Practice (DoE, 1989a) on the agricultural utilization of sewage. Examples of effective sludge treatment processes described in the Code of Practice are listed in Table 11.2. In the UK, national surveillance of human and animal disease is of a very high standard (Pike, 1986a) and has shown these measures to be very effective in preventing infection from sludge-borne pathogens (Pike and Carrington, 1986).

Table 11.1. Some of the more significant water-borne pathogens and their effects on health.

Group	Genus[1]	Effects on human health[2]
Bacteria	*Salmonella*	Typhoid fever, paratyphoid fever, enteritis, salmonellosis, food poisoning
	Shigella	Dysentery, paradysentery
	Escherichia	Enteritis (pathogenic strains)
	Vibrio	Cholera, enteritis, food poisoning
	Clostridium	Gas gangrene, tetanus, botulism
	Leptospira	Leptospirosis
Viruses	Poliovirus	Fever, poliomyelitis, enteritis
	Coxsackievirus A	Headache, muscular pain
	Coxsackievirus B	Nausea, meningitis
	Echovirus	Diarrhoea, hepatitis
	Adenovirus	Fever, respiratory infection, enteritis, inflammation of eyes (conjunctivitis), involvement of central nervous system
	Reovirus	Common cold, respiratory tract infections, diarrhoea, hepatitis
	Hepatitis A virus	Infectious hepatitis (fever, nausea, jaundice)
	Rotavirus	Diarrhoea, enteritis
Protozoa	*Entamoeba*	Amoebic dysentery
	Giardia	Diarrhoea, enteritis
	Cryptosporidium	Diarrhoea, enteritis
Trematodes	*Schistosoma*	Schistosomiasis (Bilharzia)
Cestodes	*Taenia*	Tapeworm infestation in man; in cattle infestation by bladderworm stage, *Cysticercus bovis* (cysticercosis)
Nematodes	*Ascaris*	Roundworm infestation
	Toxocara	
	Ancylostoma	Hookworm infestation
	Trichuris	Whipworms
	Globodera[3]	Potato cyst nematode

(1) Not all species within a genus and not all types within a species need be pathogenic.
(2) Not all symptoms are produced by one species and not all symptoms may be present at the same time.
(3) Plant pathogen
Adapted from Carrington (1978)

The WHO Working Group on health risks of sludge spreading on land (WHO, 1981) considered that because sludge regularly contains a variety of pathogens it inevitably poses a potential hazard, but that these pathogens only present a risk if they are ingested in sufficient numbers to cause infection. With good food hygiene, the risk will be small. Nevertheless there is inevitably an element of risk of infection when sludge is spread on farmland because it is not economically or practically feasible to achieve an absolute no-risk level even though it may be technologically possible (WHO, 1981).

Table 11.2. Examples of effective sludge treatment processes.

Process	Descriptions
Sludge pasteurization	Minimum of 30 minutes at 70 °C or minimum of 4 hours at 55 °C (or appropriate intermediate conditions), followed in all cases by primary mesophilic anaerobic digestion.
Mesophilic anaerobic digestion	Mean retention period of at least 12 days primary digestion in temperature range 35 °C ± 3 °C or of at least 20 days primary digestion in temperature range 25 °C ± 3 °C followed in each case by a secondary stage which provides a mean retention period of at least 14 days.
Thermophilic aerobic digestion	Mean retention period of at least 7 days digestion. All sludge to be subject to a minimum of 55 °C for a period of at least 4 hours.
Composting (windrow or aerated piles)	The compost must be maintained at 40 °C for at least 5 days and for 4 hours during this period at a minimum of 55 °C within the body of the pile followed by a period of maturation adequate to ensure that the compost reaction process is substantially complete.
Lime stabilization of liquid sludge	Addition of lime to raise pH to greater than 12.0 and sufficient to ensure that the pH is not less than 12 for a minimum period of 2 hours. The sludges can then be used directly.
Liquid storage	Storage of untreated liquid sludge for a minimum period of 3 months.
Dewatering and storage	Conditioning of untreated sludge with lime or other coagulants followed by dewatering and storage of the cake for a minimum period of 3 months. If sludge has been subject to primary mesophilic anaerobic digestion storage to be for a minimum period of 14 days.

Source: DoE (1989a)

Occurrence in Sewage Sludge

Bacteria, Viruses and Other Agents

Pathogenic bacteria and viruses of greatest significance are those which are faecal-borne with no alternate host in the dispersal cycle from human to human. Their incidence in sewage sludge has been discussed by Lund (1975), Pahren *et al.* (1977), Burge and Marsh (1978), Carrington (1978) and Engelbrecht (1978). Of those that are listed, *Shigella,* which causes dysentery, is usually transmitted by person to person contact or in infected food: the carrier state is virtually unknown. The cholera vibrio is endemic in the valleys of the Ganges and the Yang-tse, being spread by infected water. This disease was once common in Europe but has now virtually disappeared there since the provision of modern sanitation.

Of the pathogenic bacteria of importance in sludge disposal, *Salmonella* spp. are of the greatest concern in the UK, but only in respect of the risk to grazing animals (WHO, 1981). A survey of *Salmonella* spp. in 17 raw sludges in the UK gave a median count of only 6.0 l^{-1}: the highest value recorded was 2.2×10^4 l^{-1} and seven samples of sludge were free of the bacteria altogether (Pike *et al.*, 1988). However, *Salmonella* spp. are so ubiquitous in the environment that the proper control of human salmonellosis is through food hygiene (Coker, 1983); sewage sludge properly used in agriculture is not involved in the transmission of human salmonellosis.

Escherichia coli is carried by man and nearly all mammals as a normal gut inhabitant and exists in about 140 serological groups of which only a few are pathogenic. They are useful indicators of faecal pollution of water. Direct infection by *Clostridium* spp. usually occurs through wounds contaminated by soil. In soil the numbers are usually so high that any additions from sewage sludge make very little difference to the likelihood of infection. *Leptospira* exists in sewer rodents, but has a large reservoir in farm and wild animals. Leptospirosis is predominantly a disease of agricultural workers and cases among sewage workers are rare. *Leptospira* in sludge is more likely to occur as a result of post-treatment contamination from the urine of infected rodents rather than from sewage such that the risk of infection from sludge is minimal. *Mycobacterium tuberculosis* is not primarily water-borne: it is very resistant to drying. The incidence of *M. tuberculosis* in developed countries has declined and in the UK is now very low indeed, both in the bovine as well as in the human population, and so is of little significance in sludge disposal. However, it is recognized that hazards may be posed by newly discovered and identified pathogens such as *Campylobacter* and *Yersinia* species which are capable of being transmitted by the water route (Bruce *et al.*, 1990).

Among viruses the Enteroviruses occur widely in sewage: a single virus type can produce a range of symptoms in man. Hepatitis A virus may be present in sewage (Bosch *et al.*, 1986): it has no other host than man. An outbreak of

Hepatitis A virus infection among sludge spreaders was apparently caused through inadequate hygiene and protection precautions being taken (Pike and Carrington, 1986). There are no reports that Hepatitis B virus is transmitted by sewage-polluted water and is probably not an important biological contaminant of sludge. Similarly, there is no record of the human immunodeficiency virus (HIV), responsible for the Acquired Immune Deficiency Syndrome (AIDS), having been isolated from faeces, and epidemiological evidence shows that sewage and water have not been implicated in the transmission of HIV (Pike, 1987). Bovine spongiform encephalopathy (BSE) is a recently recognized member of a group of progressive fatal neurological disorders of man and animals. However, there is no evidence to suggest that the BSE or related scrapie agents would be present in sewage. Recent legislation was introduced in the UK to ensure that BSE-infected animals and their products are destroyed, which should reduce any risk of transmittal to sewage.

Human and Animal Parasites

These vary widely in their incidence from one geographical area to another. In tropical and sub-tropical irrigated areas *Bilharzia* is rife, but it is not present in southern USA: in this area the alternate host (a snail) is absent (Pahren *et al.*, 1977). The presence of the intermediate snail host and exposure of skin to waters populated with infected snails is essential for completion of the infection cycle. Consequently, *Bilharzia* occurs only in situations similar to paddy fields and is not a concern in the UK. The protozoan parasites *Entamoeba histolytica* and *Giardia lamblia* are commonly reported in the USA, and in N. Africa and E. Mediterranean countries. *Taenia saginata* is very common where beef is eaten uncooked or undercooked, as in the Lebanon and E. Africa. It is not regarded as a serious medical problem in the UK. Lund (1975) reported that ova of *Diphyllobothrium latum* (fish tapeworm) may be present in untreated sludge produced in the countries of N. Europe.

Ova of the nematodes *Trichuris trichiuria* (whipworm), *Necator* spp., *Ancylostoma* spp. (hookworms) and *Toxocara* spp. (dog and cat ascarids) are commonly found in sewage sludge (Hays, 1977). *Ascaris lumbricoides* is of major concern in many countries including USA and South Africa. Where it is important, because of its extreme resistance to adverse conditions and the long period of survival of its eggs, *A. lumbricoides* has been considered as the standard by which the safety of sludge disinfection measures and agricultural applications are assessed (Nell *et al.*, 1980). Theis *et al.* (1978) reported that up to 74% of sludge and sewage samples in some USA urban areas contained ova of *Ascaris* and that the infestation was commonly of the order of 50 g^{-1} of sludge. Liebmann (1964) reported that *Ascaris lumbricoides, Enterobius vermicularis,* and *Taenia saginata* were the organisms most commonly found in sludge in Central and Eastern Europe.

There are a number of important socio-agronomic factors which provide barriers to the potential transfer of animal parasites *via* sewage sludge. Over 95% of households in the UK discharge human wastes to sewers and most of the remainder discharge to septic tanks which in turn are emptied into sewage treatment works. Sludge applications to land have been largely diverted from the growing of vegetable crops on to land used for growing grass, grains, or other agricultural crops (DoE, 1989a). The cultivation of these is highly mechanized so operators are away from close contact with sludged soil. The structure of farms in the UK means that most are well-fenced and the general public are unlikely to walk over sludge-treated land. Furthermore, there are restrictions on the timing of sludge applications and the harvesting or planting of vegetable crops which may be eaten raw to ensure that the human-to-human cycle of infection with animal parasites is effectively broken (DoE, 1989a). These are described later in this chapter.

A few studies have been made of the incidence of the ova of parasitic worms in UK sewage sludge. In a survey by the Northumbrian Water Authority of 29 treatment works sludges, no parasitic worm ova were detected: the method had a sensitivity of 10 ova g^{-1} sludge (Argent *et al.*, 1977). However, Watson *et al.* (1980) (quoted by Coker, 1983) studied sludges from seven STWs in NW England over three years and on one or more occasions identified *Ascaris* and *Trichuris* ova in sludges from all seven works. *Taenia* eggs were found in sludge from all works except one. Compared with US sludges, however, the numbers present were very low. The mean concentrations were below 5 l^{-1} of sludge: by contrast Hays (1977) concluded that in the US, mean concentrations of ova of parasitic worms in sludge were 300-400 l^{-1}. Watson (1982) (personal communication, quoted by Coker, 1983) later provided data which showed that, in 411 samples taken during 1979-81, the percentages of sludge samples showing zero counts were:

Ascaris	29%
Trichuris	39%
Taenia	87%

Thus ova of these parasites were present in UK sludges but at a low level. The results seem consistent with the view that most ova present are discharged by persons who have travelled in or immigrated from countries where infestations are widespread, and have become carriers of the organisms and that there is no transmission in the UK *via* food or water as a result of the disposal of sewage sludge to agricultural land. Some of the infections may also be due to person to person contact caused by poor hygiene (E.G. Carrington, WRc Medmenham, personal communication). In the UK, infection by the beef tapeworm (*T. saginata* in man, and the larval or cysticercus stage in cattle) is regarded as the principal risk to human health (WHO, 1981), although other emerging parasites such as the protozoan *Cryptosporidium* may be carried in sludge and present a potential hazard (Bruce *et al.*, 1990). However, a recent report (Whitmore, 1993) has indicated that the risk to human health from *Cryptosporidium* is minimal for sewage sludge applied to agricultural land. Oocysts of *Cryptosporidium* may be sensitive to environmental stress increasing their susceptibility to damage during wastewater treatment and disinfection (Ransome and Carrington, 1993).

Plant Pathogens

In addition to animal pathogens, sewage sludge may also represent a vector for the potential transfer of plant pathogenic organisms. In particular, potato cyst nematodes (*Globodera rostochiensis* and *G. pallida*), which are major pests of the potato crop and are endemic in Europe (MAFF, 1977), could be introduced into the sewer system in effluent discharges from vegetable processing factories and domestic sources. Gram (1960) noted that water and sedimented soil from potato starch and sugar factories may carry cysts and spread the pest if discharged to agricultural land. Beet necrotic yellow vein virus (BNYVV) is a causal agent of rhizomania disease in sugar beet which has only recently been reported in Britain (Hill and Torrance, 1989) and could also potentially occur in sewage receiving discharges from infected crops. However, little is known about the survival of these and other plant pathogens during sludge treatment and disposal, although it can be predicted from the behaviour of other pathological groups of similar organisms.

Fate of Pathogens in Sewage and Sewage Treatment

Sewage and Primary Treatment

The efficiency of primary sedimentation in removing micro-organisms depends on the extent to which they are associated with solid matter. The rate of settlement of ova of animal parasites is determined by size and relative density. Viruses are bound to sludge particles (Lund and Ronne, 1973) and primary sludge contains higher virus concentrations than the sewage from which it is derived. However, sedimentation is a physical process and the removal of suspended viral contaminants in water occurs only to a limited extent (Shuval, 1970). The settlement of organisms either directly or by their association with sewage solids, increases the pathogen burden of sewage sludge.

The ova of *Ascaris* settle quickly, but *Taenia,* hookworm ova and cysts of the protozoan *Entamoeba histolytica* have a lower specific gravity and are not all removed by 2.5-3 hours of settlement (Cram, 1943). Therefore, Rudolphs *et al.* (1950) concluded that primary settlement was an unreliable method of removing the ova of parasitic worms, amoebic cysts, bacteria or viruses from sewage effluent. Other authors (Berg, 1973; Varma *et al.*, 1974; Pike, 1975; Whitmore, 1993) similarly agree that primary treatment has little effect on bacterial and viral pathogens.

Storage of raw sludges can effect an appreciable reduction in coliform bacteria and viruses (Carrington, 1980) (Table 11.3). For example, Bradley (1973) reported 81, 98, 99.92 and 98.82% removal of *Escherichia coli* in storage lagoons with retention periods of 1, 2.6, 8.9 and 20 days, respectively. From a review of literature, Carrington (1980) concluded that the numbers of bacteria and viruses in the liquid phase will be reduced by >90% in a pond or lagoon producing a satisfactorily stabilized effluent. Approximately 13% of the sludge dry solids used in agriculture in the UK is stabilized by storing for a minimum period of three months (CES, 1993).

Table 11.3. Relative efficacies of various methods of sludge treatment in reducing numbers of different pathogens or their period of survival.

	Relative reduction		
Process	Poor	Moderate	Good
Raw sludge storage	*Ascaris* ova *Taenia* ova *Cryptosporidium* oocysts	Viruses Bacteria	
Digestion	*Ascaris* ova[1]	Hookworm ova Bacteria *Taenia* ova	Viruses *Entamoeba* cysts *Heterodera* cysts *Cryptosporidium* oocysts
Composting			Viruses Bacteria Fungi Helminth ova
Lime treatment	*Ascaris* ova		Bacteria
Heat treatment[2]			Viruses Bacteria Helminth ova *Cryptosporidium* oocysts
Irradiation		*Ascaris* ova	Bacteria Viruses

[1] Anaerobic digestion at temperatures >36 °C will inactivate depending on exposure time
[2] Includes thermal drying and lime treatment
Adapted from Carrington (1980)

Sludge Treatments

Importance of temperature

Temperatures above 45 °C bring about denaturation of protein and hence death of all thermotolerant species and bacterial spores on a time of exposure-dependent basis (Pike and Carrington, 1986). The rate of mortality increases with increasing temperature as shown in Fig. 11.1 for various organisms in heat-treated sludges. On this basis, the time-temperature combinations (i.e. 30 min at 70 °C; or 4 h at 55 °C) given in the Code of Practice (DoE, 1989a) for some processes (Table 11.2), and those intermediate combinations, will achieve at least a 90% destruction of the principal pathogenic agents, provided that all organisms are subjected to this degree of treatment. This level of destruction indicates a 'significant reduction' in population size (Bruce *et al.*, 1990). Such reductions cannot be equated precisely with a lower risk of infection, but they do indicate that a barrier to transmission is provided.

Anaerobic digestion

Mesophilic anaerobic digestion is the method of sludge stabilization preferred by the Water Industry in the UK and approximately 44% of UK sludge (dry solids) used on agricultural land is treated by this method (CES, 1993). Anaerobic digestion of sludge significantly reduces some pathogens (Table 11.3), but does not eliminate them (Coker, 1983). Burd (1966) (quoted by Coker, 1983) reported a 99.9% reduction in pathogenic bacteria after 30 days anaerobic digestion. In contrast, Heukelekian and Albanese (1956) showed that *Mycobacterium* was very resistant and could survive 35 days of digestion. Free ammonia in digested sludge has been implicated as a viral toxin (Ward and Ashley, 1977). Sanders *et al.* (1979) estimated that 99.9% inactivation of poliovirus in digesting sludge took six days at 37 °C , although Lund and Ronne (1973) detected the virus even after 50-60 days digestion which was probably due to cross contamination by short-circuiting with fresh input of untreated sludge (Leclerc and Brouzes, 1973).

The cysts of the protozoans *Entamoeba histolytica* and *Giardia lamblia* are eliminated by anaerobic digestion (Coker, 1983). Mesophilic (35 °C) anaerobic digestion for a period of four days followed by storage of the digested sludge resulted in the complete inactivation of *Cryptosporidium* oocysts (Whitmore, 1993). However, the ova of parasitic worms are more resistant. Ova of *Ascaris* are unaffected by mesophilic digestion after three months (Cram, 1943). In contrast, Carrington (1980) considered that anaerobic digestion at temperatures >36 °C would inactivate these ova. Silverman and Guiver (1960) reported the destruction of the eggs of *Taenia saginata* by anaerobic digestion at 35 °C in one to five days in batch experiments.

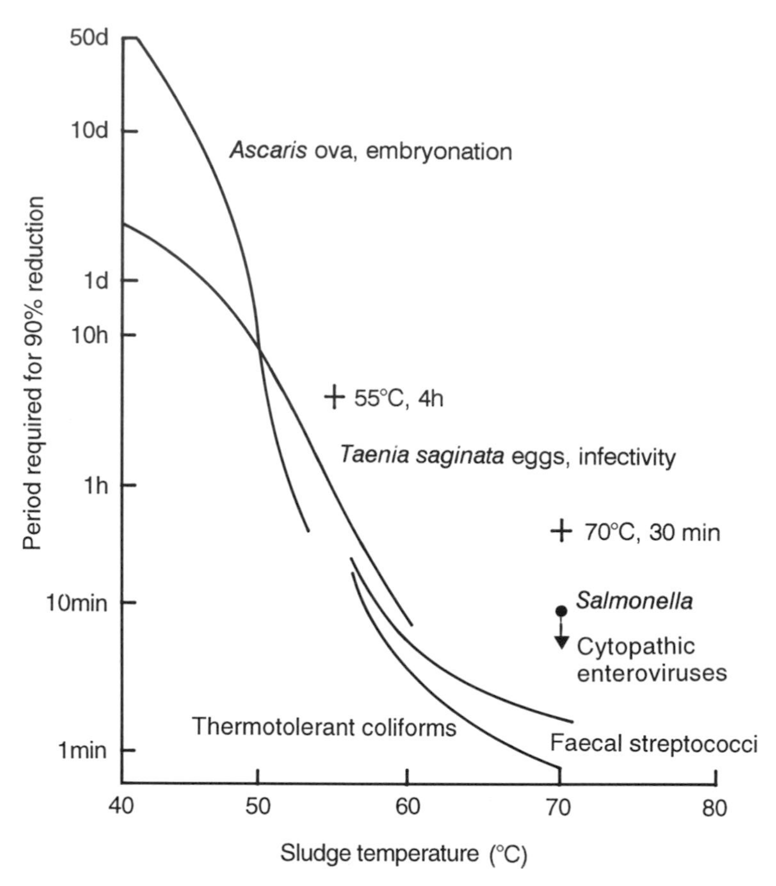

Fig. 11.1. Effect of sludge treatment temperature upon time for 95% destruction of various pathogens. Crosses indicate temperature/time conditions specified in Code of Practice (DoE, 1989a) (Bruce *et al.*, 1990).

The efficiency of pathogen removal in anaerobic digesters can be improved by drawing down treated sludge before feeding to prevent contamination of digested sludge with untreated material, or by introducing a secondary stage of treatment also known to be effective in reducing pathogens, such as lagoon storage (Pike and Carrington, 1986). The reduction in pathogen load of sludge through anaerobic digestion is sufficient for its application to agricultural land to be acceptable where agricultural practices minimize pathogen dissemination (Coker 1983; Stukenberg *et al.*, 1994). The recommended practices in the UK are described in the Code of Practice (DoE, 1989a) and are also considered later in this chapter.

Composting

Composting effectively eliminates pathogenic organisms (Table 11.3) provided the temperature reaches 55-60 °C for a minimum period of three consecutive days (Pereira Neto *et al.*, 1987). Aerated static pile methods are more effective in pathogen destruction than windrow turning techniques. This is because the surface layers of windrows remain at or near ambient temperatures and so provide a source for reinfection of the material when the windrow is turned. A recent survey indicated that only 0.5% of the sludge dry solids utilized on agricultural land in the UK was treated by composting (CES, 1993).

Lime treatment

Hall (1993) has described the various lime treatment processes which are currently available for sewage sludge. Lime treatment usually involves adding calcium hydroxide (slaked lime) to raise the pH value to 10.5-11.5. At a pH of 11.5 vegetative bacteria, including salmonellas, and most viruses are inactivated in several hours and are undetectable within 24 h (Farrell *et al.*, 1974; Sattar *et al.*, 1976; Pike and Carrington, 1986). In general, lime treatment is ineffective against the ova of parasitic worms unless pH is high (pH >12.5 for *Ascaris* spp. and pH >12 for *T. saginata*) (Table 11.3). However, the N-Viro process, which involves blending raw sludge cake with calcium oxide (quicklime) and waste cement kiln dust, also produces an exothermic reaction. Temperatures in excess of 52 °C with a pH of 12 are achieved for a minimum period of 12 hours which effectively destroys the pH insensitive pathogens (Hampton, 1992). Lime stabilization is relatively unimportant as a sludge treatment process in the UK accounting for only 0.05% of the sludge dry solids recycled to agricultural land (CES, 1993).

Pasteurization and thermal drying

Pasteurization of sludge is achieved by heating to 70 °C for 30 minutes using steam (Obrist, 1979; Pike *et al.*, 1988). Methane gas from anaerobic digestion of sludge has been used as a heat source (Kügel, 1982). This technique is not widely employed as a sludge treatment process in the UK (Frost *et al.*, 1990). Thermal drying involves heating the sludge to temperatures of at least 100 °C to produce a dried, typically granular product with 80-90% dry solids (Hall, 1993). Currently, there is only one sludge thermal drying plant operating in the UK and over 100 in Europe (Chabrier, 1994) although this number is likely to increase as the benefits of sludge minimization are realized. Pathogenic organisms are effectively eliminated from sludge by both heat treatment and thermal drying processes (Table 11.3). The risks of transmitting disease from thermally dried sludge in particular is probably negligible (Hall, 1993).

Effects of Treatments on *Salmonella* spp. and *Taenia saginata*

The effects of various sludge treatment processes upon the viability of *Salmonella* spp. have been summarized by Bruce *et al.* (1990) and are shown in Fig. 11.2. A 90% reduction in *Salmonella* spp. is indicated by the vertical axis in Fig. 11.2. Pasteurization of sludge is very effective at reducing the number of salmonellae (99.999% reduction), so are soil injection of raw sludge and lime dewatering, which achieve approximately a 99.9% reduction in numbers. The numbers of *Salmonella* spp. in sludge are reduced by 90-99% by mesophilic anaerobic digestion. Pike and Carrington (1986) noted that an extensive survey of sewage sludges from different sewage treatment works showed mesophilic anaerobic digestion in practice was the least effective of the stabilization treatments in removing salmonellae. This is principally because of short-circuiting caused by operational practice when recharging with raw sludge under 'fill and draw' operation. New plant and operational conditions are designed to avoid this problem by holding the digested sludge in a separate secondary digestion stage for a minimum of 14 days (DoE, 1989a).

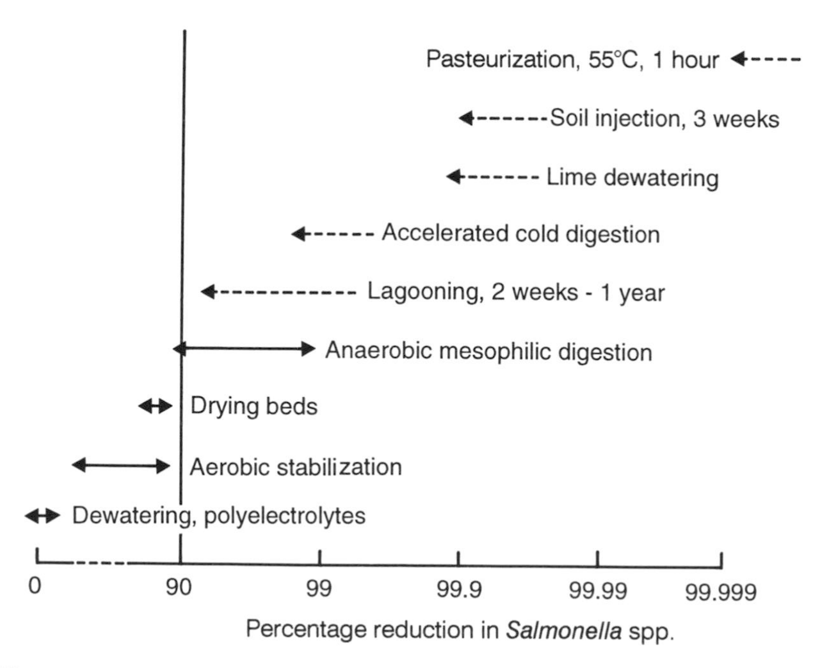

Fig. 11.2. Reductions in numbers of *Salmonella* spp. brought about by various sludge treatments (Bruce *et al.*, 1990).

Reduction in infectivity of *T. saginata* eggs for calves is shown in Fig. 11.3 for various sludge treatment processes (Bruce *et al.*, 1990). This has been calculated by comparing the number of cysts produced in infected calves three months following ingestion of either treated or untreated eggs, fed in equal numbers in sludge. As noted earlier, laboratory studies indicated the destruction of *T. saginata* eggs by mesophilic anaerobic digestion. As with salmonellae, however, the efficiency of full-scale single-stage digesters in destroying this parasite is limited by their mode of operation. However, if secondary treatment is given by lagooning (Fig. 11.3) the efficiency of destruction is greatly increased (Pike and Carrington, 1986). It appears that long-term sludge storage treatments are particularly highly effective at reducing salmonellae to low levels and also at destroying infectivity of *T. saginata* eggs, although neither organism is completely eliminated.

Effects of Treatments on Potato Cyst Nematodes and Other Plant Pathogens

Heat treatment is highly effective at reducing viability of *Globodera* spp. cysts. For example, Spaull and McCormack (1989) found that the viability of *G. rostochiensis* and *G. pallida* was reduced almost to zero by heating sludge to 70 °C for 30 minutes.

Very high levels of destruction can also be achieved by mesophilic anaerobic digestion and cold digestion (Southey and Glendinning, 1966; Turner *et al.*, 1983; Spaull and McCormack, 1989). However, Turner *et al.* (1983) suggested that batch operation of digesters would maximize the level of pathogen control by reducing the extent of contamination of treated sludge with untreated material which occurs under normal 'fill and draw' operation. Other data suggest that fungal and bacterial plant pathogens are also eliminated, or at least substantially reduced, by anaerobic digestion (Turner *et al.*, 1983). In addition, mesophilic anaerobic digestion is probably effective in the destruction of plant viruses (Carrington, 1978).

Windrow composting destroys *G. rostochiensis* in three weeks provided that a stable temperature of 50 °C is maintained by regular turning (Goffeng *et al.*, 1978), although lower temperatures (32-34 °C) may also be effective at cyst destruction (Bollen, 1985). Indeed, the majority of plant pathogens (fungi, bacteria, nematodes and viruses) do not survive efficient composting treatment (Bollen, 1985). Inactivation is caused by (a) heat generated during the thermophilic stage, (b) microbial antagonism, and (c) toxicity of conversion products formed during decomposition of organic material.

Lime treatment of sludge is not very effective at controlling the potato cyst nematodes. Spaull and McCormack (1989) reported a reduction in viability of *G. rostochiensis* and *G. pallida* cysts of only 40-45% by lime conditioning of sludge at pH 11.5. In general, lime treatment of sewage sludge is unlikely to kill all plant pathogens which may be present. However, the N-Viro process may be expected to give a high level of destruction through the combination of heat inactivation and high pH.

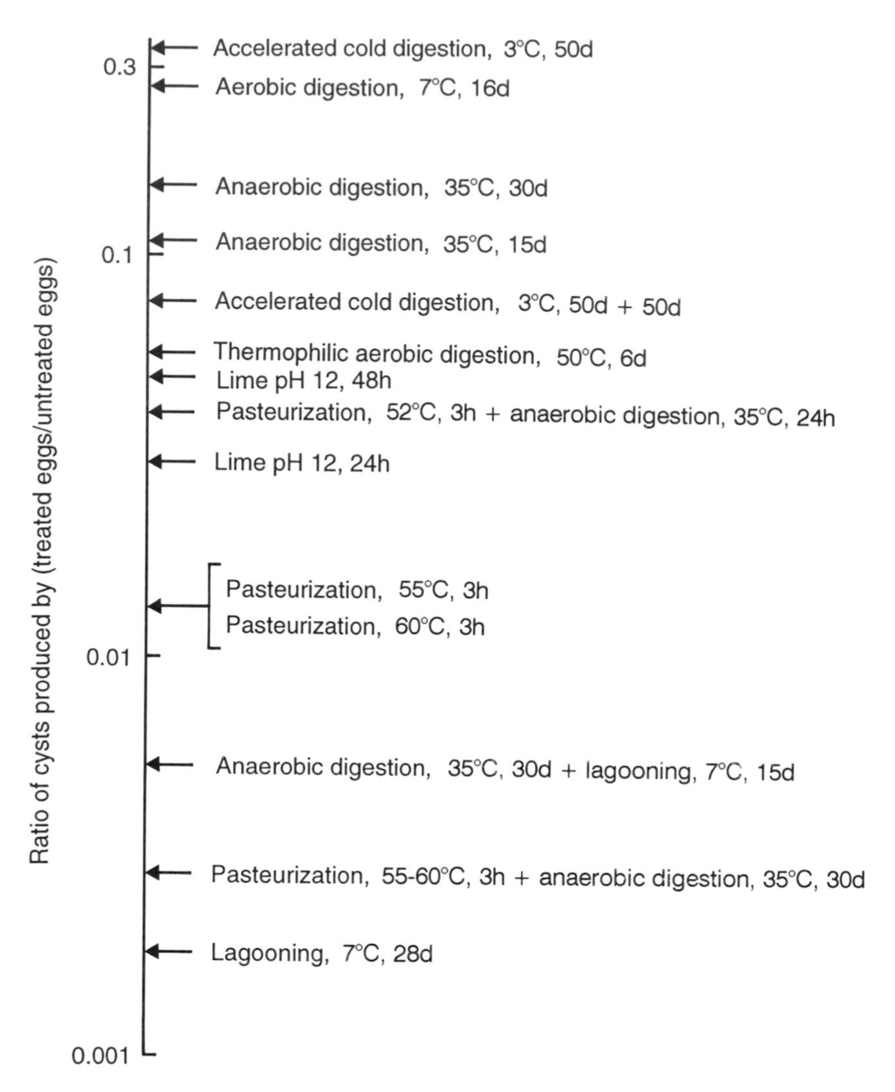

Fig. 11.3. Ranking of sludge treatment processes by their ability to reduce infectivity of *T. saginata* eggs for calves (Bruce *et al.*, 1990).

Survival in Soil and on Vegetation

Survival in Soil

The factors influencing the survival of bacteria and retention of viruses in soil have been reviewed by Gerba *et al.* (1975) and are listed in Table 11.4. In a more recent review of literature, however, Sorber and Moore (1987) concluded that temperature was the only factor, of a range of soil and environmental parameters, which influenced pathogen survival significantly in sludge-amended soils. Increasing survival time was clearly shown as a function of decreasing temperature. Faecal coliforms apparently were the only exception as their survival increased with increasing temperature at depths of 5 to 15 cm in the soil profile. This was attributed to the enrichment of sludge-amended soil with nutrients, and the protection from other environmental stresses occurring nearer to the soil surface. In addition, the response may be explained because faecal coliforms can tolerate and grow at temperatures up to 45 °C (E.G. Carrington, WRc Medmenham, personal communication). Unless examined and tested carefully, some plant coliforms can also be identified incorrectly as faecal types (E.G. Carrington, WRc Medmenham, personal communication). Nevertheless, survival times of faecal coliform bacteria remain relatively short in soil.

Bacteria are among the most sensitive of pathogens to environmental conditions outside the host gut and numbers rapidly decline on exposure to light, desiccation and antagonism when applied to soil in sludge (Coker, 1983). Decay rate of *Salmonella* is particularly dependent on local conditions and Pike and Carrington (1986) indicated survival decreases in well drained and dry soils compared with water-saturated conditions. Furthermore, light and infrequent applications of sludge permit *Salmonella* destruction whereas survival is prolonged by heavy dressings. Very long survival times often cited for *Salmonella* persistence arise from inoculation experiments in which high levels of bacteria ranging from 10^6 to 10^{10} l^{-1} were added to sludge before it was spread. Under these conditions, and assuming a maximum time to 99% destruction (T_{99}) of 45 days, persistence times in excess of five months could be anticipated (Sorber and Moore, 1987). Typically, the occurrence of *Salmonella* in sludge is much lower than this. It is worthy of note that in the majority of instances the minimum infective dose required to give clinical disease symptoms is large; Moore (1971) reports that, on average, the infective dose of *Salmonella* for humans is more than 10^4 organisms. At typical levels of contamination in sludge, a 90% reduction in *Salmonella* is normally anticipated within three weeks following sludge application (Pike and Carrington, 1986; Sorber and Moore, 1987). Indeed, Citterio and Frassinetti (1989) could not detect *S. typhimurium* after 2-3 weeks in soil treated with sludge.

Table 11.4. Factors affecting the survival of bacteria and retention of viruses in soil.

Factor	Remarks
BACTERIA	
Moisture content	Greater survival in moist soils and during wet weather
Permeability	Survival times less in sandy soils
Temperature	Longer survival times at low temperature
pH value	Shorter survival times in acid soils
Sunlight	Shorter survival times at soil surface
Organic matter	Increased survival and possible regrowth when sufficient amounts of organic matter are present
Antagonism of soil microflora	Increased survival in sterile soil
VIRUSES	
Flow rate	Virus retention decreases proportionally as flow rate increases
Cations	Cations can reduce repulsive electrostatic potential between virus and soil particles, allowing adsorption to proceed
Clays	High ion-exchange capacity and large surface area give high virus retention
Soluble organics	Compete for adsorption sites
pH value	Low pH value favours virus adsorption, high pH favours elution
Chemical composition of soil	Some metal complexes (e.g. magnetic iron oxide) readily adsorb virus particles

Source: Gerba *et al.* (1975)

The survival of apparently viable but non-culturable salmonellae in sludge-treated soil has recently raised concerns about possible health risks to grazing animals and man (Turpin *et al.*, 1993). It is doubtful whether the identification of non-culturable viable *Salmonella* raises serious questions about risks to health as these have been unable to infect mice (E.B. Pike, WRc Medmenham, personal communication). However, non-culturable *Campylobacter jejuni* found in water by Jones *et al.* (1991) were able to infect mice, although other work with *Campylobacter* has not recorded infection with non-culturable forms (E.B. Pike, WRc Medmenham, personal communication).

Furthermore, the culture medium used by Turpin *et al.* (1993) is inhibitory to *Samonella* to some extent and could underestimate the numbers of salmonellae apparently present in soil.

Despite the apparent increased survival of faecal coliforms with depth in the soil profile, these bacteria are also generally inactivated relatively quickly in sludge-treated soil. For example, Braids (1970) reported that 99% of faecal coliforms were killed in 30 days, and Bell and Bole (1978) detected 99.9% mortality of faecal coliform bacteria in liquid sludge within 35 days when applied to soil. With dried sludge, however, 99.9% destruction was achieved in 12 days. In an overall assessment of the survival of these pathogens, Sorber and Moore (1987) estimated a T_{90} value (i.e. time to 90% destruction) of <6 weeks. Faecal streptococci displayed a comparable loss in <4 weeks.

Viruses are generally inactivated rapidly near the soil surface depending principally on temperature conditions. Sorber and Moore (1987) calculated median T_{90} and T_{99} values of 3 and 6 days for viruses in warm surface soils (≈30 °C). However, at low temperatures, T_{90} values of approximately 30 days were observed. The $T_{99.99}$ for poliovirus was 100 days after a winter application of sludge inoculated with the virus: after a summer application the same decline in viability required only 7-10 days of exposure in soil (Tierney *et al.*, 1977). Viruses are obligate intracellular parasites and do not replicate outside of the appropriate host. Consequently, their numbers eventually decrease in the soil environment, irrespective of conditions, because they are not adapted to survive outside of the host animal for long periods.

Protozoan cysts are very sensitive to desiccation and only survive a few days on soil or vegetation under drying conditions (Coker, 1983; Sorber and Moore, 1987). Kowal (1983) concluded that survival times for protozoa in soil were typically about two days, possibly extending to a maximum period of ten days. Whitmore (1993) concluded for *Cryptosporidium* that oocyst viability decreased in soil, but that sludge treatment was effective in ensuring levels of oocysts applied to soil in sludge should be extremely low. In contrast, *T. saginata, Ascaris* and hookworm ova are very resistant to desiccation when present in sludge or soil. However, these parasites are also unable to replicate outside of their appropriate hosts. Pike and Carrington (1986) suggested that the loss of infectivity of the eggs of *T. saginata* arises because infection must occur within a finite period of being shed from the ripe proglottid.

Pike and Carrington (1986) estimated the survival period of *T. saginata* in soil would probably vary from a few months to a year or more depending on climatic conditions with the rate of decay slowing under cool conditions. Pike *et al.* (1983) incubated eggs of *T. saginata* under laboratory conditions which showed 90% loss of viability in 200 days. Viability overestimates infectivity, and storage of eggs for 25 and 50 days at 4 °C caused a reduction in infectivity of 70 and 95%, respectively (Pike and Carrington, 1986). Investigating the survival of *T. saginata* eggs throughout the soil profile, Storey and Phillips (1985) showed that eggs could still be found on the soil surface after 200 days and that survival increased down the soil profile. As explained earlier, this is probably because the pathogen is protected by the soil from environmental stresses including solar radiation and desiccation.

Survival times for helminth ova such as *Ascaris* are very long and persistence in soil for up to 15 years has been reported (Sorber and Moore, 1987). Schwartzbrod *et al.* (1987) estimated survival times of *Ascaris* ova in soil could range from two to seven years. However, as with all other microbial pathogens, desiccation can reduce survival substantially and a more realistic maximum survival time appears to be two years (Kowal, 1983). The reports reviewed by Sorber and Moore (1987) indicated longer survival times for these parasites than for any of the microbial groups with T_{90} values ranging from 17 to 270 days with a median of 77 days. Seasonal effects were again observed on ova survival. Summer applications of sludge gave an apparent T_{90} value of 17 days, which extended to 65 days in autumn and to over 200 days in winter.

The beneficial effect of organic matter application to soil on the activity of antagonists to soil-borne plant pathogens has been studied to a limited extent (Hadar, 1986). Soils high in organic matter tend to be naturally suppressive to soil-borne pathogens due to high populations of antagonistic micro-organisms. Soil microbial activity is greatly increased by the addition of sludge organic matter to soil so that the suppression of plant pathogenic agents may be anticipated from the agricultural use of sludge (Smith, 1991). Furthermore, Turpin *et al.* (1992) recently reported that soil amendment with sludge promoted the antagonistic effects of soil micro-organisms increasing the rate of *Salmonella* die-off in soil.

Survival on Vegetation

Desiccation and sunlight cause a high mortality among bacterial pathogens adhering to the leaves of crop plants treated with sewage sludge (Coker, 1983). Rudolphs *et al.* (1951) detected no *Salmonella* seven days after being sprayed onto vegetation. Bell and Bole (1978) showed that 90% of faecal coliforms applied to vegetation in sludge were inactivated within 48 hours and none were detected after 14 days.

Entamoeba histolytica cysts do not survive more than three days. Viruses may be eliminated from vegetation within 24 days (Larkin *et al.*, 1976). Immature *Ascaris* ova do not survive on vegetation more than 35 days (Rudolphs *et al.*, 1951), although mature or hatching ova, which occur frequently in sludge, are more able to survive adverse conditions (Engelbrecht, 1978). *Taenia* ova are susceptible to desiccation, and viability of both *Taenia* and *Ascaris* is reduced through fungal attack and degeneration (Silverman, 1955; Jones *et al.*, 1979). Various estimates for the survival time of *Taenia* on pastures, reviewed by Coker (1983), suggest that ova are killed within two to six months. Shorter survival times are associated with high temperatures and drying conditions. However, there is minimal risk of infecting grazing animals because sludge treatment is effective in pathogen destruction and surface application of raw sludge to grassland is not permitted under the regulations (SI, 1989a).

Movement in Soil and Risks to Water Sources

Few assessments have been made of the presence of micro-organisms in run-off from sludge-amended soils. However, available information suggests that sludge application to soil has no significant pathogen impact on surface waters (Edmonds, 1976; Lue-Hing *et al.*, 1979; Melanen *et al.*, 1985). The low incidence of pathogenic agents in run-off water can be anticipated given that only treated sludges can be surface-applied to soils whereas untreated sludges are incorporated which reduces run-off problems (Chapter nine). Nevertheless, elevated levels of pathogens have been reported in run-off from sludge-amended soil containing viable organisms (Theis *et al.*, 1978; Dunigan and Dick, 1980). Soil injection of sludge effectively eliminates the problem of surface run-off and pathogen contamination of surface water supplies (Hall, 1986c).

While run-off may be considered largely as the physical transport of micro-organisms associated with particulate material, vertical microbial transport is more complex. In general, micro-organisms are retained in the surface soil layers by colloidal matter in soil (Coker, 1983). The greatest risk that pathogens might enter drainage or groundwater would occur on a shallow, fine textured soil over fissured rock or land drains where the soil had dried forming fissures down to the underlying rock or drains. In such special circumstances it is possible that mass flow of liquid sludge could penetrate through and below the soil, and such applications should be avoided (Coker, 1983). The *Code of Good Agricultural Practice for the Protection of Water* (MAFF, 1991a) gives similar advice to protect water sources from pollution by organic manures.

Removal of bacteria from wastewater percolating through soil is due to both mechanical action (i.e. straining or sieving at the soil surface) and adsorption to soil particles. However, adsorbed cells do not remain permanently fixed, and release and movement would be expected since physical adsorption to particulates is a reversible process. Indeed, Gerba *et al.* (1975) noted movement of bacteria in a range of soils for distances ranging from 0.9 to 456 m. In most circumstances, however, pathogens are retained in the surface soil layers. Romero (1970) estimated that 92-97% of wastewater bacteria are retained in the top 1.0 cm of soil, and the concentrations of coliforms fell to very low levels in wastewater which had percolated through 2 m of soil. Even where high rate irrigation was practised on a coarse textured soil, the percolate was almost entirely free of faecal coliforms at a depth of 1.3 m (Glantz and Jacks, 1967). Other studies reviewed by Coker (1983) found that passage of sewage through a 3 cm soil column removed 99% of *Salmonella* and 99.9% of *E. coli*; virus particles were not detected in drainage water from sludge-treated soil.

Viral agents are effectively retained in soil by adsorption to soil particles, and the level of adsorption increases with increasing clay and organic matter content (Sorber and Moore, 1987). Drewery and Eliassen (1968) reported that most viruses are retained in the top 2 cm of columns of a range of soils, although a small proportion penetrated more deeply. The movement of viruses through soils in field studies has been demonstrated to depths of 67 m (Keswick, 1984). However, studies with lysimeters or soil cores amended with virus contaminated sludge exposed to natural rainfall showed limited or no movement below depths of 0.5-1.25 m (Sorber and Moore, 1987). When compared with the movement of 'free' viruses, sludge-bound viruses are much more effectively retained within the sludge-soil matrix at the point of application (Damgaard-Larsen *et al.*, 1977; Lue-Hing *et al.*, 1979).

The transport of protozoa and helminths in soils appears to be more limited than for bacteria or viruses, probably arising through the large size differences between these groups of micro-organisms. For example, protozoa are up to 20 times larger than bacteria and up to 2000 times larger than enteroviruses (Sorber and Moore, 1987). *Ascaris* eggs are even larger. More than 30 cm, but less than 60 cm, depth of coarse sand are necessary to remove *Ascaris* and hookworm ova from sewage effluent (Cram, 1943). Sorber and Moore (1987) considered that mechanical straining was the most important factor governing parasite transport through soil. Indeed, available literature strongly favours the view that the ova of parasites are retained at the point of sludge introduction. Groundwater monitoring for the ova of parasites has not been routinely conducted and seems unnecessary given the relative size of most parasite ova and their observed retention in the upper layers of the soil profile.

Relatively few studies conducted at sludge application sites have assessed the vertical transport of micro-organisms under field conditions. Limited monitoring has shown no measurable impact of sludge application on faecal coliform levels in groundwater (Edmonds, 1976; Lue-Hing *et al.*, 1979). In a study of the effects of effluent recharge on groundwater quality, Baxter and Clark (1984) reported the complete removal of bacteria and viruses from the infiltrating effluent by the soil and unsaturated zone beneath a recharge site in spite of very high levels of pathogen contamination in the effluent. In the groundwater outside the sewage recharge area, all viruses were removed and faecal bacteria were reduced to insignificant levels within 500 m. However, the potential risk from sewage sludge applied to agricultural soil will be considerably smaller compared with land treatment of effluents since agronomic rates of sludge addition are very much lower than rates of effluent discharge and treated sludges contain a significantly smaller pathogen load.

The available information suggests that in overall assessment the impact of pathogens in sewage sludge on ground and surface water quality is likely to be minimal and is not limiting to the agricultural utilization of sewage sludge.

Constraints on Planting, Grazing and Harvesting

The constraints on cropping specified within the CEC Directive (CEC, 1986a), implemented in the UK through the *Sludge (Use in Agriculture) Regulations 1989* (SI, 1989a), and the additional measures given in the UK *Code of Practice for Agricultural Use of Sewage Sludge* (DoE, 1989a), provide a second level of protection against transmission of infection after sludge treatment. This is achieved by allowing sufficient time for pathogens to decay or to disappear by dilution and dispersion within the soil before the land is used. The acceptable uses of treated and raw sludges in agriculture are listed in Tables 11.5 and 11.6, respectively.

The infective dose of *Salmonella* is very large in grazing livestock (Coker, 1983). For example, Hall *et al.* (1978) fed 10^5 salmonellae day^{-1} in raw sludge to calves for a period of 28 days. Salmonellae were not isolated from faeces or from their tissues at post-mortem. In view of this, and also other data from the sludge *Salmonella* surveys (Pike, 1981), the apparent risk of sludge-borne infection is very low indeed. The decay of salmonellae in sludge-treated soils is rapid, and a significant level of destruction occurs within three weeks of sludge application. Consequently, the minimum no-grazing or harvesting period for grassland stipulated in the Code of Practice (Tables 11.5 and 11.6; DoE, 1989a) ensures a very high level of protection against infection from *Salmonella.*

Sludge may not be applied within ten months before harvest to land intended for fruit growing and vegetables in contact with soil and eaten raw. This time interval is sufficient to ensure a considerable barrier to transmission of the most resistant parasites. Sludge application is not permitted to land used for growing basic seed potatoes, seed potatoes for export, or for growing basic nursery stock (including bulbs for export). These crops must be raised in land which is free of potato nematode. Sludge must not be used because the absence of cysts cannot be guaranteed, although the potential risk of infection is very small.

The use of untreated sludge on agricultural land is permitted provided it is incorporated immediately after application or injected into the soil in accordance with the WRc Code of Practice for sludge injection (WRc, 1989). Salmonellae in injected or buried sludges reach undetectable levels after about a week irrespective of soil type or temperature conditions (Andrews *et al.*, 1983). The decay in infectivity of *T. saginata* ova is slower, but is probably complete under UK conditions in soil within 3-6 months. Since properly injected sludge is effectively removed from contact with animals and man, the restrictions on use can be similar to those adopted for treated sludge with no additional risk to the environment.

Table 11.5. Acceptable uses of treated sludge in agriculture.

When applied to growing crops	When applied before planting crops
Cereals, oil seed rape Grass[(1)] Turf[(2)] Fruit trees[(3)]	Cereals, grass, fodder, sugar beet, oil seed rape, etc. Fruit trees Soft fruit[(3)] Vegetables[(4)] Potatoes[(4),(5)] Nursery stock[(6)]

(1) No grazing or harvesting within 3 weeks of application
(2) Not to be applied within 3 months before harvest
(3) Not to be applied within 10 months before harvest
(4) Not to be applied within 10 months before harvest if crops are normally in direct contact with soil and may be eaten raw
(5) Not to be applied to land used or to be used for a cropping rotation that includes the following:
(a) basic seed potatoes
(b) seed potatoes for export
(6) Not to be applied to land used or to be used for a cropping rotation that includes the following:
(a) basic nursery stock
(b) nursery stock (including bulbs) for export
Source: DoE (1989a)

Table 11.6. Acceptable uses of untreated sludge in agriculture.

When applied to growing crops by injection[(5)]	When cultivated or injected[(5)] into the soil before planting crops
Grass[(1)] Turf[(2)]	Cereals, grass, fodder, sugar beet, oil seed rape, etc. Fruit trees Soft fruit Vegetables[(3)] Potatoes[(3),(4)]

(1) No grazing or harvesting within 3 weeks of application
(2) Not to be applied within 6 months before harvest
(3) Not to be applied within 10 months before planting if crops are normally in direct contact with soil and may be eaten raw
(4) Not to be applied to land used or to be used for a cropping rotation that includes seed potatoes
(5) Injection carried out in accordance with WRc publication FR 0008 1989. *Soil Injection of Sewage Sludge - A Manual of Good Practice (2nd Edition)* (WRc, 1989)
Source: DoE (1989a)

National surveillance of human and animal disease is of a high standard in the UK and has demonstrated the effectiveness of the management practices developed against the potential transfer of pathogenic agents from sludge utilization on agricultural land (Bruce *et al.*, 1990). The minimal environmental impact of sludge-borne pathogens is emphasized further since the incidence of infectious diseases arising from the occupational exposure of wastewater treatment plant workers to sewage and sludge, which represent potentially a high risk exposure group, has not increased compared with levels normally found in the population (Langeland, 1986). Surveillance and epidemiology continue to confirm that the risks to health are extremely small, since very few cases or outbreaks of disease have been positively attributed to the use of sludge. Pike and Carrington (1986) concluded that this position would be maintained provided that sludge was applied to agricultural soil responsibly and in accordance with the Sludge Regulations and Code of Practice.

Natural Ecosystems

Sludge-borne pathogens are unlikely to represent a significant impact on natural ecosystems (Carrington, 1978). This is principally because the major sludge pathogens, such as *T. saginata* and *Ascaris*, are host specific obligate parasites, whereas other important agents like *Salmonella* would probably only contribute to background levels already present in the population. Indeed, Reasoner (1976) showed that enteric pathogens are widespread in both wild and domestic animals and that these probably act as a reservoir of infection. However, salmonellae were isolated from only eight of 1269 small wild mammals representing 16 species during an investigation in three southern English counties (Jones and Twigg, 1976). The infected animals, all of which were house mice, had been in contact with experimentally inoculated cattle. In another survey, Twigg *et al.* (1973) only identified two of 345 deer caught throughout the British Isles as having been infected in the past with *Leptospira* characterized by high levels of antibodies. This proportion is considerably lower than that found in cattle in the south of England (Carrington, 1978). Consequently, there is no available evidence suggesting that natural ecosystems are potentially at risk from biological contaminants in sewage sludge applied to agricultural soils.

Chapter twelve:

Environmental Assessment

General Comments

Recycling sewage sludge to agricultural land, to gain benefit from the essential plant nutrients and organic matter it contains, would seem a reasonable and rational method of managing a material which would otherwise need disposing of by some other non-beneficial route. Sludge also contains inorganic, organic and biological contaminants and therefore requires careful and responsible management to avoid potential environmental problems. However, in the UK, as in many countries, in spite of the regulatory basis to the agricultural use of sludge (SI, 1989a), and the additional measures required by general and specific supporting codes of practice (DoE, 1989a; MAFF, 1991a), there is concern over the potential impact sludge recycling to agriculture may have on the environment. Agricultural recycling of sewage sludge is highly sensitive to the effects of adverse publicity on sludge application operations. In particular, the misinterpretation by the media of often isolated reports of potential problems are frequently taken out of context of the overall minimal risk to the environment from sludge recycling to agricultural land. Such adverse publicity may lead to unnecessary restrictions being imposed as the regulators adopt a highly precautionary approach when these issues are raised to avoid undue confrontation from an alarmed, but scientifically uninformed public.

An overall assessment of the potential environmental impacts caused by recycling sewage sludge to agricultural land has been attempted by ascribing a qualitative description of the apparent risk or benefit, based on the scientific information discussed in the earlier chapters of this book, for each of the principal sludge components, or groups of components, against potentially impacted environmental parameters. These descriptions are listed in Table 12.1. It has been assumed in this analysis that sludge is applied according to the regulations (SI, 1989a) and codes of practice (DoE, 1989a; WRc, 1989). The restrictions on the timing of sludge applications in designated nitrate vulnerable zones (NVZs) required by the CEC Directive on nitrates (CEC, 1991b; MAFF, 1994c) have also been considered. It is assumed, therefore, that the application of sludge is avoided during the stipulated period in the autumn within the statutory NVZs. Whilst subject to codes of good agricultural practice outside of these areas (SOAFD, 1991; MAFF, 1991a), the application of sewage sludge, as well as farm wastes, is likely to be unavoidable in practice during the critical autumn period at least in the short-term. Indeed, moving the timing of manure application on agricultural land from the traditional period in autumn, required by the codes of practice, needs a complete rethink of how these materials are to be utilized in terms of application technologies and management. This cannot happen quickly and will probably require investment in extended storage facilities, new and alternative application equipment, as well as research to define best operational practices. It is emphasized, however, that ascribing an apparent risk does not necessarily imply that there is an immediate problem, only that a potentially detrimental impact could possibly occur under certain, often extreme or 'worst-case' circumstances.

Table 12.1. Environmental impact risk and benefit assessment for sewage sludge recycling to agricultural land ([1]B = beneficial effect, [2]L = low risk, [3]P = possible risk, na = not applicable).

Environmental parameter	PTEs	Organic contaminants	Pathogens	Nitrogen	Phosphorus	Organic matter
Human health	L	P(L)	L	B	B	B
Crop yields	L	L	L	B	B	B
Animal health	L	L	L	B	B	B
Groundwater quality	L	L	L	P[4]	L	L
Surface water quality	L	L	L	P(L)	P(L)	B
Air quality	L	L	L	P(L)	na	na
Soil fertility	P	L	L	B	B	B
Natural ecosystems	P	P	L	P	P	B

[1]A 'beneficial effect' (B) is attributed where a component of sludge is reported to potentially enhance an aspect of human health and environmental quality
[2] Risk is designated as 'low' (L) where environmental effects are minimized by current operational practice
[3] Risk is designated as 'possible' (P) where there is some reported evidence that current operational practice may result in a potential impact on the environment on the basis that one or more of the following conditions apply:
(a) published evidence of effects is contradictory
(b) current recommendations for environmental protection may be difficult to implement operationally in the short-term
(c) effects may occur under certain extreme 'worst-case' conditions, given the current regulations and codes of practice
(d) there is uncertainty about the environmental implications of particular sludge contaminants
[4] Would be designated as 'L' with appropriate autumn management practices and spring applications to growing crops
Letter in brackets denotes impact arising from injected or incorporated sludge

Human health is certainly the principal parameter of concern when considering the environmental impact of sewage sludge recycling to agricultural land. Soil fertility and natural ecosystems, though of importance, are probably of less concern than any potential human health implications. Differentiating the relative importance of the remaining environmental parameters is more difficult, although effects on crop yields and on animal health are clearly of concern where sludge is spread on farmland. Attention has also recently focused on impacts of organic manures on water quality, so this topic is a high priority on the political agenda within the EU. Effects on air quality, and the extent of their environmental impacts, are being realized and are likely to become increasingly relevant in the near future.

Human Health

Detrimental effects on human health arising from the agricultural use of sludge are unlikely. The principal PTEs of concern are Cd, Pb and Hg (Chapters four and six). Lead and Hg are not absorbed to any extent by crops and consequently do not pose a risk through the dietary intake of plant foods grown in sludge-amended soil. Cadmium, on the other hand, is not subject to the 'soil-plant barrier' and can accumulate in crops to concentrations which may be potentially hazardous to animals consuming them without appropriate control. However, dietary models of Cd intake from crops grown in sludge-treated soils, and in other soils highly contaminated with Cd from past industrial activities, indicate health problems are unlikely. Given the high margin of safety against potential dietary intake of Cd, which is intrinsic in the maximum permissible concentration of Cd in sludge-treated soil of 3 mg kg^{-1} (SI, 1989a), it is very unlikely that a detrimental accumulation of ingested Cd would occur. Because dietary models have assumed that all the vegetable plant foods in the diet come from sludge-treated soil at the numerical limit value for Cd, change of land use from agriculture to urban gardening, for example, would not present a potential health risk. Indeed, individuals consuming a well balanced diet, including fresh vegetable produce, actually represent a low risk group because improved mineral intake reduces Cd absorption.

Risk assessment models developed in the US (US EPA, 1992a) suggest that the direct ingestion of sludge by children is the most critical pathway of exposure to Cd, Pb and Hg. Whilst there is a remote possibility that young children may come into direct contact with sludge on agricultural land, the greatest risk is from domestic marketing of sludge in the US to home gardeners. The potential risk of PTE intake by this route is considerably smaller in the UK because very little sludge is supplied to the domestic market (Hall, 1993). Nevertheless, it is conceivable that changes in land use could increase the potential exposure of young children to PTEs if sludge-treated agricultural land were to be developed with urban housing and gardens. However, comparison of the risk-derived pollutant concentration values for sludge (US EPA, 1993) with UK limit values for soil suggests that this is unlikely to cause concern.

Livestock ingestion of PTEs from sludge surface-applied to pasture soils is another possible pathway of food chain exposure. However, of crucial importance to the human diet is that Cd and Pb do not accumulate in muscle tissue (carcass meat) entering the food chain even under 'worst-case' conditions of PTE intake by grazing livestock from sludge-treated grassland. By contrast, Cd and Pb may potentially accumulate in the kidney and liver of livestock principally through the direct ingestion of soil and surface-applied sludge (Chapter six). Nevertheless, recent feeding trials have shown that the concentrations of Cd and Pb in offal are unlikely to place the human diet at risk at the mandatory soil limit for Cd and the concentration of Pb achieved in sludge-treated soil relative to other elements, particularly Cu and Zn, and their limit values. For example, animal ingestion of soil with Cd at up to twice the statutory limit will not itself lead to unacceptable Cd concentrations in offal. In any case, offal represents only a minor component of the diet (0.5% fresh weight), compared with staple plant foods such as potatoes or cereal crops, and is therefore unlikely to have much impact overall on total dietary intake of Cd (Davis *et al.*, 1983). In contrast to Cd, Pb accumulation in the kidney of sheep may lead to exceedence of the food quality standard (1 mg Pb kg^{-1} fresh weight) under certain extreme 'worst-case' exposures considered unlikely to occur under normal agricultural management of sewage sludge. On the other hand, the concentration of Pb accumulated in the liver under these 'worst-case' conditions of exposure would remain below the considered acceptable level for food given that the maximum limit for Pb in liver is twice that allowed in kidney. It is emphasized, however, that the risk of Pb entering the human food chain is minimized further due to the marked reduction in the Pb content of sewage sludges recycled on agricultural land which has occurred in recent years. The marketing and distribution of meat and meat products locally and on a national basis will also result in a substantial market dilution factor, further reducing the dietary intake of PTEs in offal.

The main concern with Cd in sludge-treated soils is the potential change in bioavailability which may occur in the long-term and in particular during the residual period when sludge application has ceased (Chapter five). Crop uptake of Cd from sludged soil increases as soil concentrations are raised, but bioavailability will remain relatively constant or possibly decline in the long-term provided soil pH remains constant. If in future, however, soil pH were allowed to decline due to changes in land management practices or soil acidification, there is the risk that availability and crop uptake of Cd could increase. Cadmium is rarely the principal limiting element to sludge applications at the current permitted soil concentrations due to the substantial improvements in sludge quality which have occurred in recent years. Effects of decreasing soil pH conditions on potential dietary intake will therefore be considerably smaller than for sludged soils treated 10 or 15 years ago.

Nevertheless, more information on the residual behaviour of Cd in sludge-amended soils, particularly in relation to changing soil pH conditions, would be desirable to assess the potential long-term risk of Cd entering the human food chain. On balance, however, recent dietary analyses (MAFF/DoE, 1993b) show that the risk to human health from Cd in sludge-amended agricultural land is likely to remain at a very low level in practice even if a downward trend in soil pH status became apparent in the future.

Long-term trends suggest that the pH of arable soils is being maintained or improved nationally so there is apparently little cause for concern. The *Code of Good Agricultural Practice for the Protection of Soil* (MAFF, 1993a) strongly recommends that the pH of soil should be maintained at an appropriate level by liming to sustain agricultural productivity and protect soil quality. However, grassland has shown a decline in pH status overall, which may increase animal intake of Cd in herbage. The statutory soil limits for Zn, Cu and Ni are adjusted according to banded ranges of soil pH value in the UK regulations, but the limit value for Cd is set at 3 mg Cd kg^{-1} for all soils with pH >5.0. However, the adjusted values for Zn and Cu, which are the principal metals constraining the agricultural use of most sludges, automatically result in smaller accumulations of Cd in soil according to the relative concentrations of PTEs in the sludge, adjusting for background metal levels in soil. Given that Cu is frequently the most limiting element (Table 2.7), and Cd only achieves 50% of the maximum permissible soil concentration when sludge is applied to land of pH 6.0-7.0, then the Cd concentration in soil will only be 37% and 30% of its limit value in soil of pH 5.5<6.0 and 5.0<5.5, respectively. In other words, soil Cd will tend towards the normal range for uncontaminated soil (Table 2.5) with decreasing pH value. In addition, less than 7% of the agricultural land treated with sludge in the UK is within the low pH band 5.0<5.5 and 13% of land receiving sludge is in the pH 5.5<6.0 range. Most of the sludge is applied to agricultural land of pH >6.0 (80% of total treated area). On this basis there is apparently minimal risk to human health from the utilization of sewage sludge on grazed pastures. However, Cd inputs to grassland from other sources will have an additive effect on Cd levels in offal. In contrast to sewage sludge, for example, phosphatic fertilizers are potentially a major source of Cd contamination in the environment and they present a widespread and long-term risk of Cd accumulation in agricultural soils.

There is evidence from a recent survey of the nutritional status of British adults (MAFF/DH, 1990) that a significant proportion of the population may be at risk from Zn deficiency in relation to recommended minimum dietary intakes of Zn (DH, 1991). Young females appear to have the lowest intakes and anorexia nervosa has been linked to Zn deficiency (Bryce-Smith and Simpson, 1984). Therefore, it could be argued that increasing the Zn status of crops through the application of Zn to soil in sludge may represent a positive benefit of sludge trace elements on human health (Chapter four).

The potential transfer to the human food chain of certain groups of organic contaminants contained in sludge has been predicted based on known physico-chemical properties of the compounds (Chapter ten). The trace organics of particular concern are PCDDs, PCDFs and PCBs because they are lipophilic and show a high propensity to transfer and accumulate in animal fat tissue and milk fat. In addition, these compounds appear highly persistent in the environment. Livestock ingesting surface-applied sludge adhering to herbage is considered the principal route of human exposure to potentially toxic organic contaminants in sewage sludge. It could be argued that more experimental data are needed to confidently establish the potential risk to the human diet of trace organic contaminants present in surface-spread sludge on grazed pastures. Indeed, certain assessments of the potential exposure suggest that the allowable intakes of some organic contaminants could be exceeded in the human diet. However, these models may have been constructed using unrealistic assumptions about the quantity of sludge adhering to grass and the length of time grazing animals may be ingesting high intakes of sludge. The assumed dietary intake of food by an individual exclusively from sludge-amended areas is also considered as an unrealistic model of the likely extent of exposure which might be reasonably anticipated in practice (MAFF/DoE, 1993b).

Under normal operational conditions the risk of organic pollutants entering the food chain from surface-applied sludge is likely to be minimal (Stark and Hall, 1992) and management practices can reduce the level of exposure further. For example, the extent of herbage contamination with sludge is minimal when applied to short swards of rapidly growing grass, irrespective of the dry solids content of the sludge. The dry solids content of liquid digested sludges is typically in the range 2-5%. The amount of herbage contamination will increase for sludges of higher dry solids content when applied to slow growing long swards. However, grazing is generally restricted when sward growth is limited during the winter period when sludge may potentially persist in herbage beyond the minimum no-grazing interval required by the UK *Code of Practice on Agricultural Use of Sewage Sludge* (DoE, 1989a). The problem of animal ingestion of contaminants is essentially eliminated by injecting sludge into grassland soils. Market dilution arising from the bulk collection, distribution and marketing of milk further minimizes the transfer of organic contaminants to the human food chain.

There is minimal risk to human health *via* dietary intake of organic contaminants from crops grown on sludge-treated soils because there is little or no plant uptake, except perhaps in the peel of carrots which is generally removed during normal food preparation.

The low incidence of disease arising from the agricultural utilization of sewage sludge demonstrates the small environmental risk associated with the pathogen content of sludge spread on farmland (Chapter eleven). Current management practices ensure minimal risk to human health from pathogens which may be present in sludge.

Concern over potential effects on human health due to increasing concentrations of nitrate in water supplies has resulted in regulatory action within the European Union to control the application on agricultural land of all N sources including sewage sludge (CEC, 1991b). Ironically, the evidence which links nitrate ingestion with health problems is tenuous. Available data indicate the incidence of gastro-intestinal cancer decreases in areas with increasing concentrations of nitrate in drinking water, and cases of methaemoglobinaemia are extremely rare and are not associated with drinking mains supplied water (Chapter nine).

Plant nutrients and organic matter applied to soil in sludge may be considered indirectly beneficial to human health by ensuring food supplies meet national requirements. A balanced intake of vegetable foods reduces the availability and absorption of trace amounts of PTEs in the diet due to improved mineral nutrition. By supplementing the inorganic nutrient requirements of crop plants the agricultural use of sewage sludge may also reduce potential impacts on human health, and on the environment generally, arising from the manufacture of artificial fertilizers through for example, high energy consumption and NO_x emissions (Jollans, 1985).

Crop Yields

Zinc, Cu and Ni are the principal phytotoxic elements applied to soil in sludge, but current soil limit values for these elements ensure uptake into crops remains below the critical toxic concentration thresholds in plant tissues (Chapter three). The maximum permissible soil concentrations were established using sensitive crop species and coarse textured soils in field and pot culture studies, and therefore the limits protect all crops grown on a range of soil types from phytotoxicity. The soil limits are adjusted according to soil pH value to further minimize the risk of phytotoxicity.

These elements do not pose a dietary risk because they are subject to the 'soil-plant barrier' since toxic concentrations in plant tissue are lower than the amounts which are potentially injurious to animals and man.

Organic contaminants have no phytotoxic activity at the concentrations found in sludge-treated soils (Chapter ten). Plant pathogens are effectively destroyed by sludge treatment processes so effects on crop yield due to plant disease infection appear unlikely (Chapter eleven). There is no evidence which suggests that plant diseases are transmitted in sewage sludge. Furthermore, application of sludge organic matter to soil beneficially promotes the activity of soil micro-organisms potentially antagonistic to soil-borne plant pathogens (Hadar, 1986).

Sewage sludge represents an important source of plant-available N, P and S and has significant and beneficial fertilizer replacement value for these major plant nutrients. Crop yield is also increased by the application to soil of sludge organic matter due to improvements in soil physical properties.

Animal Health

Direct ingestion of sludge and sludge-treated soil represents the principal risk to grazing livestock from contaminants in sewage sludge (Chapter six). For example, a comparison of Cu intakes by sheep from sludge-treated soil with published tolerances to Cu (estimated from highly available salt-amended diets) suggests that the maximum permissible soil limit concentration for this element may result in a potentially toxic intake, although other grazing animals are much less sensitive to Cu. However, important antagonistic effects between elements in complex diet mixtures containing sludge and soil which reduce absorption by the animal are not considered in these dietary models for Cu. Consequently, the risk of Cu toxicity is actually much smaller than that anticipated from simple calculations of total intake, and feeding trials indicate that Cu toxicity to sheep under field conditions is very unlikely. Indeed, there is evidence suggesting that Cu deficiency in grazing animals is potentially widespread in England and Wales. The problem of Cu deficiency is particularly acute in Scotland (Reaves and Berrow, 1984; Berrow and Reaves, 1985). Under these conditions it could be argued that the application of Cu in sludge to grassland soils actually represents a positive benefit. Intakes of the other sludge PTEs are unlikely to impair the health of grazing animals even under 'worst-case' conditions of sludge and soil ingestion. Grazing intakes of PTEs from sludge are reduced when sludge is injected into grassland soils.

There is no evidence that organic contaminants applied to soil in sludge are toxic to grazing animals. The principal concern here is from human dietary intake resulting from accumulations in animal tissues and milk (Chapter ten). The two-level barrier against the potential transfer of animal pathogens from sludge, obtained by matching the level of sludge treatment with appropriate soil application practices, and the minimum no-grazing period, ensure the potential risk of infection from sludge-borne pathogens is minimal. The nutrients and organic matter contained in sludge increase animal production through increased pasture productivity.

Groundwater Quality

Contamination of groundwater by leaching of nitrate from sludge-treated soil, within the catchment of direct public supply abstractions on exposed aquifers, is probably the most important impact arising from the agricultural utilization of sludge in the context of current environmental legislation (CEC, 1991b). The principal problem is because sludge application to agricultural land occurs largely during the autumn period when crop uptake of the supplied N is low (Chapter nine). Nitrate consequently accumulates in soil through continued nitrification of applied and mineralized ammonium-N, and accumulated nitrate-N is susceptible to leaching loss by winter rainfall. Arable soils present the largest risk and autumn applications of sludge to land that is not cropped until the following spring should be avoided altogether because much of the applied available-N may be leached to groundwater. Furthermore, winter crops drilled according to current agronomic practices may not be very effective at absorbing and retaining the applied N. In contrast, sludge application to grassland is much less prone to leaching because grass is more effective than arable crops at absorbing mineral N in the autumn.

Management practices, such as early applications of sludge in late summer/early autumn followed by immediate sowing of winter crops may limit nitrate leaching losses from sludge-treated soils. Nitrification inhibitors may also have some potential in restricting the extent of nitrate accumulation in soil from autumn applications of sludge. However, spring applications of sludge to growing crops provide the greatest opportunity for matching N supply with crop requirements, thereby minimizing the level of residual nitrate in soil during the autumn and winter period. If this approach were extensively adopted, the impact of sludge utilization in agriculture on groundwater quality would be classified as low in Table 12.1.

Furthermore, the risk of impacting groundwater nitrate concentrations could be reduced by tactically matching sludge and fertilizer N applications more closely to crop requirements than occurs at present. Suggested rates of sludge application (WRc, 1985) to meet crop needs for nutrients can only provide a general guide to the N fertilizer replacement value of the different types of sewage sludge that are available for land application. However, it is difficult to provide accurate and site specific recommendations for the additional mineral N requirement because of the varied available N contents of sewage sludges. Field monitoring of sludges for ammonium-N content could provide a quantitative approach to improving advice for farmers on fertilizer applications on a site and sludge specific basis. However, the uniformity of sludge products as N fertilizers is likely to increase in future with improved treatment practice and in particular with increasing use of secondary digestion.

Relative to the vast quantities of N applied to farmland from fertilizers and livestock wastes, and the associated large risk to groundwater contamination with nitrate, inputs of N from sewage sludge are relatively insignificant (Chapter nine). In addition, sludge applied to agricultural land is predominantly treated by mesophilic anaerobic digestion to stabilize the organic fraction as well as to reduce the odour and pathogen content. Uniformly stabilized organic matter applied to soil in treated sludges may resist further decomposition and generally contributes little additional nitrate to the residual soil pool released through the mineralization of native soil organic N. By contrast, animal wastes are spread on farmland in untreated and potentially highly variable forms so that the mineralization of the unstabilized organic fraction can represent a major and generally unpredictable source of nitrate potentially at risk of leaching from soil. However, regulations limiting the rate of N application to soil in organic manures do not consider differences in N availability between stabilized and untreated sludges and slurries.

The N restrictions placed on certain treated sludges, and particularly on dewatered digested cakes and composted sludge products, are such that the fertilizer value of the sludge is substantially reduced, making the material potentially less attractive to farmers as a nutrient source. A principal agronomic benefit gained from applying dewatered sludges and sludge compost materials is an improvement in soil physical properties. The deteriorating physical condition and fertility of certain soils due to the destruction of soil organic matter is a recognized problem of modern intensive crop production methods, which cannot be rectified by increasing the use of inorganic fertilizers. Ironically, it is the soils which are potentially most vulnerable to nitrate leaching which are also highly susceptible to serious long-term structural decline. However, large inputs of organic manures (20-30 t dry solids) are necessary to achieve a measurable improvement in soil structural conditions, which will exceed the recommended maximum rate of N application (MAFF, 1991a). The organic matter in composted sewage sludges, in particular, is highly stabilized and large rates of application of this material are unlikely to cause much immediate or additional risk of nitrate leaching. Nevertheless, improving soil quality by adding large amounts of organic matter may raise the long-term leakage of residual nitrate from agricultural land.

In principle, constraints on the application of sludge to agricultural land to protect drinking water supplies from contamination with nitrate are justified within the catchments of direct or potential public water supply abstraction, or an exposed aquifer. Outside of these areas, restrictions on sludge N applications to levels which are less than crop requirements for N are unnecessarily restrictive of sludge application practice.

Impacts of sludge contaminants (inorganic, organic, biological), P and organic matter on groundwater quality are very unlikely. All available information indicates these constituents of sludge are retained in the surface layers of the soil and exhibit only very limited downward movement through the soil profile. The suggestion that P may be leached from sludge-treated soil, due to soil saturation with phosphate, has not been proven experimentally.

Surface Water Quality

Surface-spread sludge is potentially susceptible to displacement by rainfall run-off, although appropriate management practices (MAFF, 1991a) minimize the risk of run-off problems from the contamination of receiving surface waters by sludge contents (Chapters seven and nine). In particular, the incorporation or injection of sludge essentially negates the short-term problems of direct surface run-off of the applied sludge. However, potential impacts of PTEs and trace organics are probably small since their bioavailability is low due to binding with sediment and organic material. On the other hand, P and to a lesser extent

N in run-off from sludge-treated soils may promote eutrophication processes in water courses. In the long term, sludge application increases the P status and metal content of soil which may enter surface water courses due to soil erosion and run-off. The proportion of P which contributes to eutrophication problems from agricultural run-off/erosion, and which can be attributed to sewage sludge applications, is likely to be insignificant in most areas and in relation to farm waste disposal activities and inorganic phosphate fertilizer applications. Pathogens are unlikely to present much of a risk because surface-applied sludges must be treated, providing a barrier to infection, and raw sludges are either injected into the soil or incorporated immediately after application. Sludge organic matter improves soil structural stability and permeability thus reducing the risk of surface water pollution generally from agricultural soils due to run-off.

Air Quality

Ammonia volatilization losses from surface-spread organic manures can be large and represent an important impact on air quality by contributing to the environmental problems of acid rain (Chapter nine). Furthermore, aerial deposition of ammonia into oligotrophic environments (aquatic and heathland) disturbs the nutritional balance of these sensitive ecosystems causing fundamental changes in species composition and diversity. In quantitative terms, however, ammonia emissions from sludge-treated soils represent a relatively small proportion of the total ammonia losses from livestock wastes.

Injection or immediate incorporation of sludge prevents ammonia volatilization. Although gaseous losses of nitrous oxide, which is an important greenhouse gas, can increase when sludge is injected, the environmental impact of this is probably minimal compared with the benefits of sludge injection. Nitrification inhibitors are effective in reducing nitrous oxide emissions. Impacts of volatile PTEs, namely Hg, and trace organic contaminants from sludge-treated soil probably have little impact on air quality compared with background levels present in the aerial environment (Chapters four and ten). Low trajectory application of sludge prevents the potential aerial dispersal of pathogens in aerosols during sludge spreading.

Soil Fertility

There is concern that the long-term sustainability of agricultural production may be impaired by the application of heavy metals to soil in sewage sludge due to effects on soil fertility (Chapter eight). However, the principal decomposition and nutrient cycling processes are unaffected by concentrations of heavy metals far exceeding the currently permitted maximum levels in sludge-treated soils

(SI, 1989a). In addition, the soil macro-fauna appears remarkably tolerant to large concentrations of PTEs present in soil and is unlikely to be affected by heavy metals at current maximum soil limits. Only Zn appears to potentially place at risk certain sensitive groups of soil micro-organisms, and *Rhizobium leguminosarum* bv. *trifolii* in particular, at the statutory soil limit values. More research is required to determine the extent of the effects of Zn on *Rhizobium* in sludge-treated soils, and to establish their agronomic significance.

The expanding array of biological and biochemical assays with the potential for assessing changes in soil microbial community structure and diversity with increasingly sensitive levels of detection apparently have given contradictory information on the possible effects of soil contamination with heavy metals within the range of the soil limit values for sludge-amended agricultural land (SI, 1989a). Rather than helping to understand the potential impacts of heavy metals on the soil ecosystem these contradictory data would appear to compound the present difficulty in assessing the long-term implications and significance for soil fertility of the current maximum permissible concentrations of PTEs in soil where sewage sludge is used in agriculture.

Organic contaminants are unlikely to have any significant toxic effect on soil microbial processes at the concentrations typically found in sludge-treated agricultural soils (Chapter ten). Earthworms also appear insensitive to the presence of organic contaminants in soil although the range of compounds which have been tested is limited and there are few data on toxicity to other soil invertebrates.

Sludge-borne organic contaminants represent a relatively minor risk to soil fertility compared with agricultural pesticides. Agricultural pesticides potentially have a much greater detrimental impact on soil fertility due to large and repeated rates of application, extensive and widespread use in the environment and higher bioavailability relative to sludge-organics. The biocidal effects of pesticides on earthworms and on symbiotic N_2-fixing *Rhizobium leguminosarum* bv *trifolii* are of particular concern (Mårtensson, 1992; Springett and Gray, 1992). Consequently the risk to soil fertility of organic contaminants in sewage sludge spread on farmland has been designated as low in Table 12.1.

Sludge pathogens have no detrimental impact on soil fertility (Chapter eleven). Indeed, pathogenic organisms in sludge are susceptible to antagonistic effects of increased soil microbial activity resulting from the application of available substrate organic matter in sewage sludge. The beneficial effects of sludge organic matter and nutrients in improving soil fertility and productivity have been extensively reported and are universally recognized.

Natural Ecosystems

The principal risk to natural ecosystems is from surface run-off of sludge nutrients resulting in eutrophication of surface waters (Chapter nine). Aerial deposition of volatilized ammonia also has a detrimental effect on oligotrophic environments (Chapter nine). However, incorporation or injection of sludge largely prevents environmental problems caused by run-off and volatilization. Pathogens also present only a minimal risk. This is because certain important pathogens such as *Taenia saginata* and *Ascaris* are host specific, whereas other organisms including *Salmonella,* would probably only contribute to background levels already present in the population.

Potential environmental consequences of the bioaccumulation and transfer of Cd (the principal zootoxic element of concern in sludge) and trace organic contaminants from sludge-treated soil through natural food chains are poorly understood. Earthworms can accumulate large concentrations of Cd and organic compounds and form an important link in the transfer of toxic compounds to higher trophic levels, being a key food source for many animal groups in natural habitats. Transfer and bioaccumulation of contaminants is probably associated with herbivore-carnivore links in the food chain, and long chains are more sensitive than are short chains. Furthermore, accumulation of Cd can vary considerably since absorption is also influenced by the presence of other dietary components. These factors may explain the apparent contradictions in Cd accumulation measured in natural ecosystems. Few data have been published of the transfer of sludge organics through natural food chains. Consequently the risk to natural ecosystems of Cd and trace organic contaminants in sludge-treated agricultural soils cannot be confidently assessed, but available information suggests the risks are probably small. Sludge organic matter indirectly has a beneficial effect on natural ecosystems by reducing surface run-off pollution of water courses through improved soil structural stability.

Appendix:

Legislation and Codes of Practice

Objective

Careful control of the use of sewage sludge in agriculture is essential so that the plant nutrients and organic matter contained in sludge can be recycled to soil without detriment to the environment and in particular to human, animal and plant life from PTEs and pathogens. Statutory controls have been developed in Europe and the US and their main features are summarized below (also see Hall and Dalimier, 1994).

The European Community

The Council of the European Communities (CEC) have established and proposed a number of Directives relating to water, wastes and the environment which impact the production and disposal of sewage sludge (see Fig. A1). Directives must be transformed into national legislation by Member States usually within three years of them being adopted.

Directive 86/278/EEC on the Protection of the Environment, and in Particular of the Soil, when Sewage Sludge is Used in Agriculture

This Directive (CEC, 1986a) was adopted in June 1986 and required all Member States to implement its provisions within three years. It sets standards for sludge use in agriculture only, and does not apply to other beneficial uses of sludge on land.

The Directive prohibits the use of untreated sludge on agricultural land unless it is injected or incorporated into the soil. Treated sludge is defined as having undergone 'biological, chemical or heat treatment, long-term storage or any other appropriate process so as significantly to reduce its fermentability and the health hazards resulting from its use.'

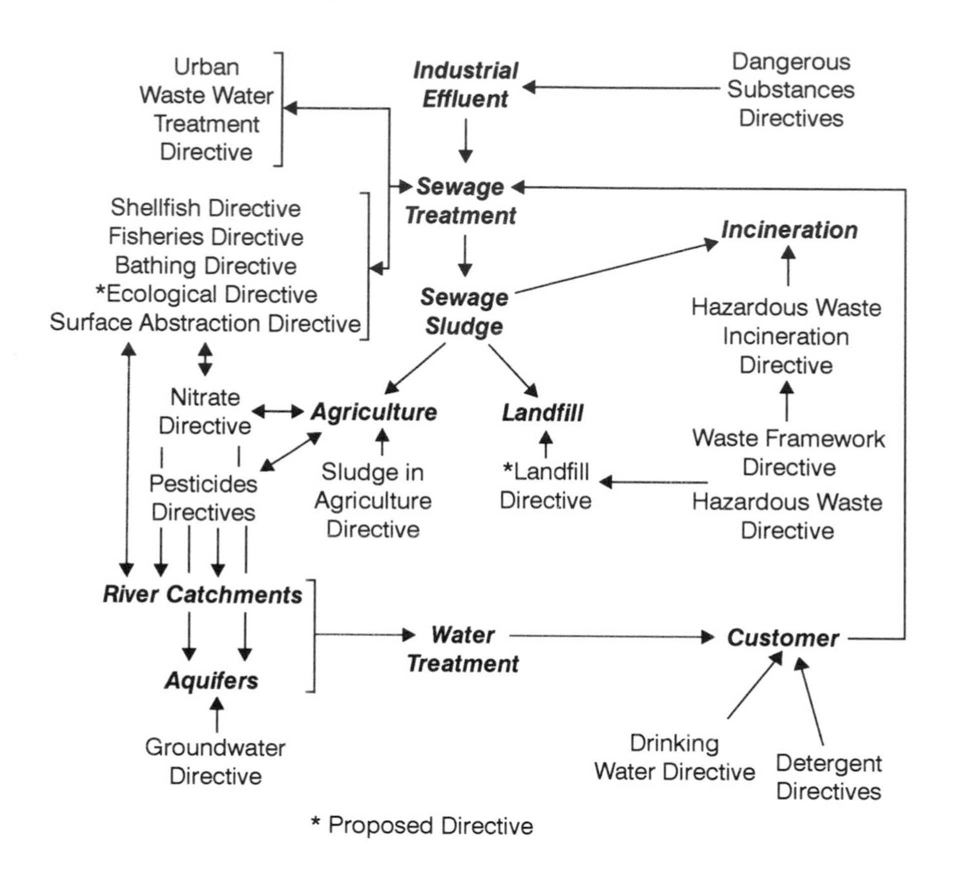

Fig. A.1. Water Directives in relation to sewage sludge.

To provide protection against potential health risks from residual pathogens, sludge must not be applied to soil in which fruit and vegetable crops are growing, or earlier than ten months before some crops are to be harvested. Grazing animals must not be allowed access to grassland or forage land earlier than three weeks after the application of sludge. The Directive also requires that sludge should be used in such a way that account is taken of the nutrient requirements of plants and that the quality of the soil and of the surface and groundwater is not impaired. The requirements on protecting water sources are in accordance with two other CEC Directives on the quality of surface water intended for the abstraction of drinking water (CEC, 1975) and on the protection of groundwater against pollution caused by certain dangerous substances (CEC, 1980a).

A major requirement of the Directive concerns the control of heavy metal accumulation in soil, and limit values for soil, sludge and rates of addition are given in the Directive Annexes 1A, 1B and 1C, respectively (see Table A1). Soil and sludge limits are given as ranges and Member States must set values below the maximum. All Member States must observe soil limit values and soil concentrations must be monitored by taking soil samples to 25 cm depth (or to the depth of the soil if less than this, but must be more than 10 cm). Member States may also choose the method of controlling the rate of addition of heavy metals to the soil (Annex 1B or 1C). The Annex 1B sets sludge quality limits and Member States must also set a sludge application rate limit in dry solids so that the rate of heavy metal addition to soil does not exceed Annex 1C. Most Member States have adopted this approach. The alternative Annex 1C sets limit values for amounts of heavy metals (as kg ha^{-1}) which may be added annually to agricultural land, based on a 10-year average. This approach has been adopted by the UK.

Table A1. Limit values for heavy metals stipulated in Directive 86/278/EEC.

Parameter	Concentrations[(1)] in soil (mg kg^{-1} dry soil)	Concentrations in sludge (mg kg^{-1} dry solids)	Annual average that may be put on soil (kg ha^{-1} $year^{-1}$)
Cadmium	1 - 3	20 - 40	0.15
Copper	50 - 140	1000 - 1750	12
Nickel	30 - 75	300 - 400	3
Lead	50 - 300	750 - 1200	15
Zinc	150 - 300	2500 - 4000	30
Mercury	1 - 1.5	16 - 25	0.1
Chromium[(2)]	-	-	-

[(1)] For soil with a pH of 6-7. Limit values may be reduced on soil with a pH <6 to account for increased mobility and availability to crops. Higher limits are permitted on soil with a pH >7 and >5% calcium carbonate, but these must not exceed the adopted pH 6-7 values by more than 50%

[(2)] No statutory limits have been set for Cr; proposed amendment to the Directive (CEC, 1990b) has been removed from the legislative process (CEC, 1993)

The Directive sets out requirements for the keeping of detailed records of the quantities of sludge produced; the quantities used in agriculture; the composition and properties of the sludge; the type of treatment given; and the places where the sludge is used. This information has to be reported to the European Commission every four years (CEC, 1994a).

Limit values for concentrations of heavy metals in sewage sludge intended for agricultural use, and in sludge-treated soils, are shown in Table A1.

Directive 91/676/EEC Concerning the Protection of Waters Against Pollution Caused by Nitrates from Agricultural Sources

The aim of this Directive (CEC, 1991b) is to prevent the pollution of water by nitrate caused by the application and storage of inorganic fertilizers and manures, and in particular to protect drinking water sources and prevent eutrophication of fresh and marine waters. Its provisions will affect the way in which sludge can be used on agricultural land.

The Directive requires Member States to designate 'vulnerable zones' which are to include:

- surface freshwaters where nitrate concentrations do or may exceed limits set by the Directive on Surface Water Intended for the Abstraction of Drinking Water (CEC, 1975);
- groundwaters containing or likely to contain more than 50 mg l^{-1} of nitrate;
- freshwater bodies, estuaries, coastal waters and marine waters which are or may become eutrophic.

Action Plans must be implemented for all vulnerable zones. The Plans must specify, at the least:

- periods when the application of certain fertilizers is prohibited;
- limits on the quantities of fertilizers which can be applied;
- a code of good agricultural practice.

The Directive therefore introduces more constraints on the use of sewage sludge than are enforced under current sludge controls. The seasonal windows available for application of sludge will be fewer and limits on total annual N application are likely to be more restrictive than those on PTEs.

Directive 91/271/EEC Concerning Urban Waste Water Treatment

The Urban Waste Water Treatment Directive (CEC, 1991a) was introduced to reduce the pollution of freshwater and estuarial and coastal waters by domestic sewage, industrial waste water and rain water run-off, collectively described as urban waste water.

The Directive sets out the following requirements with respect to the generation and disposal of sewage sludge:

- that the disposal of sewage sludge from waste water treatment is to be subject to regulation by the end of 1998;
- a requirement that the recycling of sludge be encouraged;
- that the disposal dumping of sewage sludge in surface waters be prohibited by the end of 1998;

- to monitor the disposal of sludge to ensure that the environment is protected.

Depending on the size of the sewered catchment (the term 'agglomeration' is used in the Directive) and the sensitivity of receiving water for the effluent, minimum treatment standards of sewage are laid down to be achieved in stages by the end of 2005. This will result in a substantial increase in sludge production (estimated at about 50% for the EU; Hall and Dalimier, 1994) which will require disposal to land-based outlets.

The Directive also places various reporting requirements on Member States, to describe current sludge disposal to surface waters, the programme of implementation of the Directive, and a general biennial report on sludge disposal.

Directive 91/689/EEC on Hazardous Waste

This Directive (CEC, 1991c) tightens the control of hazardous waste in the EU first introduced in the Directive on Toxic and Dangerous Waste (CEC, 1978).

Its impact on sewage sludge disposal would be *via* its definition of hazardous waste, although it is now clear that sewage sludges will generally not be regarded as hazardous. For this purpose three annexes are included in the Directive:

- Annex I lists categories or generic types of waste;
- Annex II lists constituents of waste which render them hazardous;
- Annex III lists properties of waste that render them hazardous.

Sewage sludges, untreated or unfit for use in agriculture are listed in Annex 1B (wastes which contain any of the constituents listed in Annex II and having any of the properties listed in Annex III).

Annex II contains a wide variety of inorganic and organic contaminants, many of which could be expected to be found in sewage sludge.

Annex III lists a wide ranging number of properties which may render waste hazardous. Included amongst these are:

- 'ecotoxic' - substances and preparations which present or may present immediate or delayed risks for one or more sectors of the environment;
- substances and preparations capable by any means, after disposal, of yielding another substance, e.g. leachate, which possesses any of the other characteristics in the annex;
- 'infectious' - substances containing viable micro-organisms or their toxins which are known or reliably believed to cause disease in man or other living organisms;
- other headings that may impact the disposal of sewage sludge include - irritant, harmful, teratogenic, mutagenic and carcinogenic.

The European Waste Catalogue was adopted in 1993 (CEC, 1994b) and the list of hazardous wastes, which does not include sewage sludge, was established in 1994 (CEC, 1994c). This list defines the scope of both Directive 91/689/EEC and the Directive on the incineration of hazardous waste (CEC, 1994d).

The United Kingdom

The Sludge (Use in Agriculture) Regulations 1989

These regulations (SI, 1989a) implement Directive 86/278/EEC. Some minor ambiguities were clarified and the regulations were printed as amended by the *Sludge (Use in Agriculture) (Amendment) Regulations 1990*, SI 1990/880. The control system adopted sets permissible limits for soil concentrations and rates of annual addition of PTEs. The allowable limits for Zn, Cu and Ni in soils vary with the pH of the soil. There are no restrictions on the concentrations of PTEs in sludge. The relevant limit values are summarized in Table A2.

Code of Practice for Agricultural Use of Sewage Sludge

This Code (DoE, 1989a) complements the statutory regulations, giving detailed recommendations on best practice. In addition to the statutory limits, it sets provisional limits for soil concentration and annual rate of addition for Cr (400 mg Cr kg^{-1} and 15 kg Cr ha^{-1} y^{-1}, respectively) as well as limits for these parameters for additional PTEs (Mo, Se, As and F). All of these limit values are given in Table A2. For the statutory PTEs, the required sampling depth is 25 cm (as per the Directive) but for advisory or monitoring purposes, the Code recommends sampling to 15 cm for the PTEs listed in Table A2. When sludge is applied to grassland, particularly permanent pasture, advisory limits are based on shallow (7.5 cm) soil samples (see Table A3), although the statutory limits to 25 cm still apply.

The Code describes effective sludge treatment processes which meet with the Directive requirement of ‘treatment’ (see Table 11.2). The Code also gives detailed lists of acceptable uses of treated and untreated sludges for different categories of crops (see Tables 11.5 and 11.6).

The UK Code of Practice will shortly be revised (DoE, 1995) adopting a lower advisory limit for Zn as a ‘precautionary measure’ in accordance with the recommendations of the Independent Scientific Committee review of soil fertility aspects of PTEs (MAFF/DoE, 1993a) (see Chapter eight). The new advisory limits for Zn will be 200 mg kg^{-1} for all soils of pH 5.0-7.0 and 300 mg kg^{-1} for soils of pH >7.0. The depth of soil sampling for advisory purposes will remain unchanged at 15 cm (L. McIntosh, Department of the Environment, personal communication). The advisory limits for Zn in soil under grass when samples are taken to a depth of 7.5 cm will be the same as the

Table A2. Limit values for potentially toxic elements in soil after application of sewage sludge and maximum annual rates of addition in the United Kingdom.

Parameter	Soil limit value[1] (mg kg^{-1} ds)				Maximum permissible average annual rate of PTE addition over a 10 year period (kg $ha^{-1}y^{-1}$)
	pH 5.0<5.5	pH 5.5<6.0	pH 6.0-7.0	pH >7.0	
Zinc[2]	200	250	300	450	15
Copper[2]	80	100	135	200	7.5
Nickel[2]	50	60	75	110	3
For pH 5.0 and above					
Cadmium[2]		3			0.15
Lead[2]		300			15
Mercury[2]		1			0.1
Chromium		400 (provisional)			15 (provisional)
Molybdenum		4			0.2
Selenium		3			0.15
Arsenic		50			0.7
Fluoride		500			20

[1] For soil samples taken to a depth of 25 cm (statutory) and 15 cm (advisory)
[2] Statutory limit values. Limit values for other PTEs are advisory
Sources: SI (1989a); DoE (1989a)

numerical values recommended for the 0-15 cm depth. The advisory limit for Cd of 5 mg kg^{-1} for grass managed in rotation with arable crops or grown only for conservation will be reduced to 3 mg kg^{-1}. There are currently no plans to change the statutory regulations for PTEs in sludge-treated agricultural land (Table A2).

Code of Good Agricultural Practice for the Protection of Water

The advice in this Code (MAFF, 1991a) is aimed primarily at ensuring the prevention of pollution of water by normal agricultural operations, but some of its provisions apply to the use of sewage sludge in agriculture. It is recommended that the amount of N applied in sludge should not exceed 250 kg ha^{-1} y^{-1}, and there is detailed advice on dealing with the problems associated with different soil types, ground slopes and weather conditions. Furthermore, it is recommended that applications of liquid sewage sludges to arable land in the autumn and early winter should be avoided whenever practicable.

Table A3. Limit values for potentially toxic elements in soil under grass after application of sewage sludge when samples taken to a depth of 7.5 cm (mg kg^{-1} ds).

Parameter	pH 5.0<5.5	pH 5.5<6.0	pH 6.0-7.0	pH >7.0
Zinc[(1)]	330	420	500	750
Copper[(1)]	130	170	225	330
Nickel[(1)]	80	100	125	180
For pH 5.0 and above				
Cadmium[(2)]	3/5			
Lead	300			
Mercury	1.5			
Chromium	600	(provisional)		
Molybdenum	4			
Selenium	5			
Arsenic	50			
Fluoride	500			

(1) The permitted concentrations of these elements will be subject to review when current research into their effects on the quality of grassland is completed. Until then, in cases where there is doubt about the practicality of ploughing or otherwise cultivating grassland, no sludge applications which would cause these concentrations to exceed the permitted levels specified in Table A2 should be made except in accordance with specialist agricultural advice.
(2) The permitted concentration of Cd will be subject to review when current research into its effect on grazing animals is completed. Until then, the concentration of this element may be raised to the permitted upper limit of 5 mg kg^{-1} as a result of sludge applications only under grass which is managed in rotation with arable crops and grown only for conservation. In all cases where grazing is permitted no sludge applications which would cause the concentration of Cd to exceed the lower limit of 3 mg kg^{-1} shall be made.
Source: DoE (1989a)

MAFF set up the Pilot Nitrate Scheme in 1989/90 to investigate means of reducing nitrate lost in Nitrate Sensitive Areas. This voluntary scheme tested under real farm conditions control measures by placing restrictions on the timing and rate of application of fertilizers and organic manures, including sewage sludge.

The Nitrate Directive (91/676/EEC) establishes a common framework throughout the EU for reducing nitrate pollution from agriculture. In the UK, Nitrate Vulnerable Zones (NVZs) have been designated where the nitrate concentrations in water exceed, or are expected to exceed, 50 mg l^{-1}. Compulsory action programmes are to be established and implemented by the end of 1999 at the latest. These will:

- limit nitrogen application to 210 kg ha^{-1} y^{-1} during the first four years;
- limit nitrogen application to 170 kg ha^{-1} y^{-1} subsequently unless a higher limit can be justified;
- set closed periods for nitrogen applications:
  - inorganic fertilizer, 1 September-1 February;
  - slurry, poultry manure and liquid digested sludge:
    grassland, 1 September-1 November
    arable, 1 August-1 November
- proscribe applications under adverse soil conditions, to steeply sloping land or within 10 m of surface water;
- require farmers to keep records of the use of nitrogen in inorganic fertilizer and organic manures.

Other European Countries

Control systems developed elsewhere in Europe are predominantly based on quality standards for sludge coupled with limits on the permissible annual rate of application of sludge. The limit values in sludge set by EU and non-EU countries are summarized in Table A4.

The limit values set by a number of European countries for permissible concentrations of PTEs in soils receiving applications of sludge are compared with the CEC limits in Table A5.

Legislation relating to sewage sludge in each of the Member States has been summarized by Hall and Dalimier (1994) and Morsing (1994).

The United States

The United States Environmental Protection Agency (US EPA) recently published revised sludge regulations under Clean Water Act, Section 503 (US EPA, 1993). These new regulations have been developed after a lengthy consultation involving publication of the Proposed Rule in 1989 (US EPA, 1989a). This was considered by a Peer Review which reported in 1989 (Page and Logan, 1989). The Peer Review Committee proposed the concept of NOAEL (No Observed Adverse Effect Level) which developed sludge quality limits for 10 PTEs based on a rigorous risk analysis (US EPA, 1989b) of 14 pollutant pathways so that 'clean' sludges can effectively be applied to land indefinitely.

The risk-based models developed for the Part 503 Regulation were designed to limit potential exposure of a highly exposed individual (HEI) to the pollutants of concern (US EPA, 1992a). The HEI is defined as an individual who remains for an extended period of time at or adjacent to the site where the maximum exposure occurs. Depending on the pathway of exposure, the HEI could be a human, plant, animal, or environmental endpoint, such as surface or groundwater. The 1989 Proposed Part 503 Rule (US EPA, 1989a) considered the exposed individual to be a 'most exposed individual' (MEI). This was a hypothetical (not actual) individual who would be expected to experience the greatest risk and, therefore, required the greatest protection (US EPA, 1989a). However, the consultation review of the Proposed Rule considered the MEI an unrealistic 'worst-case' of environmental exposure. Therefore, the US EPA changed the exposed individual in the risk assessment from the MEI to the HEI to protect individuals and populations that are 'highly exposed to reasonably anticipated adverse conditions'. The 14 exposure assessment pathways of concern, evaluated in the Final Part 503 Rule, are listed in Table A6.

The Final 503 Regulations (US EPA, 1993) have set 'clean' sludge quality limits (Pollutant Concentration), which correspond to the NOAEL (these values have also been described as 'Alternative Pollutant Limits' (APL) - R.L. Chaney, USDA-Agricultural Research Service, personal communication), and upper Ceiling Concentrations for 10 inorganic pollutants of principal concern in sludge (Table A7). Annual Pollutant Loading Rates have also been established for sludges which do not meet the high quality sludge standards.

The use of risk assessment in the development of the standards for recycling sewage sludge has been summarized by Ryan (1993). For the land application risk assessment the limits were presented generally in units of mass loading (kg ha^{-1}). Pollutant concentrations in sludge (mg kg^{-1}) were calculated from the cumulative pollutant loading rate (kg ha^{-1}). For this conversion it was assumed that the sludge was applied at a rate of 10 t ha^{-1} y^{-1} for 100 years (1000 t ha^{-1} total sludge application). It is important to recognize that the same time frame will be required before the exposure will be as high as the estimated level. The risk assessment derived pollutant sludge concentrations were then compared with the 99th percentile concentrations from the National Sewage Sludge Survey (NSSS) and the most restrictive values were taken as the 'clean' sludge pollutant concentration limits.

Ceiling concentration limits are less stringent than the pollutant concentration limit values. The ceiling limits were established by using the higher of the risk assessment derived pollutant sludge concentration or the NSSS 99th percentile concentration. These ceiling concentrations were established to prevent land application of sewage sludges containing high concentrations of pollutants.

Current US regulations on PTEs for the use of sewage sludge on agricultural land are listed in Table A7. Earlier proposed application limits for PTEs and the NOAEL sludge quality limits are also presented.

The 503 Rule has established two principal classifications for pathogens in sludge. Class A requires the quantification of, and has placed limits on, the density of pathogenic agents present in sludge. The sludge must also have met certain treatment and physical quality criteria. Class B has less stringent limits on the density of pathogenic agents compared with Class A, and requires the sludge to be treated by a process to significantly reduce pathogens (PSRP). Pathogen treatment processes are listed in Appendix B to Part 503 stipulating appropriate minimum temperature or pH conditions and time requirements for significant pathogen reductions to occur. Site restrictions on the application of sludge meeting the Class B requirements specifying the timing of harvesting of treated crops, do not apply to Class A sludges. One of a number of specified vector attraction reduction requirements must also be met for both classes of sludge. Vector attraction is described as the characteristic of sewage sludge that attracts rodents, flies or other organisms capable of transporting infectious agents. Vector attraction reduction requirements include, for example: injection or incorporation of sludge into the soil; minimum limits on volatile solids loss; maximum limits on the proportion of unstabilized solids in sludge and biochemical oxygen demand; and minimum temperature, pH and time requirements for treatment processes. The US requirements on pathogen reduction and the provision of additional barriers against the transfer of infection are very similar in essence to those established in the UK (DoE, 1989a).

The Final Part 503 Rule specifies requirements for the frequency of monitoring the PTEs and pathogens in sewage sludge. The regulation also stipulates the keeping of records on sludge composition, and of descriptions of how the pathogen, site restriction, vector attraction reduction requirements and management practices were met. If sludge is applied in accordance with the cumulative PTE loading (Table A7) additional records are required of when and where the sludge was applied, the area of treated land, cumulative PTE additions and the quantity of sludge spread.

Table A4. Summary of limit values for potentially toxic elements and organic contaminants in sewage sludge in EC member states, Scandinavia, Switzerland and the USA.

Parameter[1]	EC Directive 86/278/EEC	Belgium		Denmark	France		Germany		Ireland
		Flanders	Wallonia		Reference	Limit	Soil pH 5-6	Soil pH >6	
Cadmium	20 - 40	12	10	0.8	20	40	5	10	20
Copper	1000 - 1750	750	600	1000	1000	2000	800	800	1000
Nickel	300 - 400	100	100	30	200	400	200	200	300
Lead	750 - 1200	600	500	120	800	1600	900	900	750
Zinc	2500 - 4000	2500	2000	4000	3000	6000	2000	2500	2500
Mercury	16 - 25	10	10	0.8	10	20	8	8	16
Chromium	-	500	500	100	1000	2000	900	900	-
Selenium	-	-	-	-	100	200	-	-	-
Arsenic	-	-	-	-	-	-	-	-	-
Fluoride	-	-	-	-	-	-	-	-	-
Molybdenum	-	-	-	-	-	-	-	-	-
Cobalt	-	-	-	-	-	-	-	-	-
Dioxin	-	-	-		-	-	100	100	-
PCBs	-	-	-	-	-	-	0.2	0.2	-
AOX	-	-	-	-	-	-	500	500	-

Table A4 continued

Parameter[1]	Italy	Luxembourg		The Netherlands	Spain		UK
		Recommended	Limit		Soil pH <7	Soil pH >7	Grassland
Cadmium	20	20	40	1.25	20	40	-
Copper	1000	1000	1750	75	1000	1750	-
Nickel	300	300	400	30	300	400	-
Lead	750	750	1200	100	750	1200	1200
Zinc	2500	2500	4000	300	2500	4000	-
Mercury	10	16	25	0.75	16	25	-
Chromium	-	1000	1750	75	1000	1500	-
Selenium	-	-	-	-	-	-	-
Arsenic	-	-	-	15	-	-	-
Fluoride	-	-	-	-	-	-	1000
Molybdenum	-	-	-	-	-	-	-
Cobalt	-	-	-	-	-	-	-
Dioxin	-	-	-	-	-	-	-
PCBs	-	-	-	-	-	-	-
AOX	-	-	-	-	-	-	-

(1) Units - potentially toxic elements, PCBs individual congener, and AOX (mg kg^{-1} ds); dioxins (ng TEQ kg^{-1} ds)

Table A4 continued

Parameter[1]	Sweden	Norway			Finland		Switzerland	USA	
		Current	Proposed Agriculture	Proposed Non-Agric.	Sludge	Amended		Ceiling	Clean
Cadmium	2	10	4	10	1.5	3	1.5	85	39
Copper	600	1500	1000	1500	600	600	600	4300	1500
Nickel	50	100	80	100	100	100	80	420	420
Lead	100	300	100	300	100	150	500	840	300
Zinc	800	3000	700	3000	1500	1500	2000	7500	2800
Mercury	2.5	7	5	7	1	2	1.5	57	17
Chromium	100	200	125	200	300	300	500	3000	1200
Selenium	-	-	-	-	-	-	-	100	36
Arsenic	-	-	-	-	-	-	-	75	41
Fluoride	-	-	-	-	-	-	-	-	-
Molybdenum	-	-	-	-	-	-	20	75	18
Cobalt	-	-	-	-	-	-	60	-	-
Dioxin	-	-	-	-	-	-	-	-	-
PCBs	-	-	-	-	-	-	-	-	-
AOX	-	-	-	-	-	-	500	-	-

(1) Units - potentially toxic elements, PCBs individual congener, and AOX (mg kg^{-1} ds); dioxins (ng TEQ kg^{-1} ds)

Table A5. Summary of limit values for potentially toxic elements in soils in some European countries (mg kg^{-1} dry soil).

Parameter	EC Directive 86/278/EEC	Belgium			Denmark	France	Germany		Ireland	Italy	Netherlands (standard soil)
		Flanders Sandy soil	Flanders Clay/silt	Wallonia			Soil pH 5-6	Soil pH >6			
Cadmium	1 - 3	1	3	1	0.5	2	1	1.5	1	1.5	0.8
Copper	50 - 140	50	140	50	40	100	60	60	50	100	36
Nickel	30 - 75	30	75	50	15	50	50	50	30	75	35
Lead	50 - 300	50	300	100	40	100	100	100	50	100	85
Zinc	150 - 300	150	300	200	100	300	150	200	150	300	140
Mercury	1 - 1.5	1	1.5	1	0.5	1	1	1	1	1	0.3
Chromium	-	100	150	100	30	150	100	100	-	-	100
Selenium	-	-	-	-	-	10	-	-	-	-	-
Arsenic	-	-	-	-	-	-	-	-	-	-	29
Fluoride	-	-	-	-	-	-	-	-	-	-	-
Molybdenum	-	-	-	-	-	-	-	-	-	-	-
Cobalt	-	-	-	-	-	-	-	-	-	-	-
Thallium	-	-	-	-	-	-	-	-	-	-	-

Table A5 continued

Parameter	Luxembourg		Spain		United Kingdom				Finland	Switzerland	
	Recommended	Limit	Soil pH <7	Soil pH >7	Soil pH 5.0<5.5	5.5<6.0	6-7	>7		Total	Soluble
Cadmium	1	3	1	3	3	3	3	3	0.5	0.8	0.03
Copper	50	140	50	210	80	100	135	200	100	50	0.7
Nickel	30	75	30	112	50	60	75	100	60	50	0.2
Lead	50	300	50	300	300	300	300	300	60	50	1.0
Zinc	150	300	150	450	200	250	300	450	150	200	0.5
Mercury	1	1.5	1	1.5	1	1	1	1	0.2	0.8	-
Chromium	100	200	100	150	400	400	400	400	200	75	-
Selenium	-	-	-	-	3	3	3	3	-	-	-
Arsenic	-	-	-	-	50	50	50	50	-	-	-
Fluoride	-	-	-	-	500	500	500	500	-	400	25
Molybdenum	-	-	-	-	4	4	4	4	-	5	-
Cobalt	-	-	-	-	-	-	-	-	-	25	-
Thallium	-	-	-	-	-	-	-	-	-	1	-

Table A6. Environmental pathways of concern identified for application of sewage sludge to agricultural land.

Pathway	Description of HEI
1. Sewage sludge→Soil→Plant→Human	Human ingesting plants grown in sewage sludge-amended soil
2. Sewage sludge→Soil→Plant→Human	Residential home gardener
3. Sewage sludge→Human	Children ingesting sewage sludge
4. Sewage sludge→Soil→Plant→Animal→Human	Farm households producing a major portion of the animal products they consume. It is assumed that the animals eat plants grown in soil amended with sewage sludge
5. Sewage sludge→Soil→Animal→Human	Farm households consuming livestock that ingest sewage sludge while grazing
6. Sewage sludge→Soil→Plant→Animal	Livestock ingesting crops grown on sewage sludge-amended soil
7. Sewage sludge→Soil→Animal	Grazing livestock ingesting sewage sludge
8. Sewage sludge→Soil→Plant	Plants grown in sewage sludge-amended soil
9. Sewage sludge→Soil→Soil organism	Soil organisms living in sewage sludge-amended soil
10. Sewage sludge→Soil→Soil organism→Soil organism predator	Animals eating soil organisms living in sewage sludge-amended soil
11. Sewage sludge→Soil→Airborne dust→Human	Tractor operator exposed to dust while ploughing large areas of sewage sludge-amended soil
12. Sewage sludge→Soil→Surface water→Human	Water Quality Criteria for the receiving water for a person who consumes 0.04 kg day^{-1} of fish and 2 l day^{-1} of water
13. Sewage sludge→Soil→Air→Human	Human breathing volatile pollutants from sewage sludge
14. Sewage sludge→Soil→Ground water→Human	Human drinking water from wells contaminated with pollutants leaching from sewage sludge-amended soil to ground water

Source: US EPA (1992a)

Table A7. US Regulations and recommendations for the use of sewage sludge on agricultural land.

Pollutant	Proposed[1] 503 Rule, 1989	Peer Review[2] recommendation, 1989		Final Part 503 Rule, 1993[3]					
	Cumulative application (kg ha^{-1})	NOAEL sludge limits (mg kg^{-1})	Cumulative application (kg ha^{-1})	Clean sludge[4] limits (mg kg^{-1})	Ceiling sludge limits (mg kg^{-1})	Annual loading rate (kg ha^{-1} y^{-1})	Cumulative application (kg ha^{-1})	Maximum soil[5][6] concentration (mg kg^{-1})	Limiting[7] pathways
Zn	170	2700	2700	2800	7500	140	2800	1500	Phytotoxicity(8)
Cu	46	1200	1200	1500	4300	75	1500	775	Phytotoxicity(8)
Ni	78	500	500	420	420	21	420	230	Phytotoxicity(8)
Cd	18	18	>18.4	39	85	1.9	39	20 (39)	Sludge ingestion(3)
Pb	125	300	600	300	840	15	300	190 (300)	Sludge ingestion(3)
Hg	15	15	20	17	57	0.85	17	9 (17)	Sludge ingestion(3)
Cr	530	2000	>2000	1200	3000	150	3000	1540	Phytotoxicity(8)
Mo	5	35	35	18	75	0.90	18	11	Livestock feed(6)
Se	32	32	32	36	100	5.0	100	50 (36)	Sludge ingestion(3)
As	14	100	1600	41	75	2.0	41	32 (41)	Sludge ingestion(3)

(1) US EPA (1989a)
(2) Chaney (1990b)
(3) US EPA (1993)
(4) Composition of a sludge which could be applied at cumulative loadings of at least 1000 t ha^{-1} (dry solids) yet not fail the Pathway Risk Assessment which protects Highly Exposed Individuals (described as Pollutant Concentrations or Alternative Pollutant Limits)
(5) Approximate values calculated from the cumulative pollutant loading rates from Final Part 503 Rule (US EPA, 1993) assuming sludge is cultivated to 0.2 m and a soil density of 1.0; background soil concentrations are taken as median values from McGrath and Loveland (1992) or as mean values from Ure and Berrow (1982)
(6) When sludge ingestion is the limiting pathway it could be argued that the maximum soil concentration may be equivalent to the clean sludge limit value given in brackets
(7) No. of limiting pathway given in brackets - see Table A6

References

Abbasi S.A. and Soni R. (1983) Stress-induced enhancement of reproduction in earthworm *Octochaetus pattoni* exposed to chromium (VI) and mercury (II) - implications in environmental management. *International Journal of Environmental Studies* 22, 43-48.

Adams T. McM. and Adams S.N. (1983) The effect of liming and soil pH on carbon and nitrogen contained in the soil biomass. *Journal of Agricultural Science* 101, 553-558.

Adams T. McM. and Sanders J.R. (1984a) The effect of pH on the release to solution of zinc, copper and nickel from metal-loaded sewage sludge *Environmental Pollution (Series B)* 8, 85-99.

Adams T. McM. and Sanders J.R. (1984b) The effect of incubation on the compositon of soil solution displaced from four soils treated with zinc, copper or nickel-loaded sewage sludge. In: Leschber R., Davis R.D. and L'Hermite P. (eds), *Chemical Methods for Assessing Bioavailable Metals in Sludge and Soils.* Elsevier Applied Science Publishers Ltd, Barking, pp. 68-80.

Adams T. McM. and Sanders J.R. (1984c) Changes with time in the zinc, copper and nickel concentrations in solutions displaced from two sludge-treated soils. In: L'Hermite P. and Ott H. (eds), *Processing and Use of Sewage Sludge.* D. Reidel Publishing Company, Dordrecht, pp. 447-450.

Addiscott T.M. (1983) Kinetics and temperature relationships of mineralization and nitrification in Rothamsted soils with differing histories. *Journal of Soil Science* 34, 343-353.

Addiscott T.M., Whitmore A.P. and Powlson D.S. (1991) *Farming, Fertilizers and the Nitrate Problem.* CAB International, Wallingford.

Alberici T.M., Sopper W.E., Storm G.L. and Yahner R.H. (1989) Trace metals in soil, vegetation and voles from mine land treated with sewage sludge. *Journal of Environmental Quality* 18, 115-120.

Alcock R.E. and Jones K.C. (1993) Polychlorinated biphenyls in digested UK sewage sludges. *Chemosphere* 26, 2199-2207.

Aldridge K. and Alloway B.J. (1993) Lead in Soils and Food Crops: The Effects of the Speciation of Lead in Soils on Levels of Lead in Food Crops. A Literature Review for Food Science Division 1, Ministry of Agriculture, Fisheries and Food. Queen Mary and Westfield College, University of London.

Alloway B.J. and Jackson A.P. (1991) The behaviour of heavy metals in sewage sludge-amended soils. *The Science of the Total Environment* 100, 151-176.

Alloway B.J. and Tills A.R. (1984) Speciation of metals in sludge amended soils in relation to potential plant uptake. In: L'Hermite P. and Ott H. (eds), *Processing and Use of Sewage Sludge.* D. Reidel Publishing Company, Dordrecht, pp. 404-411.

Alloway B.J., Jackson A.P. and Morgan H. (1990) The accumulation of cadmium by vegetables grown on soils contaminated from a variety of sources. *The Science of the Total Environment* 91, 223-236

Amberger A. (1983) Ways to control the availability, turnover and losses of mineral fertilizer N in soils. In: *Efficient Use of Fertilizers in Agriculture. Developments in Plant and Soil Sciences*, Volume 10. Martinus Nijhoff/Dr W. Junk Publishers, The Hague, pp. 145-169.

Amberger A. (1991) Ammonia emissions during and after land spreading of slurry. In: Nielsen V.C., Voorburg J.H. and L'Hermite P. (eds), *Odour and Ammonia Emissions from Livestock Farming*. Elsevier Science Publishers Ltd, Barking, pp. 126-131.

Ames B.N., McCann J. and Yamasaki E. (1975) Methods for detecting carcinogens and mutagens with the *Salmonella*/mammalian-microsome mutagenicity test. *Mutagenic Research* 31, 347-364.

Andersen C. (1979) Cadmium, lead and calcium content, number and biomass, in earthworms (*Lumbricidae*) from sewage sludge treated soil. *Pedobiologia* 19, 309-319.

Anderson T.H. and Domsch K.H. (1993) The metabolic quotient for CO_2 (qCO_2) as a specific activity parameter to assess the effects of environmental conditions, such as pH, on the microbial biomass of forest soils. *Soil Biology and Biochemistry* 25, 393-395.

Anderson T.J. and Barrett G.W. (1982) Effects of dried sewage sludge on meadow vole (*Microtus pennsylvanicus*) populations in two grassland communities. *Journal of Applied Ecology* 19, 759-772.

Anderson, T.J., Barrett G.W., Clark C.S., Elia V.J. and Majeti V.A. (1982) Metal concentrations in tissues of meadow voles from sewage sludge-treated fields. *Journal of Environmental Quality* 11, 272-277.

Andrews D.A., Mawer S.L. and Matthews P.J. (1983) Survival of *Salmonellae* in sewage sludge injected into soil. *Effluent and Water Treatment Journal* 23, 72-74.

Angle J.S. and Baudler D.M. (1984) Persistence and degradation of mutagens in sludge-amended soil. *Journal of Environmental Quality* 13, 143-146.

Angle J.S. and Heckman J.R. (1986) Effect of soil pH and sewage sludge on VA mycorrhizal infection of soybeans. *Plant and Soil* 93, 437-441.

Angle J.S., McGrath S.P., Chaudri A.M., Chaney R.L. and Giller K.E. (1993) Inoculation effects on legumes grown in soil previously treated with sewage sludge. *Soil Biology and Biochemistry* 25, 575-580.

Anthony R.G. and Kozlowski R. (1982) Heavy metals in tissues of small mammals inhabiting waste-water-irrigated habitats. *Journal of Environmental Quality* 11, 20-22.

Apsimon H.M. and Kruse-Plass M. (1991) The role of ammonia as an atmospheric pollutant. In: Nielsen J.C., Voorburg J.H. and L'Hermite P. (eds), *Odour and Ammonia Emissions from Livestock Farming.* Elsevier Science Publishers Ltd, Barking, pp. 17-20.

Arah J.R.M. and Smith K.A. (1989) Modelling denitrification in aggregated soils: Relative importance of moisture tension, soil structure and oxidizable organic matter. In: Hansen J.A. and Henriksen K. (eds), *Nitrogen in Organic Wastes Applied to Soils.* Academic Press Limited, London, pp. 271-286.

Aranda J.M., O'Connor G.A. and Eiceman (1989) Effects of sewage sludge on di-(2-ethylhexyl) phthalate uptake by plants. *Journal of Environment Quality* 18, 45-50.

ARC; Agricultural Research Council (1980) The Nutrient Requirements of Ruminant Livestock. Commonwealth Agricultural Bureaux, Farnham Royal.

Archer J.R. (1992) UK nitrate policy implementation. In: *Nitrate and Farming Systems. Aspects of Applied Biology* 30, 11-18.

Argent V.A., Bell J.C. and Emslie-Smith M. (1977) Animal disease hazards of sludge disposal to land: Occurrence of pathogenic organisms. *Water Pollution Control* 76, 511-516.

Arnold G.W., McManus W.R. and Bush I.G. (1966) Studies in the wool production of grazing sheep. 5. Observations on teeth wear and carry-over effects. *Australian Journal of Experimental Agriculture and Animal Husbandry* 6, 101-107.

Ash C.P.J. and Lee D.L. (1980) Lead, cadmium, copper and iron in earthworms from roadside sites. *Environmental Pollution (Series A)* 22, 59-67.

Atlas R.M., Pramer D. and Bartha R. (1978) Assessment of pesticide effects on non-target soil micro-organisms. *Soil Biology and Biochemistry* 10, 231-239.

Bååth E. (1989) Effects of heavy metals in soil on microbial processes and populations (a review). *Water, Air, and Soil Pollution* 47, 335-379.

Bååth E. (1992) Measurement of heavy metal tolerance of soil bacteria using thymidine incorporation into bacteria extracted after homogenization-centrifugation. *Soil Biology and Biochemistry* 24, 1167-1172.

Bache C.A., Gutenmann W.H., St. John Jr., L.E., Sweet R.D., Hatfield H.H. and Lisk D.J. (1973) Mercury and methylmercury content of agricultural crops grown on soils treated with various mercury compounds. *Journal of Agricultural and Food Chemistry* 21, 607-613.

Barea J.M. and Azcon-Agiular C. (1983) Mycorrhizas and their significance in nodulating nitrogen-fixing plants. *Advances in Agronomy* 36, 1-54.

Baxter J.C., Barry B., Johnson P.E. and Kienholz E.W. (1982) Heavy metal reduction in cattle tissues from ingestion of sewage sludge. *Journal of Environmental Quality* 11, 616-620.

Baxter K.M. and Clark L. (1984) *Effluent Recharge. The Effects of Effluent Recharge on Groundwater Quality.* Technical Report TR 199. WRc Medmenham, Marlow.

Beauchamp E.G. (1983) Nitrogen loss from sewage sludges and manures applied to agricultural lands. In: Freney J.R. and Simpson J.R. (eds), *Gaseous Loss of Nitrogen from Plant-Soil Systems.* Martinus Nijhoff/Dr W. Junk Publishers, The Hague, pp. 181-194.

Beauchamp E.G., Kidd G.E. and Thurtell G. (1978) Ammonia volatilization from sewage sludge applied in the field. *Journal of Environmental Quality* 7, 141-146.

Beckett P.H.T. and Brindley P. (1983) Changes in the extractabilities of the heavy metals in water-logged sludge-treated soils. *Water Pollution Control* 82, 107-113.

Beckett P.H.T. and Davis R.D. (1977) Upper critical levels of toxic elements in plants. *New Phytologist* 79, 95-106.

Beckett P.H.T. and Davis R.D. (1979) The disposal of sewage sludge onto farmland: the scope of the problem of toxic elements. *Water Pollution Control* 78, 419-445.

Beckett P.H.T. and Davis R.D. (1982) Heavy metals in sludge - are their toxic effects additive? *Water Pollution Control* 81, 112-119.

Beckett P.H.T., Davis R.D., Milward A.F. and Brindley P. (1977) A comparison of the effect of different sewage sludges on young barley. *Plant and Soil* 48, 129-141.

Beckett P.H.T., Warr E. and Davis R.D. (1983) Cu and Zn in soils treated with sewage sludge: Their 'extractability' to reagents compared with their 'availability' to plants. *Plant and Soil* 70, 3-14.

Bell R.G. and Bole J.B. (1978) Elimination of faecal coliform bacteria from soil irrigated with municipal sewage farm effluent. *Journal of Environmental Quality* 7, 193-196.

Bell P.F., James B.R. and Chaney R.L. (1991) Heavy metal extractability in long-term sewage sludge and metal salt-amended soils. *Journal of Environmental Quality* 20, 481-486.

Bellin C.A. and O'Connor G.A. (1990) Plant uptake of pentachlorophenol from sludge-amended soils. *Journal of Environmental Quality* 19, 598-602.

Bellin C.A., O'Connor G.A. and Jin Y. (1990) Sorption and degradation of

pentachlorphenol in sludge-amended soils. *Journal of Environmental Quality* 19, 603-608.
Bengtsson G. and Tranvik L. (1989) Critical metal concentrations for forest soil invertebrates. *Water, Air, and Soil Pollution* 47, 381-417.
Berg G. (1973) Re-assessment of the virus problem in sewage and in surface and renovated waters. *Progress in Water Technology* 3, 87-94.
Berg R.C., Morse W.J. and Johnson T.M. (1987) Hydrogeologic Evaluation of the Effects of Surface Application of Sewage Sludge to Agricultural Land Near Rochton, Illinois. Environmental Geology Notes 119. Department of Energy and National Resources, Illinois State Geological Survey.
Bergback B., Anderberg S. and Lohm U. (1994) Accumulated environmental impact: The case of cadmium in Sweden. *The Science of the Total Environment* 145, 13-28.
Berglund S. and Hall J.E. (1988) Sludge and slurry disposal techniques and environmental problems - a review. In: Nielsen V.C., Voorburg J.H. and L'Hermite P. (eds), *Volatile Emissions from Livestock Farming and Sewage Operations.* Elsevier Applied Science Publishers Ltd, Barking, pp. 60-72.
Bergstrom L. and Brink N. (1986) Effects of differentiated applications of fertilizer N on leaching losses and distribution of inorganic N in the soil. *Plant and Soil* 93, 333-345.
Berrow M.L. (1986) An overview of soil contamination problems. In: Lester J.M., Perry R., Sterritt R.M. (eds), *Chemicals in the Environment.* Selper, London, pp. 543-552.
Berrow M.L. and Burridge J.C. (1980) Trace element levels in soils: effects of sewage sludge. In: *Inorganic Pollution and Agriculture.* MAFF Reference Book 326. HMSO, London, pp. 159-183.
Berrow M.L. and Burridge J.C. (1981) Persistence of metals in available form in sewage sludge treated soils under field conditions. In: *International Conference Heavy Metals in the Environment,* Amsterdam. CEP Consultants Ltd, Edinburgh, pp. 202-204.
Berrow M.L. and Burridge J.C. (1984) Persistence of metals in sewage sludge treated soils. In: L'Hermite P. and Ott H. (eds), *Processing and Use of Sewage Sludge.* D. Reidel Publishing Company, Dordrecht, pp. 418-422.
Berrow M.L. and Burridge J.C. (1991) Uptake, distribution and effects of metal compounds on plants. In: Merian E. (ed), *Metals and Their Compounds in the Environment.* VCH Verlagsgesellschaft mbH, Weinheim, pp. 399-410.
Berrow M.L. and Reaves G.A. (1985) Extractable copper concentrations in Scottish soils. *Journal of Soil Science* 36, 31-43.
Berrow M.L. and Webber J. (1972) Trace elements in sewage sludge. *Journal of the Science of Food and Agriculture* 23, 93-100.
Berrow M.L., Morrison A.R., Park J.S. and Sharp B.L. (1990) The long term partitioning of copper in waters extracted from polluted soils using high performance/size reduction exclusion liquid chromatography. In: *Proceedings 4th International Conference - Environmental Contamination,*

Barcelona. CEP Consultants Ltd, Edinburgh, pp. 85-87.
Bertilsson G. (1988) Lysimeter studies of nitrogen leaching and nitrogen balances as affected by agricultural practices. *Acta Agriculturae Scandinavica* 38, 3-11.
Bertrand J.E., Lutrick M.C., Breland H.C. and West R.L. (1980) Effects of dried digested sludge and corn grown on soil treated with liquid digested sludge on performance, carcass quality and tissue residues in beef steers. *Journal of Animal Science* 50, 35-40.
Bertrand J.E., Lutrick M.C., Edds G.T. and West R.L. (1981) Metal residues in tissues, animal performance and carcass quality with beef steers grazing pensacola bahiagrass pastures treated with liquid digested sludge. *Journal of Animal Science* 53, 146-153.
Beveridge T.J. (1989) Role of cellular design in bacterial metal accumulation and mineralization. *Annual Review of Microbiology* 43, 147-171.
Beyer W.N., Chaney R.L. and Mulhern B.M. (1982) Heavy metal concentrations in earthworms from soil amended with sewage sludge. *Journal of Environmental Quality* 11, 381-385.
Bidwell A.M. and Dowdy R.H. (1987) Cadmium and zinc availability to corn following termination of sewage sludge applications. *Journal of Environmental Quality* 16, 438-442.
Biederbeck V.O., Campbell C.A. and Smith A.E. (1987) Effects of long-term 2,4-D field applications on soil biochemical processes. *Journal of Environmental Quality* 16, 257-262.
Bingham F.T., Page A.L., Mitchell G.A. and Strong J.E. (1979) Effects of liming an acid soil amended with sewage sludge enriched with Cd, Cu, Ni and Zn on yield and Cd content of wheat grain. *Journal of Environmental Quality* 8, 202-207.
Bittell J.E. and Miller R.J. (1974) Lead, cadmium and calcium selectivity coefficients on a montmorillonite, illite and kaolinite. *Journal of Environmental Quality* 3, 250-253.
Bitzer C.C. and Sims J.T. (1988) Estimating the availability of nitrogen in poultry manure through laboratory and field studies. *Journal of Environmental Quality* 17, 47-54.
Bjerre G.K. and Schierup H.H. (1985) Uptake of six heavy metals by oats as influenced by soil type and additions of cadmium, lead, zinc and copper. *Plant and Soil* 88, 57-69.
Blake L., Johnston A.E. and Goulding K.W.T. (1994) Mobilization of aluminium in soil by acid deposition and its uptake by grass cut for hay - a chemical time bomb. *Soil Use and Management* 10, 51-55.
Block J.C. (1986) Biological health risks of sludge disposal. In: Block J.C., Havelaar A.H. and L'Hermite P. (eds), *Epidemiological Studies of Risks Associated with the Agricultural Use of Sewage Sludge: Knowledge and Needs.* Elsevier Applied Science Publishers Ltd, Barking, pp. 123-134.
Bloomfield C. and McGrath S.P. (1982) A comparison of the extractabilities of

Zn, Cu, Ni and Cr from sewage sludges prepared by treating raw sewage with metal salt before or after anaerobic digestion. *Environmental Pollution (Series B)* 3, 193-198.

Boawn L.C. and Rasmussen P.E. (1971) Crop response to excessive zinc fertilization of alkaline soil. *Agronomy Journal* 63, 874-876.

Bollen G.J. (1985) The fate of plant pathogens during composting of crop residues. In: Gasser J.K.R. (ed), *Composting of Agricultural and Other Wastes.* Elsevier Applied Science Publishers Ltd, Barking, pp. 282-290.

Bolton J. (1975) Liming effects on the toxicity to perennial ryegrass of a sewage sludge contaminated with zinc, nickel, copper and chromium. *Environmental Pollution* 9, 295-304.

Bosch A., Lucena F. and Jofre J. (1986) Fate of human enteric viruses (rotaviruses and enteroviruses) in sewage after primary sedimentation. *Water Science and Technology* 18, 47-52.

Bossert I., Kachel W.M. and Bartha R. (1984) Fate of hydrocarbons during oily sludge disposal in soil. *Applied and Environmental Microbiology* 47, 763-767.

Bouche M.B. (1972) *Lumbriciens de France.* Institut Nationale de la Recherche Agronomique.

Boyle M. (1989) The environmental microbiology of chlorinated aromatic decomposition. *Journal of Environmental Quality* 18, 395-402.

Bradley R.M. (1973) Chlorination of effluents and the Italian concept. *Effluent and Water Treatment Journal* 13, 683-689.

Brady N.C. (1990) *The Nature and Properties of Soils*, 10th Edition. Macmillan Publishing Company, New York.

Braids O.C. (1970) Liquid digested sludge gives field crops necessary nutrients. *Illinois Research* 12, 6-10.

Brams E., Anthony W. and Weatherspoon L. (1989) Biological monitoring of an agricultural food chain: Soil cadmium and lead in ruminant tissues. *Journal of Environmental Quality* 18, 317-323.

Bray B.J., Dowdy R.H., Goodrich R.D. and Pamp D.E. (1985) Trace metal accumulations in tissues of goats fed silage produced on sewage sludge-amended soil. *Journal of Environmental Quality* 14, 114-118.

Bremner I. (1981) Effects of the disposal of copper-rich slurry on the health of grazing animals. In: L'Hermite P. and Dehandtschutter J. (eds), *Copper in Animal Wastes and Sewage Sludge.* D. Reidel Publishing Company, Dordrecht, pp. 245-255.

Bremner J.M. and Blackmer A.M. (1978) Nitrous oxide: emission from soils during nitrification of fertilizer nitrogen. *Science* 199, 295-296.

Bremner I. and Mills C.F. (1979) Effects of diet on the toxicity of heavy metals. In: *Management and Control of Heavy Metals in the Environment.* CEP Consultants Ltd, Edinburgh, pp. 139-146.

Briggs D.J. and Courtney F.M. (1989) *Agriculture and Environment: The Physical Geography of Temperate Agricultural Systems.* Longman

Scientific and Technical, Harlow.

Brockman J.S. (1988) Grassland. In: Halley, R.J. and Soffe, R.J. (eds), *The Agricultural Notebook*, 18th Edition. Blackwell Scientific Publications, Oxford, pp. 177-208.

Brookes P.C. and McGrath S.P. (1984) Effects of metal toxicity on the size of the soil microbial biomass. *Journal of Soil Science* 35, 341-346.

Brookes P.C., McGrath S.P., Klein D.A. and Elliott E.T. (1984) Effects of heavy metals on microbial activity and biomass in field soils treated with sewage sludge. In: *Environmental Contamination*, International Conference, London. CEP Consultants Ltd, Edinburgh, pp. 574-583.

Brookes P.C., Powlson D.S. and Jenkinson D.S. (1984) Phosphorus in the soil microbial biomass. *Soil Biology and Biochemistry* 16, 169-175.

Brookes P.C., McGrath S.P. and Heijnen C.E. (1986) Metal residues in soils previously treated with sewage sludge and their effects on growth and nitrogen fixation by blue-green algae. *Soil Biology and Biochemistry* 18, 345-353.

Brown, K.W., Thomas J.C. and Slowey J.F. (1983) The movement of metals applied to soils in sewage effluent. *Water, Air, and Soil Pollution* 19, 43-54.

Brown K.W., Donnolly K.C., Thomas J.C., Davol P. and Scott B.R. (1986) Mutagenic activity of soils amended with two refinery wastes. *Water, Air, and Soil Pollution* 29, 1-13.

Bruce A.M. and Davis R.D. (1983) Utilization of sewage sludge in agriculture - maximising benefits and minimising risks. In: *International Symposium on Biological Reclamation and Land Utilization of Urban Wastes*, Naples. WRc Report No. 228-S. WRc Medmenham, Marlow, pp. 11-14.

Bruce A.M., Pike E.B. and Fisher W.J. (1990) A review of treatment process options to meet the EC Sludge Directive. *Journal of the Institution of Water and Environmental Management* 4, 1-13.

Brums E.A., Anthony W. and Weatherspoon L. (1989) Biological monitoring of a biological food chain: Soil, cadmium and lead in ruminant tissues. *Journal of Environmental Quality* 18, 317-323.

Bryce-Smith D. (1986) Environmental chemical influences on mentation and behaviour. *Chemical Society Reviews* 15, 93.

Bryce-Smith D. and Simpson R.I.D. (1984) Case of anorexia nervosa responding to zinc sulphate. *The Lancet* (August 11), 350.

Bumpus J.A. and Aust S.D. (1987) Biodegradation of environmental pollutants by the white root fungus. *Bioessays* 6, 166-170.

Bumpus J.A.. Tien M., Wright D. and Aust S. (1985) Oxidation of persistent environmental pollutants by a white rot fungus. *Science* 228, 1434-1436.

Burd R.G. (1966) A study of sludge handling and disposal. Federal Water Pollution Control Administration, Report No. WP-20-4. US Dept. of Interior, Cincinnatti.

Burge W.D. and Marsh P.B. (1978) Infectious disease hazards of landspreading sewage wastes. *Journal of Environmental Quality* 7, 1-9.

Burns I.G. and Greenwood D.J. (1982) Estimation of the year-to-year variations in nitrate leaching in different soils and regions of England and Wales. *Agriculture and Environment* 7, 35-45.

Butler G.W. and Jones D.I.H. (1973) Mineral biochemistry of herbage. In: Butler G.W. and Bailey R.W. (eds), *Chemistry and Biochemistry of Herbage.* Academic Press, pp. 127-162.

Buttigieg A.D., Klessa D.A. and Hall D.A. (1990) The Contamination of Herbage Following the Application of Sewage Sludge to Pasture. Report No. FR 0078. Foundation for Water Research, Marlow.

Cabrera D., Young S.D. and Rowell D.L. (1988) The toxicity of cadmium to barley plants as affected by complex formation. *Plant and Soil* 105, 195-204.

Cabrera C., Ortega E., Gallego C., Lopez M.C., Lorenzo M.L. and Asensio C. (1994) Cadmium concentration in farmlands in southern Spain: Possible sources of contamination. *The Science of the Total Environment* 153, 261-265.

Calabrese, E.J., Barnes, R., Stanek III E.J., Pastides H., Gilbert C.E., Veneman P., Wang X., Lasztity A. and Kostecki P.T. (1989) How much soil do young children ingest: An epidemiologic study. *Regulatory Toxicology and Pharmacology* 10, 123-137.

Carlton-Smith C.H. (1987) *Effects of Metals in Sludge-treated Soils on Crops.* Technical Report TR 251. WRc Medmenham, Marlow.

Carlton-Smith C.H. and Davis R.D. (1983) Comparative uptake of heavy metals by forage crops grown on sludge-treated soil. In: *International Conference on Heavy Metals in the Environment*, 6-9 September, Heidelberg. CEP Consultants Ltd, Edinburgh.

Carlton-Smith C.H. and Stark J.H. (1985) Mobility of Metals in Sludged Soil (EI 9316 SLD). Final Report to the Department of the Environment: June 1982 - May 1985. WRc Report No. 1047-M/2. WRc Medmenham, Marlow.

Carlton-Smith C.H. and Stark J.H. (1987) Sites with a History of Sludge Deposition. Interim Report of Field Trials: April 1983-March 1986 (SDA 9166 SLD). WRc Report No. DoE 1376-M. WRc Medmenham, Marlow.

Carlton-Smith C.H., Stark J.H. and Thien Y.J. (1985) Sites With a History of Sludge Disposal: Pot trial in Relation to soil Microbial Biomass. Progress Report to Department of the Environment: March-October 1985. WRc Report No. 1104-M. WRc Medmenham, Marlow.

Carlton-Smith C.H., Stark J.H., Carrington E.G. and Kichenside N.J. (1987) Sites with a Long History of Sludge Disposal. Supplementary Pot and Field Trial Studies in Relation to Soil Microbial Biomass (SDA 9166 SLD). WRc Report No. DoE 1375-M. WRc Medmenham, Marlow.

Carlton-Smith C.H., Kichenside N.J. and Thomas B.A. (1988) Effect of Soil pH on Metal Availability to Crops (LDTS 9325 SLD). Final Report to the Department of the Environment. WRc Report No. DoE 1619-M/1. WRc Medmenham, Marlow.

Carrington E.G. (1978) *The Contribution of Sewage Sludges to the Dissemination of Pathogenic Micro-organisms in the Environment.* Technical Report TR 71. WRc Medmenham, Marlow.

Carrington E.G. (1980) *The Fate of Pathogenic Micro-organisms During Waste-water Treatment and Disposal.* Technical Report TR 128. WRc Medmenham, Marlow.

Carroll W.D. and Ross R.D. (1981) A Study to Assess the Environmental Impact of Injecting Raw Sewage Sludge into Agricultural Soil. Final Report to Environmental Protection Service, Environment Canada, Department of Supply and Services, Contract: ISS80-00232. Waterworks, Waste and Disposal Department, City of Winnipeg.

Carter M.R. (1986) Microbial biomass and mineralizable nitrogen in solonetzic soils: Influence of gypsum and lime amendments. *Soil Biology and Biochemistry* 18, 531-537.

Cartwright N.G. and Painter H. (1991) An Assessment of the Environmental Quality Standards for Inorganic Nutrients Necessary to Prevent Eutrophication (Nuisance Growth of Algae). WRc Report No. NR 2397. WRc Medmenham, Marlow.

CAST; Council for Agricultural Science and Technology (1976) Application of Sewage Sludge to Cropland: Appraisal of Potential Hazards of the Heavy Metals to Plants and Animals. Department of Commerce, Report PB-264 015-US, National Technical Information Service, Springfield, Virginia.

Catt J.A., Christian D.G., Goss M.J., Harris G.L. and Howse K.R. (1992) Strategies to reduce nitrate leaching by crop rotation, minimal cultivation and straw incorporation in the Brimstone Farm Experment, Oxfordshire. In: *Nitrate and Farming Systems. Aspects of Applied Biology* 30, 255-262.

Cavallaro N. and McBride M.B. (1978) Copper and cadmium adsorption characteristics of selected acid and calcareous soils. *Soil Science Society of America Journal* 42, 550-556.

Cavallaro N. and McBride M.B. (1980) Activities of Cu^{2+} and Cd^{2+} in soil solutions as affected by pH. *Soil Science Society of America Journal* 44, 729-732.

CEC; Council of the European Communities (1975) Council of European Communities Directive on Drinking Water 16 June 1975. Concerning the quality required of surface water intended for the abstraction of drinking water in the Member States (75/440/EEC). *Official Journal of the European Communities* No. L 194, 25 July 1975.

CEC; Council of the European Communities (1976) Council Directive of 4 May 1976 on pollution caused by certain dangerous substances discharged into the aquatic environment of the Community (76/464/EEC). *Official Journal of the European Communities* No. L 129/23-29.

CEC; Council of the European Communities (1978) Council Directive of 20 March 1978 on toxic and dangerous waste (78/319/EEC). *Official Journal of the European Communities* No. L 84/43-48.

CEC; Council of the European Communities (1980a) Council Directive of 17 December 1979 on the protection of groundwater against pollution caused by dangerous substances (80/68/EEC). *Official Journal of the European Communities* No. L 20/43-48.

CEC; Council of the European Communities (1980b) Council Directive of 15 July 1980 relating to the quality of water intended for human consumption (80/778/EEC). *Official Journal of the European Communities* No. L 229/11-29.

CEC; Council of the European Communities (1986a) Council Directive of 12 June 1986 on the protection of the environment, and in particular of the soil, when sewage sludge is used in agriculture (86/278/EEC). *Official Journal of the European Communities* No. L 181/6-12.

CEC; Council of the European Communities (1986b) Council Directive of 12 June 1986 on limit values and quality objectives for discharges of certain dangerous substances included in List I of the Annex to Directive 76/464/EEC, (86/280/EEC). *Official Journal of the European Communities* No. L 181/16-27.

CEC; Council of the European Communities (1988) Council Directive of 16 June 1988 amending Annex II to Directive 86/280/EEC on limit values and quality objectives for discharges of certain dangerous substances included in List I of the Annex to Directive 76/464/EEC (76/347/EEC). *Official Journal of the European Communities* No. L 158/35-41.

CEC; Council of the European Communities (1990a) Council Directive of 27 July 1990 amending Annex II to Directive 86/280/EEC on limit values and quality objectives for discharges of certain dangerous substances included in List I of the Annex to Directive 76/464/EEC (90/415/EEC). *Official Journal of the European Communities* No. L 219/49-57.

CEC; Council of the European Communities (1990b) Amendments to the Proposal for a Council Directive amending, in respect of chromium, Directive 86/278/EEC on the protection of the environment, and in particular of soil, when sewage sludge is used in agriculture. Com (90) 85 Final, Brussels, 27 March 1990.

CEC; Council of the European Communities (1991a) Council Directive of 21 May 1991 concerning urban waste water treatment (91/271/EEC). *Official Journal of the European Communities* No. L 135/40-52.

CEC; Council of the European Communities (1991b) Council Directive of 12 December 1991 concerning the protection of waters against pollution caused by nitrates from agricultural sources (91/676/EEC). *Official Journal of the European Communities* No. L 375/1-8.

CEC; Council of the European Communities (1991c) Council Directive of 12 December 1991 on hazardous waste (91/689/EEC). *Official Journal of the European Communities* No. L 377/20-27.

CEC; Commission of the European Communities (1992) *Towards Sustainability: A European Community Programme of Policy and Action in*

Relation to the Environment and Sustainable Development. Com (92) 23 Final - Vol. II, 2 March, 1992.

CEC; Commission of the European Communities (1993) Withdrawal of certain proposals and drafts from the Commission to the Council (93/C 228/04). *Official Journal of the European Communities* No. C 228/13.

CEC; Commission of the European Communities (1994a) Commission Decision of 24 October 1994 concerning questionnaires for Member States reports on the implementation of certain Directives in the waste sector (implementation of Council Directive 91/692/EEC) (94/741/EC). *Official Journal of the European Communities* No. L 296/42-55.

CEC; Commission of the European Communities (1994b) Commission Decision of 20 December 1993 on the European Waste Catalogue. *Official Journal of the European Communities* No. L 5.

CEC; Council of the European Communities (1994c) Council Decision of 22 December 1994 establishing a list of hazardous waste pursuant to Article 1(4) of Council Directive 91/689/EEC on hazardous waste (94/904/EC). *Official Journal of the European Communities* No. L 356/14-22.

CEC; Council of the European Communities (1994d) Council Directive 94/67/EC of 16 December 1994 on the incineration of hazardous waste. *Official Journal of the European Communities* No. L 365/34-45.

CES, Consultants in Environmental Sciences Limited (1993) UK Sewage Sludge Survey. Final Report. CES, Gateshead.

Chabrier J.P. (1994) Sludge drying in Europe. Paper presented to European Conference on Sludge and Organic Waste, 12-15 April, Wakefield, UK. *Aqua. Enviro.*, Dept. of Civil Engineering, University of Leeds.

Chalk P.M. and Smith C.J. (1983) Chemodenitrification. In: Freney J.R. and Simpson J.R. (eds), *Gaseous Loss of Nitrogen from Plant-Soil Systems.* Martinus Nijhoff/Dr W. Junk Publishers, The Hague, pp. 65-89.

Chambers B.J. and Smith K.A. (1992) Soil mineral nitrogen arising from organic manure applications. In: *Nitrate and Farming Systems. Aspects of Applied Biology* 30, 135-143.

Chambers B.J., McGrath S.P. and Hall J.E. (1994) Effects of Sewage Sludge Applications to Agricultural Soils on Soil Microbial Activity and the Implications for Agricultural Productivity and Long Term Soil Fertility. Contract Ref: CSA 2566. Ministry of Agriculture, Fisheries and Food, London.

Chander K. (1991) The effects of heavy metals from past applications of sewage sludge on the soil microbial biomass and activity. Ph.D. thesis, University of Reading.

Chander K. and Brookes P.C. (1991a) Effects of heavy metals from past applications of sewage sludge on microbial biomass and organic matter accumulation in a sandy loam and a silty loam UK soil. *Soil Biology and Biochemistry* 23, 927-932.

Chander K. and Brookes P.C. (1991b) Microbial biomass dynamics during decomposition of glucose and maize in metal-contaminated and non-contaminated soils. *Soil Biology and Biochemistry* 23, 917-925.

Chander K. and Brookes P.C. (1991c) Plant inputs of carbon to metal-contaminated soil and effects on the soil microbial biomass. *Soil Biology and Biochemistry* 24, 1169-1177.

Chaney R.L. (1973) Crop and food chain effects of toxic elements in sludges and effluents. In: *Recycling Municipal Sludges and Effluents on Land.* National Association State University and Land Grant Colleges, Washington D.C., pp. 129-141.

Chaney R.L. (1980) Health risks associated with toxic metals in municipal sludge. In: Bitton G. (ed), *Sludge-Health Risks of Land Application.* Ann Arbor Science Publishers, Ann Arbor, pp. 59-83.

Chaney R.L. (1983) Potential effects of waste constituents on the food chain. In: Parr J.F., Marsh P.B. and Kla J.M. (eds), *Land Treatment of Hazardous Wastes.* Noyes Data Corporation, Park Ridge, New Jersey, pp. 152-240.

Chaney R.L. (1988) Effective utilization of sewage sludge on cropland in the United States and toxicological consideration for land application. In: *Proceedings Second International Symposium Land Application of Sewage Sludge*, Tokyo. Association for the Utilization of Sludge, Tokyo, pp. 77-105.

Chaney R.L. (1990a) Twenty years of land application research. Part I. *Biocycle* 31, (9), 54-59.

Chaney R.L. (1990b) Public health and sludge utilization. Part II. *Biocycle* 31, (10), 68-73.

Chaney R.L. and Lloyd C.A. (1979) Adherence of spray-applied liquid digested sewage sludge to tall fescue. *Journal of Environmental Quality* 8, 407-411.

Chaney R.L. and Ryan J.A. (1993) Heavy metals and toxic organic pollutants in MSW-composts: Research results on phytoavailability, bioavailability, fate, etc. In: Hoitink, H.A.J. and Keener, H.M. (eds), *Science and Engineering of Composting: Design, Environmental, Microbiological and Utilization Aspects.* Renaissance Publications, Worthington, Ohio, pp. 451-506.

Chaney R.L., Stoewsand G.S., Furr A.K., Bache C.A. and Lisk D.J. (1978) Elemental content of tissues of guinea pigs fed Swiss chard grown on municipal sewage sludge-amended soil. *Journal of Agricultural and Food Chemistry* 26, 994-997.

Chaney R.L., Bruins R.J.F., Baker D.E., Korcak R.F., Smith J.E. and Cole D. (1987) Transfer of sludge-applied trace elements to the food chain. In: Page A.L., Logan T.J. and Ryan J.A. (eds), *Land Application of Sludge Food Chain Implications.* Lewis Publishers Inc., Chelsea, Michigan, pp. 67-99.

Chang A.C., Page A.L., Foster K.W. and Jones T.E. (1982a) A comparison of cadmium and zinc accumulation by four cultivars of barley grown in sludge-amended soils. *Journal of Environmental Quality* 11, 409-412.

Chang A.C., Page A.L. and Bingham F.T. (1982b) Heavy metal absorption by winter wheat following termination of cropland sludge applications.

Journal of Environmental Quality 11, 705-708.
Chang A.C., Page A.L., Warneke J.E., Resketo M.R. and Jones T.E. (1983) Accumulation of cadmium and zinc in barley grown on sludge-treated soils: A long-term field study. *Journal of Environmental Quality* 12, 391-397.
Chang A.C., Page A.L., Warneke J.E. and Grgurevic E. (1984a) Sequential extraction of soil heavy metals following a sludge application. *Journal of Environmental Quality* 13, 33-38.
Chang A.C., Warneke J.E., Page A.L. and Lund L.J. (1984b) Accumulation of heavy metals in sewage sludge-treated soils. *Journal of Environmental Quality* 13, 87-91.
Chang A.C., Page A.L. and Warneke J.E. (1987a) Long-term sludge applications on cadmium and zinc accumulation in Swiss chard and radish. *Journal of Environmental Quality* 16, 217-221.
Chang A.C., Hinesly T.D., Bates T.E., Doner H.E., Dowdy R.H. and Ryan J.A. (1987b) Effects of long-term sludge application on accumulation of trace elements by crops. In: Page A.L., Logan T.J. and Ryan J.A. (eds), *Land Application of Sludge Food Chain Implications.* Lewis Publishers Inc., Chelsea, Michigan, pp. 53-66.
Chang A.C., Page A.L., Pratt P.F. and Warneke J.E. (1988) Leaching of nitrate from freely drained-irrigated fields treated with municipal sludges. In: *Planning Now for Irrigation and Drainage in the 21st Century.* Lincoln, Nebraska, pp. 455-467.
Chang A.C., Granato T.C. and Page A.L. (1992) A methodology for establishing phytotoxicity criteria for chromium, copper, nickel and zinc in agricultural land application of municipal sewage sludges. *Journal of Environmental Quality* 21, 521-536.
Chang A.C., Page A.L. and Asano T. (1993) Developing Human Health-Related Chemical Guidelines for Reclaimed Wastewater and Sewage Sludge Applications in Agriculture. Submitted to World Health Organization, Geneva. Technical Service Agreement Nos. GL/GLO/CWS/058/RB90.300 and GL/GLO/DGP/449/RB/90.30.
Chaudri A.M., McGrath S.P. and Giller K.E. (1992a) Survival of the indigenous population of *Rhizobium leguminosarum* biovar *trifolii* in soil spiked with Cd, Zn, Cu and Ni salts. *Soil Biology and Biochemistry* 24, 625-632.
Chaudri A.M., McGrath S.P. and Giller K.E. (1992b) Metal tolerance of isolates of *Rhizobium leguminosarum* biovar *trifolii* from soil contaminated by past applications of sewage sludge. *Soil Biology and Biochemistry* 24, 83-88.
Chaudri A.M., McGrath S.P., Giller K.E., Rietz E. and Sauerbeck D.R. (1993) Enumeration of indigenous *Rhizobium leguminosarum* biovar *trifolii* in soils previously treated with metal-contaminated sewage sludge. *Soil Biology and Biochemistry* 25, 301-309.
Chaussod R. (1981) Valeur fertilisante ozote des boues residuaires. In: L'Hermite P. and Ott H. (eds), *Characterisation, Treatment and Use of Sewage Sludge.* D. Reidel, Dordrecht, pp. 449-465.

Chaussod R., German J.C. and Catroux R. (1978) Determination de la valeur fertilisante des boues residuaires - Aptitude a libere l'azote. Rapport Minstere de l'Environnements. INRA, Laboratoire de Microbiologie des Sols, Dijon.

Chaussod R., Gupta S.K., Hall J.E., Pommel B. and Williams J.H. (1985) *Nitrogen and Phosphorus Value of Sewage Sludge.* Concerted Action Treatment and Use of Sewage Sludge Revision of Document NR. SL/82/82-XII/ENJ/35/82. Commission of the European Communities, Luxembourg.

Chaussod R., Catroux G. and Juste C. (1986) Effects of anaerobic digestion of organic wastes on carbon and nitrogen mineralization rates: Laboratory and field experiments. In: Dam Kofoed A., Williams J.H. and L'Hermite P. (eds), *Efficient Land Use of Sludge and Manure.* Elsevier Applied Science Publishers Ltd, Barking, pp. 24-36.

Christensen T.H. (1984a) Cadmium soil sorption at low concentrations: I. Effect of time, cadmium load, pH and calcium. *Water, Air and Soil Pollution* 21, 105-114.

Christensen T.H. (1984b) Cadmium soil sorption at low concentrations: II. Reversibility, effect of changes in solute composition and effect of soil ageing. *Water, Air, and Soil Pollution* 21, 115-125.

Christian D.G., Crees R. and Dowdell R.J. (1985) Yield and uptake of fertilizer nitrogen by direct-drilled winter barley growing on a chalk soil. *Soil Use and Management* 1, 74-79.

Christian D.G., Goodlass G. and Powlson D.S. (1992) Nitrogen uptake by cover crops. In: *Nitrate and Farming Systems. Aspects of Applied Biology* 30, 291-300.

Christie P. and Beattie J.A.M. (1989) Grassland soil microbial biomass and accumulation of potentially toxic elements from long-term slurry application. *Journal of Applied Ecology* 26, 597-612.

Christie P. and Kilpatrick D.J. (1992) Vesicular-arbuscular mycorrhiza infection in cut grassland following long-term slurry application. *Soil Biology and Biochemistry* 24, 325-330.

Chumbley C.G. (1971) *Permissible Levels of Toxic Metals in Sewage Used on Agricultural Land.* Agricultural and Development Advisory Paper No. 10. Ministry of Agriculture, Fisheries and Food, Pinner.

Chumbley C.G. and Unwin R.J. (1982) Cadmium and lead content of vegetable crops grown on land with a history of sewage sludge application. *Environmental Pollution (Series B)* 4, 231-237.

Citterio, B and Frassinetti S. (1989) Sopravvivenga di *Salmonella typhimurium* nel trattato con fanghi anaerobi provenienti de deprazione di acque urbane. *Annali di Microbiologia ed Enzimologia* 39, 111-118.

Clark R.G. and Stewart D.J. (1983) Fluorine (F). In: Grace N.D. (ed), *The Mineral Requirements of Grazing Ruminants*, Occasional Publication 9. New Zealand Society of Animal Production, pp. 124-134.

Clark R.G. and Towers N.R. (1983) Introduction. In: Grace N.D. (ed) *The Mineral Requirements of Grazing Ruminants*, Occasional Publication 9. New Zealand Society of Animal Producton, pp. 9-12.

Clark S.A. and Mahanty H.K. (1991) Influence of herbicides on growth and nodulation of white clover, *Trifolium repens. Soil Biology and Biochemistry* 28, 725-730.

Cline G.R. and O'Connor G.A. (1984) Cadmium sorption and mobility in sludge-amended soil. *Soil Science* 138, 248-254.

Coker E.G. (1983) Biological aspects of the disposal - utilization of sewage sludge on land. *Advances in Applied Biology* 9, 257-322.

Coker E.G. and Carlton-Smith C.H. (1986) Phosphorus in sewage sludges as a fertilizer. *Waste Management Research* 4, 303-319.

Coker E.G. and Davis R.D. (1979) A Protocol for Experiments on the Effects of Sludge-Borne Metals on the Growth and Composition of Crops Grown in Soils with a History of Sludge Application. WRc Report No. LR 1081. WRc Medmenham, Marlow.

Coker E.G., Hodgson D.R. and Smith A.T. (1984) The Effects of Undigested Primary Sewage Sludge on the Growth and Nitrogen Uptake of Barley and Permanent Grass. WRc Report No. 652-M. WRc Medmenham, Marlow.

Coker E.G., Hall J.E., Carlton-Smith C.H. and Davis R.D. (1987a) Field investigations into the manurial value of lagoon-matured digested sewage sludge. *Journal of Agricultural Science, Cambridge* 109, 467-478.

Coker E.G., Hall J.E., Carlton-Smith C.H. and Davis R.D. (1987b) Field investigations into the manurial value of liquid undigested sewage sludge when applied to grassland. *Journal of Agricultural Science, Cambridge* 109, 479-494.

Comber S.D.W. and Gunn A.M. (1994) Diffuse Sources of Heavy Metals to Sewers. Report No. FR 0470. Foundation for Water Research, Marlow.

Cooke G.W. (1982) *Fertilizing for Maximum Yield*, Third edition. Granada Publishing Limited, St Albans.

Coppoolse J. (1992) Pollution by heavy metals from diffuse sources in the Netherlands. RIZA Document. The Netherlands Workshop in Diffuse Sources of Pollution by Heavy Metals, European Institution for Water, May 1992.

Corey R.B., Fujii R. and Hendrickson L.L. (1981) Bioavailability of heavy metals in soil-sludge systems. In: *Proceedings of the Fourth Annual Madison Conference Application Research and Practice Municipal and Industrial Waste*. University of Wisconsin-Extension, Madison, pp. 449-465.

Corey R.B., King L.D., Lue-Hing C., Fanning D.S., Street J. and Walker J.M. (1987) Effects of sludge properties on accumulation of trace elements by crops. In: Page A.L., Logan T.J. and Ryan J.A. (eds), *Land Application of Sludge Food Chain Implications*. Lewis Publishers Inc., Chelsea, Michigan, pp. 25-51.

Cornfield A.H., Beckett P.H.T. and Davis R.D. (1976) Effect of sewage sludge on mineralization of organic carbon. *Nature* 260, 518-520.

Cottenie A., Dhaese A. and Camerlynck R. (1976) Plant quality response to uptake of polluting elements. *Qualitas Plantarum* 26, 293-319.

Cottenie A., Kiekens L. and van Landschoot G. (1984) Problems of the mobility and predictability of heavy metal uptake by plants. In: L'Hermite P. and Ott H. (eds), *Processing and Use of Sewage Sludge*. D. Reidel Publishing Company, Dordrecht, pp. 124-131.

Couillard D. and Zhu S. (1992) Bacterial leaching of heavy metals from sewage sludge for agricultural application. *Water, Air, and Soil Pollution* 63, 67-80.

Cram E.B. (1943) The effect of various treatment processes on the survival of helminth ova and protozoan cysts in sewage. *Sewage Works Journal* 15 1119-1138.

Critchley R.F. and Agg A.R. (1986) Sources and Pathways of Trace Metals in the United Kingdom. WRc Report No. ER 822-M. WRc Medmenham, Marlow.

Crush J.R. (1987) Nitrogen fixation. In: Baker, M.J. and Williams, W.M. (eds), *White Clover*. CAB International, Wallingford, pp. 185-201.

Cunningham J.D., Keeney D.R. and Ryan J.A. (1975) Phytotoxicity and uptake of metals added to soils as inorganic salts or in sewage sludge. *Journal of Environmental Quality* 4, 460-462.

Curry J.P. and Cotton D.C.F. (1980) Effects of heavy pig slurry contamination on earthworms in grassland. In: *Proceedings of the 7th International Conference of Soil Biology*. US Environmental Protection Agency Report No. EPA 560/13-80-038. Office of Pesticide and Toxic Substances, Washington, pp. 336-343.

Damgaard-Larsen S., Jensen K.O., Lurd E. and Nissen B. (1977) Survival and movement of enterovirus in connection with land disposal of sludges. *Water Research* 11, 503-508.

Darmody R.G., Foss J.E., McIntosh M. and Wolf D.C. (1983) Municipal sewage sludge compost-amended soils: Some spatiotemporal treatment effects. *Journal of Environmental Quality* 12, 231-236.

Davies B.E. (1992) Inter-relationships between soil properties and the uptake of cadmium, copper, lead and zinc from contaminated soils by radish (*Rhaphanus sativus* L.). *Water, Air, and Soil Pollution* 63, 331-342.

Davies B.E. (1995) Lead. In: Alloway B.J. (ed), *Heavy Metals in Soils*, Second edition. Blackie Academic and Professional, Glasgow, pp. 206-223.

Davies B.E., Conway D. and Holt S. (1979) Lead pollution of London soils: A potential restriction on their use for growing vegetables. *Journal of Agricultural Science, Cambridge* 93, 749-752.

Davis R.D. (1979) Uptake of copper, nickel and zinc by crops growing in contaminated soils. *Journal of the Science of Food and Agriculture* 30, 937-947.

Davis R.D. (1980) *Control of Contamination Problems in the Treatment and*

Disposal of Sewage Sludge. Technical Report TR 156. WRc Medmenham, Marlow.

Davis R.D. (1981) Uptake of molybdenum and copper by forage crops growing on sludge-treated soils and its implications for the health of grazing animals. In: *Heavy Metals in the Environment*, International Conference, Amsterdam. CEP Consultants Ltd, Edinburgh, pp. 194-197.

Davis R.D. (1984a) Crop uptake of metals (cadmium, lead, mercury, copper, nickel, zinc and chromium) from sludge-treated soil and its implications for soil fertility and for the human diet. In: L'Hermite P. and Ott H. (eds), *Processing and Use of Sewage Sludge*. D. Reidel Publishing Company, Dordrecht, pp. 349-357.

Davis R.D. (1984b) Cadmium - a complex environmental problem. Part II. Cadmium in sludges used as fertilizer. *Experientia* 40, 117-126.

Davis R.D. (1989) Agricultural utilization of sewage sludge: A review. *Journal of the Institution of Water and Environmental Management* 3, 351-355.

Davis, R.D. and Beckett P.M.T. (1978) Upper critical levels of toxic elements in plants. Part II. Critical levels of copper in young barley, wheat, rape, lettuce and ryegrass and of nickel and zinc in young barley and ryegrass. *New Phytologist* 80, 23-32.

Davis R.D. and Carlton-Smith C.H. (1980) *Crops as Indicators of the Significance of Contamination of Soil by Heavy Metals*. Technical Report TR 140. WRc Medmenham, Marlow.

Davis R.D. and Carlton-Smith C.H. (1981) The preparation of sewage sludges of controlled metal content for experimental purposes. *Environmental Pollution (Series B)* 2, 167-177.

Davis R.D. and Carlton-Smith C.H. (1984) An investigation into the phytotoxicity of zinc, copper and nickel using sewage sludge of controlled metal content. *Environmental Pollution (Series B)* 8, 163-185.

Davis R.D. and Coker E.G. (1980) *Cadmium in Agriculture, with Special Reference to the Utilization of Sewage Sludge on Land*. Technical Report TR 139. WRc Medmenham, Marlow.

Davis R.D. and Dalimier F. (1994) Waste Management - Sewage Sludge. Part 2 - Quality Criteria, Classification and Strategy Development. WRc Report No. EC 3757. WRc Medmenham, Marlow.

Davis R.D., Beckett P.H.T. and Wollan E. (1978) Critical levels of twenty potentially toxic elements in young spring barley. *Plant and Soil* 49, 395-408.

Davis R.D., Howell K., Oake R.J. and Wilcox P. (1984) Significance of organic contaminants in sewage sludges used on agricultural land. In: *International Conference on Environmetal Contamination*, 10-13 July, Imperial College, London. CEP Consultants Ltd, Edinburgh, pp. 73-79.

Davis R.D., Stark J.H. and Carlton-Smith C.H. (1983) Cadmium in sludge-treated soil in relation to potential human dietary intake of cadmium. In: Davis R.D., Hucker G. and L'Hermite P. (eds), *Environmental Effects of*

Organic and Inorganic Contaminants in Sewage Sludge. D. Reidel Publishing Company, Dordrecht, pp. 137-146.

Davis R.D., Carlton-Smith C.H., Stark J.H. and Campbell J.A. (1988) Distribution of metals in grassland soils following surface applications of sewage sludge. *Environmental Pollution* 49, 99-115.

Dean R.B. and Suess M.J. (eds) (1985) The risk to health of chemicals in sewage sludge applied to land. *Waste Management and Research* 3, 251-278.

de Bertoldi M., Rambelli A., Giovannetti M. and Griselli M. (1978) Effects of benomyl and captan on rhizosphere fungi and the growth of *Allium cepa. Soil Biology and Biochemistry* 10, 265-268.

Decker A.M., Davidson J.P., Hammond R.C., Mohanty S.B., Chaney R.L. and Rumsey T.S. (1980) Animal performance on pastures topdressed with liquid sewage sludge and sludge compost. In: *Proceedings of the National Conference on Municipal and Industrial Sludge Utilization and Disposal.* Information Transfer Inc., Silver Spring, Maryland, pp. 37-41.

de Haan S. (1981) Effect of phosphorus in sewage sludge on phosphorus in crops and drainage water. In: Hucker T.W.G. and Catroux G. (eds), *Phosphorus in Sewage Sludge and Animal Waste Slurries.* D. Reidel Publishing Company, Dordrecht, pp. 241-253.

Dehne H.W. (1982) Interaction between vesicular arbuscular mycorrhizal fungi and plant pathogens. *Phytopathology* 72, 1115-1119.

Demirjian Y.A., Westman T.R., Joshi A.M., Rop D.J., Buhl R.V. and Clark W.R. (1984) Land treatment of contaminated sludge with wastewater irrigation. *Journal of the Water Pollution Control Federation* 56, 370-377.

Demirjian Y.A., Joshi A.M. and Westman T.R. (1987) Fate of organic compounds in land application of contaminated municipal sludge. *Journal of the Water Pollution Control Federation* 59, 32-38.

DH; Department of Health (1991) *Dietary Reference Values for Food, Energy and Nutrients for the United Kingdom.* Report on Health and Social Subjects 41. Report of the Panel on Dietary Reference Values of the Committee on Medical Aspects of Food Policy. HMSO, London.

Diercxsens P. and Tarradellas J. (1983) Presentation of the analytical and sampling methods and of results on organo-chlorine in soils improved with sewage sludges and compost. In: Davis R.D., Hucker G. and L'Hermite P. (eds), *Environmental Effects of Organic and Inorganic Contaminants in Sewage Sludge.* D. Reidel Publishing Company, Dordrecht, pp. 59-68.

Dijkshoorn W., Lampe J.E.M. and van Broekhoven L.W. (1981) Influence of soil pH on heavy metals in ryegrass from sludge-amended soil. *Plant and Soil* 61, 277-284.

Dirkzwager A.H. (1991) National sewage sludge policy in the Netherlands (state of affairs, June 1990). *European Water Pollution Control* 1 13-18.

DoE; Department of the Environment (1980) *Cadmium in the Environment and its Significance to Man.* Pollution Paper 17. HMSO, London.

DoE; Department of the Environment (1986) *Nitrate in Water*. Pollution Paper No. 26. HMSO, London.
DoE; Department of the Environment (1989a) Code of Practice for Agricultural Use of Sewage Sludge, HMSO, London.
DoE; Department of the Environment (1989b) *Dioxins in the Environment*. Pollution Paper No. 27. HMSO, London.
DoE; Department of the Environment (1991) The Use of Herbicides in Non-Agricultural Situations in England and Wales. Report No. FR/D0002. Foundation for Water Research, Marlow.
DoE; Department of the Environment (1993) The Examination of Sewage Sludges for Polychlorinated Dibenzo-*p*-dioxins and Polychlorinated Dibenzo-furans. Final Report. Report No. FR/D 0009. Foundation for Water Research, Marlow.
DoE; Department of the Environment (1995) Government response to the recommendations from the review of current rules on sewage sludge use in agriculture. Department of the Environment News Release 283, 22 May 1995.
DoE/NWC; Department of the Environment/National Water Council (1983) Sewage sludge survey 1980 data. DoE/NWC Standing Committee on the Disposal of Sewage Sludge.
Doelman P. (1986) Resistance of soil microbial communities to heavy metals. In: Jensen V., Kjoller A. and Sorensen C.H. (eds), *Microbial Communities in Soil*, FEMS Symposium No. 33. Elsevier Applied Science, Barking, pp. 369-384.
Donnelly K.C., Brown K.W. and Thomas J.C. (1989) Mutagenic potential of municipal sewage sludge amended soils. *Water, Air, and Soil Pollution* 48, 435-449.
Donovan W.C. and Logan T.J. (1983) Factors affecting ammonia volatilization from sewage sludge applied to soil in a laboratory study. *Journal of Environmental Quality* 12, 584-590.
Dowdy R.H., Larson W.E., Titrud J.M. and Latterell J.J. (1978) Growth and metal uptake of snapbeans grown on sewage sludge-amended soil. A four year study. *Journal of Environmental Quality* 7, 252-257.
Dowdy R.H., Latterell J.J., Hinesly T.D., Grossman R.B. and Sullivan D.L. (1991) Trace metal movement in an aeric ochraqualf following 14 years of annual sludge applications. *Journal of Environmental Quality* 20, 119-123.
Drescher-Kaden U., Brüggeman R., Matthes B. and Matthies M. (1992) Contents of organic pollutants in German sewage sludges. In: Hall J.E., Sauerbeck D.R. and L'Hermite P. (eds), *Effects of Organic Contaminants in Sewage Sludge on Soil Fertility, Plants and Animals*. Commission of the European Communities, Luxembourg, pp. 14-34.
Dressler R.L., Storm G.L., Tzilkowski W.M. and Sopper W.E. (1986) Heavy metals in cottontail rabbits on mined lands treated with sewage sludge. *Journal of Environmental Quality* 15, 278-281.

Drewery W.A. and Eliassen R. (1968) Virus movement in groundwater. *Journal of the Water Pollution Control Federation* 40, 257-271.

Drewes C.D., Vining E.P. and Callahan C.A. (1984) Non-invasive electrophysiological monitoring: A sensitive method for detecting sublethal neurotoxicity in earthworms. *Environmental Toxicology and Chemistry* 3, 599-607.

Duah-Yentumi S. and Johnson D.B. (1986) Changes in soil microflora in response to repeated applications of some pesticides. *Soil Biology and Biochemistry* 18, 629-635.

Duarte-Davidson R. and Jones K.C. (1994) Polychlorinated biphenyls (PCBs) in the UK population: Estimated intake, exposure and body burden. *The Science of the Total Environment* 151, 131-152.

Ducommun A. and Matthey W. (1989) The use of sewage sludge on agricultural land - impact on soil fauna. In: Dirkzwager A.H. and L'Hermite P. (eds), *Sewage Sludge Treatment and Use: New Developments, Technological Aspects and Environmental Effects.* Elsevier Science Publishers Ltd, Barking, pp. 440-444.

Dudley L.M., McNeal B.L. and Baham J.E. (1986) Time-dependent changes in soluble organics, copper, nickel and zinc from sludge amended soils. *Journal of Environmental Quality* 15, 188-192.

Dudley L.M., McNeal B.L., Baham J.E., Coray C.S. and Cheng H.H. (1987) Characterization of soluble organic compounds and complexation of copper, nickel and zinc in extracts of sludge-amended soils. *Journal of Environmental Quality* 16, 341-348.

Duncomb D.R., Larson W.E., Clapp C.E., Dowdey R.H., Linden D.R. and Johnson W.K. (1982) Effect of liquid wastewater sludge application on crop yield and water quality. *Journal of the Water Pollution Control Federation* 54, 1185-1193.

Dunigan E.P. and Dick R.P. (1980) Nutrient and coliform losses in run-off from fertilized and sewage sludge-treated soil. *Journal of Environmental Quality* 9, 243-250.

Duynisveld W.H.M., Strebel O. and Böttcher J. (1988) Are nitrate leaching from arable land and nitrate pollution of groundwater avoidable? *Ecological Bulletins* 39, 116-125.

Eaton D.C. (1985) Mineralization of polychlorinated biphenyls by *Phanerochaete chrysosporium*: a ligninolytic fungus. *Enzyme and Microbial Technology* 7, 194-196.

Edmonds R.L. (1976) Survival of coliform bacteria in sewage sludge applied to a forest clearcut and potential movement into groundwater. *Applied Environmental Microbiology* 32, 537-546.

Edwards C.A. and Thompson A.R. (1973) Pesticides and the soil fauna. *Residue Reviews* 45, 1-79.

Edwards R., Lepp N.W. and Jones K.C. (1995) Other less abundant elements of potential environmental significance. In: Alloway, B.J. (ed), *Heavy Metals in Soils,* Second edition. Blackie Academic and Professional, Glasgow, pp. 306-352.

El-Aziz R., Angle J.S. and Chaney R.L. (1991) Metal tolerance of *Rhizobium meliloti* isolated from heavy-metal contaminated soils. *Soil Biology and Biochemistry* 23, 795-798.

Elliott H.A. and Denneny C.M. (1982) Soil adsorption of cadmium from solutions containing organic ligands. *Journal of Environmental Quality* 11, 658-663.

Elliott H.A., Liberati M.R. and Huang C.P. (1986) Competitive adsorption of heavy metals by soils. *Journal of Environmental Quality* 15, 214-219.

Elseewi A.A., Page A.L. and Bingham F.T. (1978) Availability of sulfur in sewage sludge to plants: A comparative study. *Journal of Environmental Quality* 7, 213-217.

Emmerich W.E., Lund L.J., Page A.L. and Chang A.L. (1982a) Predicted solution phase forms of heavy metals in sewage sludge-treated soils. *Journal of Environmental Quality* 11, 182-186.

Emmerich W.E., Lund L.J., Page A.L. and Chang A.C. (1982b) Solid phase forms of heavy metals in sewage sludge-treated soils. *Journal of Environmental Quality* 11, 178-181.

Emmerich W.E., Lund L.J., Page A.L. and Chang A.C. (1982c) Movement of heavy metals in sewage sludge-treated soils. *Journal of Environmental Quality* 11, 174-178.

ENDS; Environmental Data Services (1992) Dioxin formation in sewage sludge calls land spreading into question. ENDS Report 213, October 1992, 10-11.

Engelbrecht R.S. (1978) Microbial hazards associated with wastewater and sludge. *Public Health Engineer* 6, 219-226.

Epstein E., Keane D.B., Meisinger J.J. and Legg J.O. (1978) Mineralization of nitrogen from sewage sludge and sludge compost. *Journal of Environmental Quality* 7, 217-221.

Eriksson J.E. (1989) The influence of pH, soil type and time on adsorption and uptake by plants of Cd added to the soil. *Water, Air, and Soil Pollution* 48, 317-335.

Estes G.O., Knoop W.E. and Houghton F.D. (1973) Soil-plant response to surface-applied mercury. *Journal of Environmental Quality* 2, 451-452.

Fåhraeus G. and Ljunggren H. (1968) Pre-infection phases of the legume symbiosis. In: Gray T.R.G. and Parkinson D. (eds), *The Ecology of Soil Bacteria: An International Symposium.* Liverpool University Press, Liverpool, pp. 396-421.

Fairbanks B.C., O'Connor G.A. and Smith S.E. (1985) Fate of di-2-(ethylhexyl) phthalate in three sludge-amended New Mexico soils. *Journal of Environmental Quality* 14, 479-483.

Fairbanks, B.C., O'Connor G.A. and Smith S.E. (1987) Mineralization and volatilization of polychlorinated biphenyls in sludge-amended soils. *Journal of Environmental Quality* 16, 18-25.

Fangmeier A., Hadwiger-Fangmeier A., Van der Eerden L. and Hans-Jürgen J. (1994) Effects of atmospheric ammonia on vegetation - A review. *Environmental Pollution* 86, 43-82.
Farrah H. and Pickering W.F. (1978) Extraction of heavy metal ions sorbed on clays. *Water, Air, and Soil Pollution* 9, 491-498.
Farrell J.B., Smith J.E., Hathaway S.W. and Dean R.B. (1974) Lime stabilization of primary sludges. *Journal of the Water Pollution Control Federation* 46, 113-121.
Fenn L.B. and Hossner L.R. (1985) Ammonia volatilization from ammonium or ammonium-forming nitrogen fertilizers. *Advances in Soil Science* 1, 123-169.
Ferrer M.R., Gonzalez-Lopez J. and Ramos-Cormenzana A. (1986) Effect of some herbicides on the biological acitvity of *Azotobacter vinelandii. Soil Biology and Biochemistry* 18, 237-238.
Field A.C. and Purves D. (1964) The intake of soil by grazing sheep. *Proceedings of the Nutrition Society* 23, 24-25.
Fielder A.G. and Peel S. (1992) The selection and management of species of cover crop. In: *Nitrate and Farming Systems. Aspects of Applied Biology* 30, 283-290.
Fleming G.A. (1986) Soil injestion by grazing animals; a factor in sludge treated grassland. In: Davis R.D., Haeni H. and L'Hermite P. (eds), *Factors Influencing Sludge Utilization Practices in Europe*. Elsevier Applied Science Publishers Ltd, Barking, pp. 43-50.
Fleming G.A. (1993) The production of animal wastes. In: *Environment Agriculture and Livestock Farming in Europe*. Proceedings of European Conference, Mantova, Italy. Commission of the European Communities.
Flemming C.A., Ferris F.G., Beveride T.J. and Bailey G.W. (1990) Remobilization of toxic heavy metals adsorbed to bacterial wall-clay composites. *Applied and Environmental Microbiology* 56, 3191-3203.
Fließbach A. and Reber H. (1991) Auswirkungen einer langjährigen Zufuhr von Klärschlammen auf Bodenmikroorganismen und ihre Liestungen. In: Sauerbeck, D.R. and Lübben, S. (eds), *Auswirkungen von Siedlungsabfallen auf Böden, Bodenorganismen and Pflanzen*, Volume 6. Berichte aus der Ökologischen Forschung, Forschung, Forschungszentrum Jülich GmbH, pp. 327-358.
Fließbach A. and Reber H. (1992) Effects of long-term sewage sludge applications on soil microbial parameters. In: Hall, J.E., Sauerbeck, D.R. and L'Hermite, P. (eds), *Effects of Organic Contaminants in Sewage Sludge on Soil Fertility, Plants and Animals*. Commission of the European Communities, Luxembourg, pp. 184-192.
Fließbach A., Martens R. and Reber H.H. (1994) Soil microbial biomass and microbial activity in soils treated with heavy metal contaminated sewage sludge. *Soil Biology and Biochemistry* 26, 1201-1205.
Forman D., Al-Dabbagh S. and Doll R. (1985) Nitrates, nitrites and gastric

cancer in Great Britain. *Nature* 313, 620-625.

Fox M.R.S. (1988) Nutritional factors that may influence bioavailability of cadmium. *Journal of Environmental Quality* 17, 175-180.

Fries G.F. (1982) Potential polychlorinated biphenyl residues in animal products from application of contaminated sewage sludge to land. *Journal of Environmental Quality* 11, 14-20.

Fries G.F. (1991) Organic contaminants in terrestrial food chains. In: Jones, K.C. (ed), *Organic Contaminants in the Environment.* Elsevier Science Publishers Ltd, Barking, pp. 207-236.

Fries G.F. and Marrow G.S. (1981) Chlorobiphenyl movement from soil to soybean plants. *Journal of Agricultural and Food Chemistry* 29, 757-759.

Froment M.A., Chalmers A.G. and Smith K.A. (1992) Nitrate leaching from autumn and winter application of animal manures to grassland. In: *Nitrate and Farming Systems. Aspects of Applied Biology* 30, 153-156.

Frost R., Powlesland C., Hall J.E., Nixon S.C. and Young C.P. (1990) Review of Sludge Treatment and Disposal Techniques. WRc Report No. PRD 2306-M/1. WRc Medmenham, Marlow.

Furrer O.J. (1981) Accumulation and leaching of phosphorus as influenced by sludge application. In: Hucker T.W.G. and Catroux G. (eds), *Phosphorus in Sewage Sludge and Animal Waste Slurries.* D. Reidel Publishing Company, Dordrecht, pp. 235-240.

Furrer O.J. and Bolliger R. (1981) Phosphorus content of sludge from Swiss sewage treatment plants. In: Hucker T.W.G. and Catroux G. (eds) *Phosphorus in Sewage Sludge and Animal Waste Slurries.* D. Reidel Publishing Company, Dordrecht, pp. 91-97.

Furrer O.J. and Gupta S.K. (1985) Phosphate balance in long-term sewage sludge and pig slurry fertilized field experiment. In: Williams J.H., Guidi G. and L'Hermite P. (eds), *Long-term Effects of Sewage Sludge and Farm Slurries Applications.* Elsevier Applied Science Publishers Ltd, Barking, pp. 146-150.

Furrer O.J. and Stauffer W. (1986) Influence of sewage sludge and slurry application on nutrient leaching losses. In: Dam Kofoed A., Williams J.H. and L'Hermite P. (eds), *Efficient Land Use of Sludge and Manure.* Elsevier Applied Science Publishers Ltd, Barking, pp. 108-115.

Gadd G.M. and White C. (1989) Heavy metal and radionuclide accumulation and toxicity in fungi and yeasts. In: Poole R.K. and Gadd G.M. (eds), *Metal-Microbe Interactions,* Special Publications of the Society for General Microbiology, Volume 26. IRL Press, Oxford, pp. 19-38.

Gaffney G.R. and Ellertson R. (1979) Ion uptake of redwinged blackbirds nesting on sludge-treated spoils. In: Sopper W. and Kerr S. (eds), *Utilization of Municipal Sewage Effluent on Forest and Disturbed Land.* Pennsylvania State University, pp. 507-516.

Galbally I.E. and Roy C.R. (1983) The fate of nitrogen compounds in the atmosphere. In: Freney J.R. and Simpson J.R. (eds), *Gaseous Loss of*

Nitrogen from Plant-soil Systems. Martinus Nijhoff/Dr W Junk Publishers, The Hague, pp. 265-284.
Garwood E.A. and Ryden J.C. (1986) Nitrate loss through leaching and surface run-off from grassland: Effects of water supply, soil type and management. In: van der Mear H.G., Ryden J.C. and Ennik G.C. (eds), *Nitrogen Fluxes in Intensive Grassland Systems.* Martinus Nijhoff Publishers, Dordrecht, pp. 99-113.
Garwood E.A. and Sinclair J. (1979) Use of water by six grass species, 2. Root distritibution and use of soil water. *Journal of Agricultural Science, Cambridge* 93, 25-35.
Gaunt J.L., Loveland P., Garroway J.L. and Doyle P.J. (1990) Evaluation of Soil Quality Including Soil Structure and Fertility (LDS 9321): Volume I. Final Report to the Department of the Environment: December 1986 to March 1989. WRc Report No. DoE 2104-M/1. WRc Medmenham, Marlow.
Gavaskar A.R., Arthur M.F., Cornaby B.W., Wickramanayake G.B. and Zwick T.C. (1990) Sewage sludge application to agricultural land. In: *Solid/Liquid Separation: Waste Management and Productivity.* Battelle Press, Columbus, pp. 429-437.
Gemmell M.A. (1986) General epidemiology of *Taenia saginata.* In: Block J.C., Havelaar A.H. and L'Hermite P. (eds), *Epidemiological Studies of Risks Associated with the Agricultural Use of Sewage Sludge: Knowledge and Needs.* Elsevier Applied Science Publishers, Barking, pp. 60-71.
Georgievskii V.I. (1982a) General information on minerals. In: Georgievskii V.I. (ed), *Mineral Nutrition of Animals.* Butterworth, London, pp. 70-91.
Georgievskii V.I. (1982b) The physiological role of microelements. In: Georgievskii V.I. (ed), *Mineral Nutrition of Animals.* Butterworth, London, pp. 171-224.
Gerba C.P., Wallis C. and Melnick J.L. (1975) The fate of wastewater bacteria and viruses in the soil. *Proceedings of the American Society of Civil Engineering Irrigation and Drainage Division* 101, IR3, 157-173.
Gerritse R.G., Vriesema R., Dalenberg J.W. and de Roos H.P. (1982) Effect of sewage sludge on trace element mobility in soils. *Journal of Environmental Quality* 11, 359-364.
Gestring W.D. and Jarrell W.M. (1982) Plant availability of phosphorus and heavy metals in soils amended with chemically treated sewage sludge. *Journal of Environmental Quality* 11, 669-675.
Gibson A.H., Curnow B.C., Bergersen F.J., Brockwell J. and Robinson A.C. (1975) Studies of field populations of *Rhizobium:* Effectiveness of strains of *Rhizobium trifolii* associated with *Trifolium subterraneum* L. pastures in south-eastern Australia. *Soil Biology and Biochemistry* 7, 95-102.
Gibson W.P. and Burns R.G. (1977) The breakdown of malathion in soil and soil components. *Microbial Ecology* 3, 219-230.
Giddens J. and Rao A.M. (1975) Effect of incubation and contact with soil on microbial and nitrogen changes in poultry manure. *Journal of*

Environmental Quality 4, 275-278.
Gilbert J. (1994) The fate of environmental contaminants in the food chain. *The Science of the Total Environment* 143, 103-111.
Gilbert W.S. and Lewis C.E. (1982) Residues in soil, pasture and grazing dairy cattle after the incorporation of dieldrin and heptachlor into soil before cropping. *Australian Journal of Experimental Agriculture and Animal Husbandry* 22, 106-115.
Gildon A. and Tinker P.B. (1983a) Interactions of vesicular-arbuscular mycorrhizal infection and heavy metals in plants. I. The effects of heavy metals on the development of vesicular-arbuscular mycorrhizas. *New Phytologist* 95, 247-261.
Gildon A. and Tinker P.B. (1983b) Interactions of vesicular-arbuscular mycorrhizal infections and heavy metals in plants. II. The effects of infection on uptake of copper. *New Phytologist* 95, 263-168.
Giller K.E. and Day J.M. (1985) Nitrogen fixation in the rhizosphere: Significance in natural and agricultural systems. In: Fitter, A.H. (ed), *Ecological Interactions in Soil*, Special Publication Number 4 of The British Ecological Society. Blackwell Scientific Publications, Oxford, pp. 127-147.
Giller K.E., McGrath S.P. and Hirsch P.R. (1989) Absence of nitrogen fixation in clover grown on soil subject to long-term contamination with heavy metals is due to survival of only ineffective *Rhizobium*. *Soil Biology and Biochemistry* 21, 841-848.
Giller K.E., Nussbaum R., Chaudri A.M. and McGrath S.P. (1993) *Rhizobium meliloti* is less sensitive to heavy-metal contamination in soil than *R. leguminosarum* bv. *trifolii* or *R. loti*. *Soil Biology and Biochemistry* 25, 273-278.
Giordano P.M. and Mortvedt J.J. (1976) Nitrogen effects on mobility and plant uptake of heavy metals in sewage sludge applied to soil columns. *Journal of Environmental Quality* 5, 165-168.
Giordano P.M., Mortvedt J.J. and Mays D.A. (1975) Effect of municipal wastes on crop yields and uptake of heavy metals. *Journal of Environmental Quality* 4, 394-399.
Giordano P.M., Mays D.A. and Behel Jr. A.D. (1979) Soil temperature effects on uptake of cadmium and zinc by vegetables grown on sludge-amended soil. *Journal of Environmental Quality* 8, 233-236.
Glantz P.J. and Jacks T.M. (1967) Significance of *Escherichia coli* serotypes in waste water effluent. *Journal of the Water Pollution Control Federation* 39, 1918-1921.
Glendining M.J., Poulton P.R. and Powlson D.S. (1992) The relationship between inorganic N in soil and the rate of fertilizer N applied on the Broadbalk Wheat Experment. In: *Nitrate and Farming Systems. Aspects of Applied Biology* 30, 95-102.
Glockemann B. and Larink O. (1989) Einfluss von klärschlammdüngung und

schwermetallbe lastung auf milben, speziell Gamasiden, in einem ackerboden. *Pedobiologia* 33, 237-246.

Goffeng G., Oeydvin J., Hammeraas B. and Loewe A. (1978) The survival of potato cysts nematode during windrow composting of municipal waste. *Kongsvingerundersoekelsene* 3(11), 14.

Goulding K.W.T. (1990) Nitrogen deposition to land from the atmosphere. *Soil Use and Management* 6, 61-63.

Goyer R.A. (1981) Lead. In: Bronner F. and Coburn J.W. (eds), *Disorders of Mineral Metabolism, Volume 1. Trace Minerals.* Academic Press, Inc. (London) Ltd, London, pp. 159-199.

Grace N.D. (1983) *The Mineral Requirements of Grazing Ruminants.* Occasional Publication No. 9. New Zealand Society of Animal Production.

Gram E. (1960) Quarantines. In: Horsfall J.G. and Dimond A.E. (eds), *Plant Pathology: An Advanced Treatise, Volume III. The Diseased Population: Epidemics and Control.* Academic Press Limited, London, pp. 313-356.

Grant R.O. and Olesen S.E. (1984) Sludge utilization in spruce plantations on sandy soils. In: Berglund S., Davis R.D. and L'Hermite P. (eds), *Utilization of Sewage Sludge on Land: Rates of Application and Long-term Effects of Metals.* D. Reidel Publishing Company, Dordrecht, pp. 79-90.

Green C.F. and Ivins J.D. (1985) Time of sowing and the yield of winter wheat. *Journal of Agricultural Science, Cambridge* 104, 235-238.

Griffin G.F. and Laine A.F. (1983) Nitrogen mineralization in soils previously amended with organic wastes. *Agronomy Journal* 75, 124-129.

Guerra A. (1994) The effect of organic matter content on soil erosion in simulated rainfall experiments in W. Sussex, UK. *Soil Use and Management* 10, 60-64.

Guidi G. and Hall J.E. (1984) Effects of sewage sludge on the physical and chemical properties of soils. In: L'Hermite P. and Ott H. (eds), *Processing and Use of Sewage Sludge.* D. Reidel Publishing Company, Dordrecht, pp. 295-306.

Hadar Y. (1986) The role of organic matter in the introduction of biofertilizers and biocontrol agents to soils. In: Chen Y. and Avnimelech Y. (eds), *The Role of Organic Matter in Modern Agriculture.* Martinus Nijhoff Publishers, Dordrecht, pp. 169-179.

Haghiri F. (1974) Plant uptake of cadmium as influenced by cation exchange capacity, organic matter, zinc and soil temperature. *Journal of Environmental Quality* 3, 180-183.

Haigh R.A. and White R.E. (1986) Nitrate leaching from a small, underdrained, grassland, clay catchment. *Soil Use and Management* 2, 65-70.

Haimi J., Salminen J., Huhta V., Knuutinen J. and Palm H. (1992) Bioaccumulation of organochlorine compounds in earthworms. *Soil Biology and Biochemistry* 24, 1699-1703.

Hall G.A., Jones P.W. and Aitken M.M. (1978) The pathogenesis of experimental intra-ruminal infections of cows with *Salmonella dublin.*

Journal of Comparative Pathology 88, 409-418.
Hall J.E. (1984) Predicting the nitrogen value of sewage sludges. In: L'Hermite P. and Ott H. (eds), *Processing and Use of Sewage Sludge.* D. Reidel Publishing Company, Dordrecht, pp. 268-278.
Hall J.E. (1985) The cumulative and residual effects of sewage sludge nitrogen on crop growth. In: Williams J.H., Guidi G. and L'Hermite P. (eds), *Long-term Effects of Sewage Sludge and Farm Slurry Applications.* Elsevier Applied Science Publishers Ltd, Barking, pp. 73-83.
Hall J.E. (1986a) The Agricultural Value of Sewage Sludge. WRc Report No. ER 1220-M. WRc Medmenham, Marlow.
Hall J.E. (1986b) Machinery spreading: soil injection as a barrier to odour dispersion. In: Nielsen V.C., Voorburg J.H. and L'Hermite P. (eds), *Odour Prevention and Control of Organic Sludge and Livestock Farming.* Elsevier Applied Science Publishers Ltd, Barking, pp. 194-206.
Hall J.E. (1986c) Soil injection of organic manures. In: Solbé, J.F. de L.G. (ed), *Effects of Land Use on Fresh Waters: Agriculture, Forestry, Mineral Exploitation, Urbanisation.* WRc Medmenham, Marlow, pp. 521-527.
Hall J.E. (1988) Environmental effects of ammonia volatilization from agriculture. In: Vetter H., Steffens G. and L'Hermite P. (eds), *Safe and Efficient Slurry Utilization.* SL/131/89. XII/ENV/3/89. Commission of the European Communities, Luxembourg, pp. 71-77.
Hall J.E. (1992) Treatment and use of sewage sludge. In: Bradshaw A.D., Southwood R. and Warner F. (eds), *The Treatment and Handling of Wastes.* Chapman & Hall, London, pp. 63-82.
Hall J.E. (1993) Alternative Uses of Sewage Sludge for Land Application (EHA 3516). Final Report to the Department of the Environment: October 1992 to March 1993. WRc Report No. 3357/1. WRc Medmenham, Marlow.
Hall J.E. (ed) (1995) *Animal Waste Management.* Proceedings of the Seventh Technical Consultation on the European Cooperative Research Network on Animal Waste Management, Bad Zwischenahn, Germany, 17-20 May. REUR Technical Series 34. Food and Agriculture Organization, Rome.
Hall J.E. and Coker E.G. (1983) Some effects of sewage sludge on soil physical conditions and plant growth. In: Catroux G., L'Hermite P. and Suess E. *The Influence of Sewage Sludge Application of Physical and Biological Properties of Soils.* D. Reidel Publishing Company, Dordrecht, pp. 43-60.
Hall J.E. and Dalimier F. (1994) Waste Management - Sewage Sludge. Part I - Survey of Sludge Production, Treatment, Quality and Disposal in the European Union. WRc Report No. EC 3646. WRc Medmenham, Marlow.
Hall J.E. and Ryden J.C. (1986) Current UK research into ammonia losses from sludges and slurries. In: Dam Kofoed A., Williams J.H. and L'Hermite P. (eds), *Efficient Land Use of Sludge and Manure.* Elsevier Applied Science Publishers Ltd, Barking, pp. 180-192.
Hall J.E. and Williams J.H. (1984) The use of sewage sludge on arable and grassland. In: Berglund S., Davis R.D. and L'Hermite P. (eds), *Utilization*

of Sewage Sludge on Land: Rates of Application and Long-term Effects of Metals. D. Reidel Publishing Company, Dordrecht, pp. 22-35.

Hall J.E., Godwin R.J., Warner N.L. and Davis J.M. (1986) Soil Injection of Sewage Sludge. WRc Report No. ER 1202-M. WRc Medmenham, Marlow.

Hall J.E., Papadopoulos P. and Carlton-Smith C.H. (1988) The effect of soil pH on the concentration of heavy metals in ryegrass and soil solution. In: Astruc M. and Lester J.N. (eds), *Heavy Metals in the Hydrological Cycle*. Selper Ltd., London, pp. 281-286.

Hall J.E., L'Hermite P. and Newman P.J. (eds) (1992) *Treatment and Use of Sewage Sludge and Liquid Agricultural Wastes. Review of COST 68/681 Programme, 1972-90*. Commission of the European Communities, Luxembourg.

Halsall C., Burnett V., Davis B., Jones P., Pettit C. and Jones K.C. (1993) PCBs and PAHs in U.K. urban air. *Chemosphere* 26, 2185-2197.

Hampton A.N. (1992) The Simon N-Viro sludge to soil sludge pasteurization process. Utilization of Sewage Sludge on Land. A Workshop in The University of Birmingham, January 7-8, 1992.

Hann M.J., Atkinson C.J., Godwin R.J. and Smith S.R. (1992) The Fate of Nitrogen Resulting from the Application of Sewage Sludge. Report No. FR 0316. Foundation for Water Research, Marlow.

Hansen J.A. and Henriksen K. (eds) (1989) Gaseous losses. In: *Nitrogen in Organic Wastes Applied to Soils*. Academic Press Ltd, London, pp. 185-286.

Haque A. and Ebing W. (1983) Toxicity determination of pesticides to earthworms in the soil substrate. *Zeitschrift für Pflanzenkrankheiten und Pflanzenschutz* 90, 395-408.

Harding S.A., Clapp C.E. and Larson W.E. (1985) Nitrogen availability and uptake from field soils five years after addition of sewage sludge. *Journal of Environmental Quality* 14, 95-100.

Harrad S.J., Sewart A.P., Alcock R., Boumphrey R., Burnett V., Duarte-Davidson R., Halsall C., Sanders G., Waterhouse K., Wild S.R. and Jones K.C. (1994) Polychlorinated biphenyls (PCBs) in the British environment: Sinks, sources and temporal trends. *Environmental Pollution* 85, 131-146.

Harrison D.C., Mol J.C.M. and Healy W.B. (1970) DDT residues in sheep from ingestion of soil. *New Zealand Journal of Agricultural Research* 13, 664-672.

Hartenstein R. (1981) Potential use of earthworms as a solution to sludge management. *Water Pollution Control* 80, 638-643.

Hartenstein R., Neuhauser E.F. and Collier J. (1980) Accumulation of heavy metals in the earthworm *Eisenia foetida*. *Journal of Environmental Quality* 9, 23-26.

Hartenstein R., Neuhauser E.F. and Narahara A. (1981) Effects of heavy metal and other elemental additives to activated sludge on growth of *Eisenia foetida. Journal of Environmental Quality* 10, 372-376.

Harter R.D. (1983) Effect of soil pH on adsorption of lead, copper, zinc and nickel. *Soil Science Society of America Journal* 47, 47-51.

Hassall K.A. (1982) *The Chemistry of Pesticides: Their Metabolism, Mode of Action and Uses in Crop Production.* Macmillan Education Ltd, Basingstoke.

Hatch D.J., Jones L.H.P. and Burau R.G. (1988) The effect of pH on the uptake of cadmium by four plant species grown in flowing solution culture. *Plant and Soil* 105, 121-126.

Hatton D. and Pickering W.F. (1980) The effect of pH on the retention of Cu, Pb, Zn and Cd by clay-humic acid mixtures. *Water, Air, and Soil Pollution* 14, 13-21.

Hayman D.S. (1983) The physiology of vesicular arbuscular endo-mycorrhizal symbiosis. *Canadian Journal of Botany* 61, 944-963.

Hays B.D. (1977) Potential for parasitic disease transmission with land application of sewage plant effluents and sludges. *Water Research* 11, 583-595.

Healy W.B. (1967) Ingestion of soil by sheep. *Proceedings of the New Zealand Society of Animal Production* 27, 109-120.

Healy W.B. (1973) Nutritional aspects of soil ingestion by grazing animals. In: Butler G.W. and Bailey R.W. (eds), *Chemistry and Biochemistry of Herbage*, Volume 1. Academic Press Inc. (London) Ltd., London, pp. 567-588.

Healy W.B. and Ludwig T.G. (1965) Wear of sheep's teeth. 1. The role of ingested soil. *New Zealand Journal of Agricultural Research* 8, 737-752.

Hedgecott S. and Rogers H.R. (1991) The Potential Impact of Domestic Chemical Use on Fresh Water Quality. Report for Royal Commission on Environmental Pollution. WRc Report No. CO 2672-M/1. WRc Medmenham, Marlow.

Hegsted D.M. (1971) Interactions in nutrition. In: Mertz W. and Cornatzer W.E. (eds), *Newer Trace Elements in Nutrition.* Marcel Dekker Inc., New York, pp. 19-32.

Heimbach F. (1985) Comparison of laboratory methods using *Eisenia foetida* and *Lumbricus terrestris*, for the assessment of the hazard of chemicals to earthworms. *Zeitschrift für Pflanzenkrankheiten und Pflanzenschutz* 92, 186-193.

Hellström T. and Dahlberg, A-G. (1994) Swedish experience in gaining acceptance for the use of biosolids in agriculture. In: *The Management of Water and Wastewater Solids for the 21st Century: a Global Perspective.* Paper presented at WEF Conference, 19-22 June 1994, Washington, ESQ. WEF Speciality Conference Series Proceedings.

Helmke P.A., Robarge W.P., Korotev R.L. and Schomberg P.J. (1979) Effects of soil-applied sewage sludge on concentrations of elements in earthworms. *Journal of Environmental Quality* 8, 322-327.

Hembrock-Heger A. (1992) Persistent organic contaminants in soils, plants and food. In: Hall J.E., Sauerbeck D.R. and L'Hermite P. (eds), *Effects of Organic Contaminants in Sewage Sludge on Soil Fertility, Plants and Animals.* Commission of the European Communities, Luxembourg, pp. 78-89.

Hemington W.S. (1992) Practical measures to reduce nitrate leaching in a NSA. In: *Nitrate and Farming Systems. Aspects of Applied Biology* 30, 393-398.

Heukelekian H. and Albanese M. (1956) Enumeration and survival of tubercle bacilli in polluted waters. 2. Effect of sewage treatment and natural purification. *Sewage and Industrial Wastes* 28, 1094-1102.

Higgins A.J. (1984) Land application of sewage sludge with regard to cropping systems and pollution potential. *Journal of Environmental Quality* 13, 441-448.

Hill S.A. and Torrance L. (1989) Rhizomania disease of sugar beet in England. *Plant Pathology* 38, 114-122.

Hinesly T.D., Ziegler E.L. and Tyler J.J. (1976) Selected chemical elements in tissues of pheasants fed corn grain from sewage sludge-amended soil. *Agro-Ecosystems* 3, 11-26.

Hinesly T.D., Redborg K.E., Ziegler E.L. and Alexander J.D. (1982) Effect of soil cation exchange capacity on the uptake of cadmium by corn. *Soil Science Society of America Journal* 46, 490-497.

Hinesly T.D., Redborg K.E., Pietz R.I. and Ziegler E.L. (1984) Cadmium and zinc uptake by corn (*Zea mays* L.) with repeated applications of sewage sludge. *Journal of Agricultural and Food Chemistry* 32, 155-163.

Hirsch P.R., Jones M.J., McGrath S.P. and Giller K.E. (1993) Heavy metals from past applications of sewage sludge decrease the genetic diversity of *Rhizobium leguminosarum* biovar *trifolii* populations. *Soil Biology and Biochemistry* 25, 1485-1490.

Hoff J.D., Nelson D.W. and Sutton A.L. (1981) Ammonia volatilization from liquid swine manure applied to cropland. *Journal of Environmental Quality* 10, 90-95.

Hogg T.J., Stewart J.W.B. and Brettany J.R. (1978) Influence of the chemical form of mercury on its adsorption and ability to leach through soils. *Journal of Environmental Quality* 7, 440-445.

Hogue D.E., Parrish J.J., Foote R.H. and Stouffer J.R. (1984) Toxicologic studies with male sheep grazing on municipal sludge amended soil. *Journal of Toxicology and Environmental Health* 14, 153-161.

Hohla G.N., Jones R.L. and Hinesly T.D. (1978) The effect of anaerobically digested sewage sludge on organic fractions of Blount silt loam. *Journal of Environmental Quality* 7, 559-563.

Holding A.J. and King J. (1963) The effectiveness of indigenous populations of *Rhizobium trifolii* in relation to soil factors. *Plant and Soil* 18, 191-198.

Holt M.S., Matthijs E. and Waters J. (1989) The concentrations and fate of linear alkylbenzene sulphonate in sludge amended soils. *Water Research*

23, 749-759.

Holtzclaw K.M., Keech D.A., Page A.L., Sposito G., Ganje T.J. and Ball N.B. (1978) Trace metal distributions among the humic acid, the fulvic acid and precipitable fractions extracted with NaOH from sewage sludges. *Journal of Environmental Quality* 7, 124-127.

Honeycutt C.W., Potaro L.J. and Halteman W.A. (1991) Predicting nitrate formation from soil, fertilizer, crop residue, and sludge with thermal units. *Journal of Environmental Quality* 20, 850-856.

Hopke P.K., Plewa K.J. and Stapleton P. (1987) Reduction of mutagenicity of municipal wastewaters by land treatment. *The Science of the Total Environment* 66, 193-202.

Hopkin S.P. (1989) *Ecophysiology of Metals in Terrestrial Invertebrates*. Elsevier Applied Science Publishers Ltd, Barking.

Horstmann M., Kaune A., McLachlan M.S., Reissinger M. and Hutzinger O. (1992) Temporal variability of PCDD/F concentrations in sewage sludge. *Chemosphere* 25, 1463-1468.

Huang W-Y., Breach E.D., Fernandez-Cornejo J. and Uri N.D. (1994) An assessment of the potential risks of groundwater and surface water contamination by agricultural chemicals used in vegetable production. *The Science of the Total Environment* 153, 151-167.

Hutton M. and Symon C. (1986) The quantities of cadmium, lead, mercury and arsenic entering the UK environment from human activities. *The Science of the Total Environment* 57, 129-150.

Huylebroeck J. (1981) Review of research projects on ground water pollution from agricultural use of sewage sludge. Treatment and Use of Sewage Sludge. Commission of the European Communities. Cost Project 68 Bis. Final Report III. Technical annexes, pp. 491-521.

Huysman F., Verstraete W. and Brookes P.C. (1994) Effect of manuring practices and increased copper concentrations on soil microbial populations. *Soil Biology and Biochemistry* 26, 103-110.

Ilic P., Knofel S. and Sach-Paulus, N. (1994) Contamination of cultivated soils with polychlorinated dioxins/furans and polychlorinated biphenyls relevant to agriculture utilization of sewage sludge in the area of the Frankfurt District Cooperative. *Korrespondenz Abwasser* 41, 1268-1275.

Insam H. and Haselwandter K. (1989) Metabolic quotient of the soil microflora in relation to plant succession. *Oecologia* 79, 174-178.

Insam H., Parkinson D. and Domsch K.H. (1989) Influence of macroclimate on soil microbial biomass. *Soil Biology and Biochemistry* 21, 211-221.

Insam H., Mitchell C.C. and Dormaar J.F. (1991) Relationship of soil microbial biomass and activity with fertilization practice and crop yield of three ultisols. *Soil Biology and Biochemistry* 23, 459-464.

Isensee A.R. and Jones G.E. (1971) Absorption and translocation of root and foliage applied 2,4-dichlorophenol, 2,7-dichlorodibenzo-*p*-dioxin, and 2,3,7,8-tetrachlorodibenzo-*p*-dioxin. *Journal of Agricultural and Food*

Chemistry 19, 1210-1214.

Istas J.R., de Borger R., de Temmerman L., Guns, Meeus-Verdinne K., Ronse A., Scokart P. and Termonia M. (1988) *Environment and Quality of Life: Effect of Ammonia on the Acidification of the Environment.* EUR 11857EN. Commission of the European Communities, Luxembourg.

Jacks G.V., Brind W.D. and Smith R. (1955) *Mulching.* Technical Communication No. 49 of the Commonwealth Bureau of Soil Science. Commonwealth Agricultural Bureaux, Farnham Royal.

Jackson A.P. and Alloway B.J. (1991) The bioavailability of cadmium to lettuce and cabbage in soils previously treated with sewage sludges. *Plant and Soil* 132, 179-186.

Jackson N.E. (1985) Tolerance of red clover to toxic metals in sewage treated soil. In: *Research and Development in the Midlands and Western Region, 1985.* Ministry of Agriculture, Fisheries and Food/Agricultural Development and Advisory Service, Wolverhampton, pp. 173-176.

Jackson W.A., Wilkinson S.R. and Leonard R.A. (1977) Land disposal of broiler litter: Changes in concentration of chloride, nitrate nitrogen, total nitrogen, and organic matter in a Cecil sandy loam. *Journal of Environmental Quality* 6, 58-62.

Jacobs L.W. and Zabik M.J. (1983) Importance of sludge-borne organic chemicals for land application programs. In: *Proc. 6th Ann. Madison Conf. of Applied Research and Practice on Municipal and Industrial Waste,* 14-15 Sept, Madison, Dept of Engineering and Applied Science. University of Wisconsin, Madison, pp. 418-426.

Jacobs L.W., O'Connor G.A., Overcash M.A., Zabik M.J. and Rygiewicz P., Machno P., Munger S. and Elseavi A.A. (1987) Effects of trace organics in sewage sludges on soil-plant systems and assessing their risk to humans. In: Page A.L., Logan T.J. and Ryan J.A. (eds), *Land Application of Sludge Food Chain Implications.* Lewis Publishers Inc., Chelsea, Michigan, pp. 101-143.

Jakubowski W. (1986) US EPA-sponsored epidemiological studies of health effects associated with the treatment and disposal of wastewater and sewage sludge. In: Block J.C., Havelaar A.H. and L'Hermite P. (eds), *Epidemiological Studies of Risks Associated with the Agricultural Use of Sewage Sludge: Knowledge and Needs.* Elsevier Applied Science Publishers Ltd, Barking, pp. 140-153.

James B.R. and Bartlett R.J. (1984) Plant-soil interactions of chromium. *Journal of Environmental Quality* 13, 67-70.

Jansson P-E., Antil R.S. and Borg G.C.H. (1989) Simulation of nitrate leaching from arable soils treated with manure. In: Hansen J.A. and Henriksen K. (eds), *Nitrogen in Organic Wastes Applied to Soils.* Academic Press Limited, London, pp. 150-166.

Jarvis S.C. (1992) Grazed grassland management and nitrogen losses: an overview. In: *Nitrate and Farming Systems. Aspects of Applied Biology* 30,

207-214.

Jarvis S.C. (1994) The pollution potential and flows of nitrogen to waters and the atmosphere from grassland under grazing. In: Ap Dewi I., Axford R.F.E., Marai I.F.M. and Omed H.M. (eds), *Pollution in Livestock Production Systems.* CAB International, Wallingford, pp. 227-239.

Jenkinson D.S. (1990a) The turnover of organic carbon and nitrogen in soil. *Philosophical Transactions of the Royal Society* B329, 361-368.

Jenkinson D.S. (1990b) An introduction to the global nitrogen cycle. *Soil Use and Management* 6, 56-61.

Jenkinson D.S. and Ladd J.N. (1981) Microbial biomass in soil: Measurement and turnover. In: Paul E.A. and Ladd J.N. (eds), *Soil Biochemistry*, Volume 5. Marcel Dekker, New York, pp. 415-471.

Jenkinson D.S. and Parry L.C. (1989) The nitrogen cycle in the Broadbalk wheat experiment: A model for the turnover of nitrogen through the soil microbial biomass. *Soil Biology and Biochemistry* 21, 535-541.

Jenkinson D.S. and Powlson D.S. (1976) The effects of biocidal treatments on metabolism in soil - V. A method for measuring soil biomass. *Soil Biology and Biochemistry* 8, 209-213.

Jin Y. and O'Connor G.A. (1990) Behaviour of toluene added to sludge-amended soils. *Journal of Environmental Quality* 19, 573-579.

Jing J. and Logan T.J. (1992) Effects of sewage sludge cadmium concentration on chemical extractability and plant uptake. *Journal of Environmental Quality* 21, 73-81.

John M.K. and van Laerhoven C.J. (1976) Effects of sewage sludge composition, application rate and lime regime on plant availability of heavy metals. *Journal of Environmental Quality* 5, 246-251.

Johnson C.B., Mannering J.V. and Moldenhauer W.C. (1979) Influence of surface roughness and clod stability on soil and water loss. *Soil Science Society of America Journal* 43, 772-777.

Johnson D.E., Kienhol E.W., Baxter J.C., Spangler E. and Ward G.M. (1981) Heavy metal retention in tissues of cattle fed high cadmium in sewage sludge. *Journal of Animal Science* 52, 108-114.

Johnson H.S. (1971) Reduction of stratospheric ozone by nitrogen oxide catalysts from supersonic transport exhaust. *Science* 173, 517-522.

Johnston A.E. (1981) Accumulation of phosphorus in a sandy loam soil from farmyard manure (FYM) and sewage sludge. In: Hucket T.W.G. and Catrouse G. (eds), *Phosphorus in Sewage Sludge and Animal Waste Slurries.* D. Reidel Publishing Company, Dordrecht, pp. 273-285.

Johnston A.E., McGrath S.P., Poulton P.R. and Lane P.W. (1989) Accumulation and loss of nitrogen from manure, sludge and compost: Long-term experiments at Rothamsted and Woburn. In: Hansen J.A. and Henriksen K. (eds), *Nitrogen in Organic Wastes Applied to Soils.* Academic Press Limited, London, pp. 126-139.

Johnston N.B., Beckett P.H.T. and Waters C.J. (1983) Limits of zinc and copper toxicity from digested sludge applied to agricultural land. In: Davis R.D., Hucket G. and L'Hermite P. (eds), *Environmental Effects of Organic and Inorganic Contaminants in Sewage Sludge*. D. Reidel Publishing Company, Dordrecht, pp. 75-81.

Joint Conference on Recycling Municipal Sludges and Effluents on Land (1973) National Association of State University and Land Grant Colleges, Washington D.C.

Jollans J.L. (1985) *Fertilizers in UK Farming*. CAS Report 9. Centre for Agricultural Strategy, University of Reading, Reading.

Jones A.E., Crewe W. and Owen R.R. (1979) Enhanced survival of eggs of *Ascaris* following ingestion by earthworms. *Transactions of the Royal Society of Tropical Medicine and Hygiene* 73, 325.

Jones D.M., Sutcliffe E.M. and Curry A. (1991) Recovery of viable but non-culturable *Campylobacter jejuni*. *Journal of General Microbiology* 137, 2477-2482.

Jones K.C. and Johnston A.E. (1989) Cadmium in cereal grain and herbage from long-term experimental plots at Rothamsted, UK. *Environmental Pollution* 57, 199-216.

Jones K.C. and Wild S.R. (1991) Organic Chemicals Entering Agricultural Soils in Sewage Sludges: Screening for their Potential to Transfer to Crop Plants and Livestock. Report No. FR 0169. Foundation for Water Research, Marlow.

Jones K.C., .Obbard J.P., Ineson P., Kichenside N. and Stark J.H. (1987) Microbial Biomass and Microbially Mediated N-transformation Processes in Historically Sludged and Metal Contaminated Grassland Soils. WRc Unpublished Report. WRc Medmenham, Marlow.

Jones P.W. (1986) Sewage sludge as a vector of salmonellosis. In: Block J.C., Havelaar A.H. and L'Hermite P. (eds), *Epidemiological Studies of Risks Associated with the Agricultural Use of Sewage Sludge: Knowledge and Needs*. Elsevier Applied Science Publishers Ltd, Barking, pp. 21-33.

Jones P.W. and Twigg G.I. (1976) Salmonellosis in wild animals. *Journal of Hygiene* 71, 51-54.

Jones S.G., Brown K.W., Deuel L.E. and Donnelly K.C. (1979) Influence of simulated rainfall on the retention of sludge heavy metals by the leaves of forage crops. *Journal of Environmental Quality* 8, 69-72.

Kabata-Pendias A. and Pendias H. (1992) *Trace Elements in Soils and Plants*, 2nd Edition. CRC Press Inc., Boca Raton.

Kampe W. (1984) Cd and Pb in the consumption of foodstuffs depending on various contents of heavy metals In: L'Hermite P. and Ott H. (eds), *Processing and Use of Sewage Sludge*. D. Reidel Publishing Company, Dordrecht, pp. 334-348.

Kampe W. and Leschber R. (1989) Occurrence of organic pollutants in soil and plants after intensive sewage sludge application. In: Quaghebeur D., Temmerman I. and Angeletti G. (eds), *Organic Contaminants in Waste*

Water, Sludge and Sediment Occurrence, Fate and Disposal. Elsevier Applied Science Publishers Ltd, Barking, pp. 35-41.

Karapanagiotis N.K., Sterritt R.M. and Lester J.N. (1991) Heavy metal complexation in sludge-amended soil. The role of organic matter in metal retention. *Environmental Technology* 12, 1107-1116.

Keefer R.F., Codling E.E. and Singh R.N. (1984) Fractionation of metal-organic components extracted from a sludge-amended soil. *Soil Science Society of America Journal* 48, 1054-1059.

Kelley W.D., Simpson T.W., Renea Jr., R.B., McCart G.D. and Martens D.C. (1984) Perception of technical personnel on research and education needs for land application of sewage sludge. *Water, Air, and Soil Pollution* 22, 181-185.

Kelling K.A., Keeney D.R., Walsh L.M. and Ryan J.A. (1977a) A field study of the agricultural use of sewage sludge: III. Effect on uptake and extractability of sludge-borne metals. *Journal of Environmental Quality* 6, 352-358.

Kelling K.A., Peterson A.E. and Walsh L.M. (1977b) Effect of wastewater sludge on soil moisture relationships and surface run-off. *Journal of the Water Pollution Control Federation* 49, 1698-1703.

Kemppainen E. (1990) Injection of cow slurry to grass and barley. In: Hall J.E. (ed), *Recent Developments in Animal Waste Utilisation.* REUR Technical Series 17. Food and Agriculture Organisation, Rome, pp. 242-248.

Kenrick M.A.P., Clark L., Baxter K.M., Fleet M., James H.A., Gibson T.M. and Turrell M.B. (1985) *Trace Organics in British Aquifers - A Baseline Survey.* Technical Report TR 223. WRc Medmenham, Marlow.

Keswick B. (1984) Sources of groundwater pollution. In: Bitton G. and Gerba C.P. (eds), *Groundwater Pollution Microbiology.* John Wiley and Sons, New York, pp. 39-64.

Khan D.H. and Frankland B. (1983) Effects of cadmium and lead on radish plants with particular reference to movement of metals through soil profile and plant. *Plant and Soil* 70, 335-345.

Kiekens L. (1984) Behaviour of heavy metals in soils. In: Berglund S., Davis R.D. and L'Hermite P. (eds), *Utilization of Sewage Sludge on Land: Rates of Application and Long-term Effects of Metals.* D. Reidel Publishing Company, Dordrecht, pp. 126-134.

Kiekens L. and Cottenie A. (1985) Principles of investigations on the mobility and plant uptake of heavy metals. In: Leschber R., Davis R.D. and L'Hermite P. (eds), *Chemical Methods for Assessing Bio-available Metals in Sludges and Soils.* Elsevier Applied Science Publishers Ltd, Barking, pp. 32-41.

Killham K. (1985) A physiological determination of the impact of environmental stress on the activity of microbial biomass. *Environmental Pollution (Series A)* 38, 283-294.

Kim S.J., Chang A.C., Page A.L. and Warneke J.E. (1988) Relative concentrations of cadmium and zinc in tissue of selected food plants grown on sludge-treated soils. *Journal of Environmental Quality* 17, 568-573.

King L.D. (1973) Mineralization and gaseous loss of nitrogen in soil-applied liquid sewage sludge. *Journal of Environmental Quality* 2, 356-358.

King L.D. (1981) Effect of swine manure lagoon sludge and municipal sewage sludge on growth, nitrogen recovery, and heavy metal content of fescuegrass. *Journal of Environmental Quality* 10, 465-472.

King L.D. (1988a) Effect of selected soil properties on cadmium content of tobacco. *Journal of Environmental Quality* 17, 251-255.

King L.D. (1988b) Retention of metals by several soils of the South-eastern United States. *Journal of Environmental Quality* 17, 239-246.

King L.D. (1988c) Retention of cadmium by several soils of the South-eastern United States. *Journal of Environmental Quality* 17, 246-250.

King L.D. and Dunlop W.R. (1982) Application of sewage sludge to soils high in organic matter. *Journal of Environmental Quality* 11, 608-616.

Kirby D.R. and Stuth J.W. (1980) Soil ingestion rates of steers following brush management in central Texas. *Journal of Range Management* 33, 207-209.

Kirchmann H., Åström H. and Jönsäll G. (1991) Organic pollutants in sewage sludge. 1. Effect of toluene, naphthalene, 2-methylnaphthalene, 4-n-nonylphenol and di-2-ethylhexyl phthalate on soil biological processes and their decompositon in soil. *Swedish Journal of Agricultural Research* 21, 107-113.

Kladivko E.J. and Nelson D.W. (1979) Surface run-off from sludge-amended soils. *Journal of the Water Pollution Control Federation* 51, 100-110.

Klessa D.A. and Desira-Buttigieg A. (1992) The adhesion to leaf surfaces of heavy metals from sewage sludge applied to grassland. *Soil Use and Management* 8, 115-121.

Kloke A. (1983) Tolerable amounts of heavy metals in soils and their accumulation in plants. In: Davis R.D., Hucker G. and L'Hermite P. (eds), *Environmental Effects of Organic and Inorganic Contaminants in Sewage Sludge*. D. Reidel Publishing Company, Dordrecht, pp. 171-175.

Koeppe D.E. (1981) Lead: Understanding the minimal toxicity of lead in plants. In: Lepp N.W. (ed), *Effect of Heavy Metal Pollution on Plants, Volume 1; Effect of Trace Metals on Plant Function.* Applied Science Publishers, London, pp. 55-76.

Kolenbrander G.J. (1981) Effect of injection of animal waste on ammonia losses by volatilization on arable land and grassland. In: Brogan J.C. (ed), *Nitrogen Losses and Surface Run-off from Landspreading of Manures.* Martinus Nijhoff/Dr W. Junk Publishers, The Hague, pp. 425-430.

Koomen I., McGrath S.P. and Giller K.E. (1990) Mycorrhizal infection of clover is delayed in soils contaminated with heavy metals from past sewage sludge applications. *Soil Biology and Biochemistry* 22, 871-873.

Korcak R.F. and Fanning D.S. (1985) Availability of applied heavy metals as a function of type of soil material and metal source. *Soil Science* 140, 23-34.

Kowal N.E. (1983) An overview of public health effects. In: Page A.L., Gleason III T.L., Smith Jr J.E., Iskander I.K. and Sommers L.E. (eds), *Proceedings of the 1983 Workshop on Utilization of Municipal Wastewater and Sludge on Land.* University of California, Riverside, pp. 329-394,

Kroeze C. (1994) Nitrous oxide and global warming. *The Science of the Total Environment* 143, 193-209.

Kruse E.A. and Barrett G.W. (1985) Effects of municipal sludge and fertilizer on heavy metal accumulation in earthworms. *Enviromental Pollution (Series A)* 38, 235-244.

Kügel G. (1982) Simultaneous pasteurisation-digestion (SPD-process). *Water Science and Technology* 14, 739-748.

Kula H. and Kokta C. (1992) Side effects of selected pesticides on earthworms under laboratory and field conditons. *Soil Biology and Biochemistry* 24, 1711-1714.

Kumper F. (1985) Corrosion problems in domestic heating. *Gas/Wasser/Warme* 39, 14-16.

Kuo S. (1986) Concurrent sorption of phosphate and zinc, cadmium or calcium by a hydrous ferric oxide. *Soil Science Society of America Journal* 50, 1412-1419.

Kuo S. and McNeal B.L. (1984) Effects of pH and phosphate on cadmium sorption by a hydrous ferric oxide. *Soil Science Society of America Journal* 48, 1040-1044.

Kuthubutheen A.J. and Pugh G.J.F. (1979) The effects of fungicides on soil fungal populations. *Soil Biology and Biochemistry* 22, 297-303.

Kwan K.H.M. and Smith S.R. (1990) Sites with a Long History of Sludge Disposal. A Study of Microbial Activity in Soil Contaminated with Heavy Metals Resulting from the Application of Sewage Sludge. WRc Report No. DoE 2508-M. WRc Medmenham, Marlow.

Lake D.L., Kirk P.W.W. and Lester J.N. (1984) Fractionation, characterization and speciation of heavy metals in sewage sludge and sludge-amended soils: A review. *Journal of Environmental Quality* 13, 175-183.

Lambert D.H., Baker D.E. and Cole H. (1979) The role of mycorrhizae in the interactions of phosphorus with zinc, copper and other elements. *Soil Science Society of America Journal* 43, 976-980.

Langeland G. (1986) Infection risk among sewage workers. In: Block J.C., Havelaar A.H. and L'Hermite P. (eds), *Epidemiological Studies of Risks Associated with the Agricultural Use of Sewage Sludge: Knowledge and Needs.* Elsevier Applied Science Publishers Ltd, London, pp. 135-139.

Larkin E.R., Tierney J.T. and Sullivan R. (1976) Persistence of virus on sewage irrigated vegetables. *Journal of the Environmental Engineering Division of the American Society of Civil Engineers* 102, 29-35.

Larsen K.E. (1984) Cadmium content in soil and crops after use of sewage sludge. In: Berglund S., Davis R.D. and L'Hermite P. (eds), *Utilization of Sewage Sludge on Land: Rates of Application and Long-term Effects of*

Metals. D. Reidel Publishing Company, Dordrecht, pp. 157-165.

Latterell J.J., Dowdy R.H. and Ham G.E. (1976) Sludge-borne metal uptake by soybeans as a function of soil cation exchange capacity. *Communications in Soil Science and Plant Analysis* 7, 465-476.

Leclerc H. and Brouzes P. (1973) Sanitary aspects of sludge treatment. *Water Research* 7, 355-360.

Lee H.J. (1975) Trace elements in animal production. In: Nicholas D.J.D. and Egan A.R. (eds), *Trace Elements in Soil-Plant Animal Systems.* Academic Press Inc. (London) Ltd, London, pp. 39-54.

le Riche H.H. (1968) Metal contamination of soil in the Woburn Market-Garden experiment resulting from the application of sewage sludge. *Journal of Agricultural Science, Cambridge* 71, 205-208.

Lester J.N. (1983) Occurrence, behaviour and fate of organic micropollutants during waste water and sludge treatment processes. In: Davis R.D., Hucker G. and L'Hermite P. (eds), *Environmental Effects of Organic and Inorganic Contaminants in Sewage Sludge.* D. Reidel Publishing Company, Dordrecht, pp. 3-18.

Leung D. and Chant S.R. (1990) Effects of sewage sludge treatment of soils on nodulation and leghaemoglobin content of clover. *Microbios* 64, 85-92.

Leung D. and Miles R.J. (1991) Heavy metal resistant strains of *Rhizobium leguminosarum* biovar *trifolii:* Isolation, characterization and root nodule induction. In: Farmer J.G. (ed), *Heavy Metals in the Environment,* Volume 2, International Conference, Edinburgh. CEP Consultants Ltd, Edinburgh, pp. 302-305.

Leung D., Miles R.J. and Smith S.R. (1993) An Investigation of the Interaction Between Heavy Metals and *Rhizobium leguminosarum* bv. *trifolii* and also of the Distribution of Nitrifying Bacteria in Sewage Sludge-Treated Agricultural Soils. Report No. FR 0342. Foundation for Water Research, Marlow.

Levesque M.P. and Mathur S.P. (1986) Soil tests for copper, iron, manganese zinc in histosols: I. The influence of soil properties, iron, manganese and zinc on the level and distribution of copper. *Soil Science* 142, 153-163.

Levine M.B., Hall A.T., Barrett G.W. and Taylor D.H. (1989) Heavy metal concentrations during ten years of sludge treatment to an old-field community. *Journal of Environmental Quality* 18, 411-418.

Lewin V.H. and Beckett P.H.T. (1980) Monitoring heavy metal accumulation in agricultural soils treated with sewage sludge: Conclusion. *Effluent and Water Treatment Journal* 20, 217-221.

Lewis J.A., Papavizas G.C. and Hora T.S. (1978) Effect of some herbicides on microbial activity in soil. *Soil Biology and Biochemistry* 10, 137-141.

Liebhardt W.C., Golt C. and Tupin J. (1979) Nitrate and ammonium concentrations of groundwater resulting from poultry manure applications. *Journal of Environmental Quality* 8, 211-215.

Liebmann H. (1964) Parasites in sewage and the possibilities of their extinction.

In: *Advances in Water Pollution Research*, Proceedings of the 2nd International Conference, Tokyo. Pergamon Press, Oxford, pp 269-276.

Lindberg S.E., Jackson D.R., Huckabee J.W., Janzen S.A., Levin M.J. and Lund J.R. (1979) Atmospheric emission and plant uptake of mercury from agricultural soils near the Almadén mercury mine. *Journal of Environmental Quality* 8, 572-578.

Lindemann W.C., Connell G. and Urquhart N.S. (1988) Previous sludge addition effects on nitrogen mineralization in freshly amended soil. *Soil Science Society of America Journal* 52, 109-112.

Lindsay D.E. (1983) Effects arising from the presence of persistent organic compounds in sludge. In: Davis R.D., Hucker G. and L'Hermite P. (eds), *Environmental Effects of Organic and Inorganic Contaminants in Sewage Sludge*. D. Reidel Publishing Company, Dordrecht, pp. 19-26.

Lloyd C.A., Chaney R.L., Hornick S.B. and Mastradone P.J. (1981) Labile cadmium in soils of long-term sludge utilization farms. *Agronomy Abstracts* 1981, 29.

Logan T.J. and Chaney R.L. (1983) Utilization of municipal wastewater and sludge on land-metals. In: Page A.L., Gleason III T.L., Smith Jr J.E., Iskander I.K. and Sommers L.E. (eds), *Proceedings of the 1983 Workshop, on Utilization of Municipal Wastewater and Sludge on Land*. University of California, Riverside, pp. 235-326.

Logan T.J. and Chaney R.L. (1987) Nonlinear rate response and relative crop uptake of sludge cadmium for land application of sludge risk assessment. In: *Proceedings Sixth International Conference Heavy Metals in the Environment*, Volume 1. CEP Consultants Ltd, Edinburgh, pp. 387-389.

Logan T.J. and Feltz R.E. (1985) Plant uptake of cadmium from acid-extracted anaerobically digested sewage sludge. *Journal of Environmental Quality* 14, 495-500.

Logan T.J., Chang A.C., Page A.L. and Ganje T.J. (1987) Accumulation of selenium in crops grown on sludge-treated soil. *Journal of Environmental Quality* 16, 349-352.

Lorenz F. and Steffens G. (1992) Agronomically efficient and environmentally careful slurry application to arable crops. In: *Nitrate and Farming Systems. Aspects of Applied Biology* 30, 109-116.

Lorenz S.E., McGrath S.P. and Giller K.E. (1992) Assessment of free-living nitrogen fixation activity as a biological indicator of heavy metal toxicity in soil. *Soil Biology and Biochemistry* 24, 601-606.

Lue-Hing C., Sedita S.J. and Rao K.C. (1979) Viral and bacterial levels resulting from land application of digested sludge. In: Sopper W.E. and Kerr S.N. (eds), *Utilization of Municipal Sewage Effluent and Sludge on Forest and Disturbed Land*. The Pennsylvania State University Press, Pennsylvania, pp. 445-462.

Lund E. (1975) Public health aspects of waste water treatment. In: *Radiation for a Clean Environment*. IAEA, Vienna, pp. 45-60.

Lund E. and Ronne V. (1973) On the isolation of virus from sewage treatment plant sludges. *Water Research* 7, 863-891.

Lund L.J., Page A.L. and Nelson C.O. (1976) Movement of heavy metals below sewage disposal ponds. *Journal of Environmental Quality* 5, 330-334.

Ma W. (1982) The influence of soil properties and worm-related factors on the concentrations of heavy metals in earthworms. *Pedobiologia* 24, 109-119.

Ma W. (1984) Sublethal toxic effects of copper on growth, reproduction and litter breakdown activity in the earthworm *Lumbricus rubellus* with observations on the influence of temperature and soil pH. *Environmental Pollution* (*Series A*) 33, 207-219.

Ma W. (1987) Heavy metal accumulation in the mole, *Talpa europea*, and earthworms as an indicator of metal bioavailability in terrestrial environments. *Bulletin of Environmental Contamination and Toxicology* 39, 933-938.

Ma W., Edelman Th., van Beersum I. and Jans Th. (1983) Uptake of cadmium, zinc, lead and copper by earthworms near a zinc-smelting complex: Influence of soil pH and organic matter. *Bulletin of Environmental Contamination and Toxicology* 30, 424-427.

Maag M. (1989) Denitrification losses from soil receiving pig slurry or fertilizer. In: Hansen J.A. and Henriksen K. (eds), *Nitrogen in Organic Wastes Applied to Soils.* Academic Press, London, pp. 235-246.

MacDonald A.J., Powlson D.S., Poulton P.R. and Jenkinson D.S. (1989) Unused fertilizer nitrogen in arable soils - its contribution to nitrate leaching. *Journal of the Science of Food and Agriculture* 46, 407-419.

Macnicol R.D. and Beckett P.H.T. (1985) Critical tissue concentrations of potentially toxic elements. *Plant and Soil* 85, 107-129.

Madariaga G.M. and Angle J.S. (1992) Sludge-borne salt effects on survival of *Bradyrhizobium japonicum*. *Journal of Environmental Quality* 21, 276-280.

MAF; Ministry of Agriculture and Fisheries (1937) *Manures and Manuring.* Bulletin No. 36. HMSO, London.

MAFF; Ministry of Agriculture, Fisheries and Food (1977) Cyst Eelworms on Potato. Advisory Leaflet 284. MAFF Publications, London.

MAFF; Ministry of Agriculture, Fisheries and Food (1981) *Lime and Liming.* MAFF Reference Book 35. HMSO, London.

MAFF; Ministry of Agricutlure, Fisheries and Food (1983) Mineral Trace Element and Vitamin Allowances for Ruminant Livestock. MAFF/ADAS Interdepartmental Working Party.

MAFF; Ministry of Agriculture, Fisheries and Food (1985) Advice on Avoiding Pollution from Manures and Other Slurry Wastes. Booklet 2200. HMSO, London.

MAFF; Ministry of Agriculture, Fisheries and Food (1990a) Agricultural Statistics United Kingdom 1988. HMSO, London.

MAFF; Ministry of Agriculture, Fisheries and Food (1990b) Pilot Nitrate Scheme: Nitrate Sensitive Areas. MAFF, London.

MAFF; Ministry of Agriculture, Fisheries and Food (1990c) Pesticide Usage Survey Report 78: Arable Farm Crops 1988. Reference Book 578. MAFF Publications, London.

MAFF; Ministry of Agriculture, Fisheries and Food (1991a) Code of Good Agricultural Practice for the Protection of Water. PB 0587. MAFF Publications, London.

MAFF; Ministry of Agriculture, Fisheries and Food (1991b) Pesticide Usage Survey Report 79: Grassland and Fodder Crops 1989. PB 0480. MAFF Publications, London.

MAFF; Ministry of Agriculture, Fisheries and Food (1992) Code of Good Agricultural Practice for the Protection of Air. PB 0618. MAFF Publications, London.

MAFF; Ministry of Agriculture, Fisheries and Food (1993a) Code of Good Agricultural Practice for the Protection of Soil. PB 0617. MAFF Publications, London.

MAFF; Ministry of Agriculture, Fisheries and Food (1993b) Pilot Nitrate Sensitive Areas Scheme: Report on the First 3 Years. MAFF, Environmental Protection Division, London.

MAFF; Ministry of Agriculture, Fisheries and Food (1994a) *Fertilizer Recommendations for Agricultural and Horticultural Crops.* Reference Book 209. HMSO, London.

MAFF; Ministry of Agriculture, Fisheries and Food (1994b) The Nitrate Sensitive Areas Scheme: Explanatory Booklet. PB 1729. MAFF, London.

MAFF; Ministry of Agriculture, Fisheries and Food (1994c) Consultation Document: Designation of Vulnerable Zones in England and Wales Under the EC Nitrate Directive (91/676). PB 1715. MAFF, London.

MAFF/DH; Ministry of Agriculture, Fisheries and Food/Department of Health (1990) *The Dietary and Nutritional Survey of British Adults.* HMSO, London.

MAFF/DoE; Ministry of Agriculture, Fisheries and Food/Department of the Environment (1993a) Review of the Rules for Sewage Sludge Application to Agricultural Land. Soil Fertility Aspects of Potentially Toxic Elements. Report of the Independent Scientific Committee. MAFF Publications, London.

MAFF/DoE; Ministry of Agriculture, Fisheries and Food/Department of the Environment (1993b) Review of the Rules for Sewage Sludge Application to Agricultural Land. Food Safety and Relevant Animal Health Aspects of Potentially Toxic Elements. Report of the Steering Group on Chemical Aspects of Food Surveillance. MAFF Publications, London.

Mahler R.J., Bingham F.T., Sposito G. and Page A.L. (1980) Cadmium-enriched sewage sludge application to acid and calcareous soils: Relation between treatment, cadmium in saturation extracts and cadmium uptake. *Journal of Environmental Quality* 9, 359-364.

Mahler R.J., Bingham F.T., Page A.L. and Ryan J.A. (1982) Cadmium-enriched sewage sludge application to acid and calcareous soils: Effect on soil and nutrition of lettuce, corn, tomato and Swiss chard. *Journal of Environmental Quality* 11, 694-700.

Mann H.H. and Patterson H.D. (1963) The Woburn Market-garden Experiment: Summary 1944-60. In: *Rothamsted Experimental Station Report for 1962.* Rothamsted Experimental Station, Harpenden, pp. 186-193.

Marcomini A., Capel P.D. and Giger W. (1988) Residues of detergent-derived organic pollutants and polychlorinated biphenyls in sludge-amended soil. *Naturwissenschaften* 75, 460-462.

Marcomini A., Capel P.D., Lichtensteiger Th., Brunner P.H. and Giger W. (1989) Behaviour of aromatic surfactants and PCBs in sludge-treated soil and landfills. *Journal of Environmental Quality* 18, 523-528.

Marks M.J., Williams J.H. and Chumbley C.G. (1980) Field experiments testing the effects of metal contaminated sewage sludges on some vegetable crops. In: *Inorganic Pollution and Agriculture.* MAFF Reference Book 326. HMSO, London, pp. 235-251.

Mårtensson A.M. (1992) Effects of agrochemicals and heavy metals on fast-growing rhizobia and their symbiosis with small-seeded legumes. *Soil Biology and Biochemistry* 24, 435-445.

Mårtensson A.M. and Witter E. (1990) The influence of various soil amendments on nitrogen fixing micro-organisms in a long-term field experiment, with special reference to sewage sludge. *Soil Biology and Biochemistry* 22, 977-982.

Marumoto T., Anderson J.P.E. and Domsch K.H. (1982) Mineralization of nutrients from soil microbial biomass. *Soil Biology and Biochemistry* 14, 469-475.

Mashhady A.S. (1984) Heavy metals extractable from a calcareous soil treated with sewage sludge. *Environmental Pollution (Series B)* 8, 51-62.

Matsumura F., Quensen J. and Tsushimoto G. (1983) Microbial degradation of TCDD in a model ecosystem. In: Tucker R.E., Young A.L. and Gray A.P. (eds), *Human and Environmental Risks of Chlorinated Dioxins and Related Compounds.* Plenum Press, New York, pp. 191-219.

Matthews M.R., Miller III, F.A. and Hyfantis Jr. G.J. (1981) Florence demonstration of fertilizer from sludge. *Industrial Engineering Chemical Production Research and Development* 20, 567-574.

McBride M.B. (1995) Toxic metal accumulation from agricultural use of sludge: Are the USEPA regulations protective? *Journal of Environmental Quality* 24, 5-18.

McGill, W.B., Shields, J.A. and Paul, E.A. (1975) Relation between carbon and nitrogen turnover in soil organic fractions of microbial origin. *Soil Biology and Biochemistry* 7, 57-63.

McGrath D. (1981) Implications of applying copper-rich pig slurry to grassland - effects on plants and soils. In: L'Hermite P. and Dehandtschutter J. (eds), *Copper in Animal Wastes and Sewage Sludge.* D. Reidel Publishing

Company, Dordrecht, pp. 144-153.

McGrath S.P. (1984) Metal concentrations in sludges and soil from a long-term field trial. *Journal of Agricultural Science, Cambridge* 103, 25-35.

McGrath S.P. (1987) Long-term studies of metal transfers following application of sewage sludge. In: Coughtrey P.J., Martin M.H. and Unsworth M.H. (eds), *Pollutant, Transport and Fate in Ecosystems.* Blackwell Scientific Publications, Oxford, pp. 301-317.

McGrath S.P. (1994) Effects of heavy metals from sewage sludge on soil microbes in agricultural ecosystems. In: Ross S.M. (ed) *Toxic Metals in Soil-Plant Systems.* John Wiley and Sons Ltd, Chichester, pp. 247-274.

McGrath S.P. (1995) Chromium and nickel. In: Alloway B.J. (ed), *Heavy Metals in Soils,* Second edition. Blackie Academic and Professional, Glasgow, pp. 152-178.

McGrath S.P. and Cegarra J. (1992) Chemical extractability of heavy metals during and after long-term applications of sewage sludge to soil. *Journal of Soil Science* 43, 313-321.

McGrath S.P. and Lane P.W. (1989) An explanation for the apparent losses of metals in a long-term field experiment with sewage sludge. *Environmental Pollution* 60, 235-256.

McGrath S.P. and Loveland P.J. (1992) *The Soil Geochemical Atlas of England and Wales.* Blackie Academic and Professional, London.

McGrath S.P., Brookes P.C. and Giller K.E. (1988) Effects of potentially toxic metals in soil derived from past applications of sewage sludge on nitrogen fixation by *Trifolium repens* L. *Soil Biology and Biochemistry* 20, 415-424.

McGrath S.P., Chaudri A.M. and Giller K.E. (1994) Long-term effects of land application of sewage sludge: Soils, micro-organisms and plants. In: *Proceedings of the 15th International Congress of Soil Science.* Acapulco Mexico, pp. 517-533.

McIlveen W.D. and Cole Jr. H. (1974) Influence of heavy metals on nodulation of red clover. *Phytopathology* 64, 583.

McIntyre A.E. and Lester J.N. (1984) Occurrence and distribution of persistent organochlorine compounds in U.K. sewage sludges. *Water, Air, and Soil Pollution* 23, 397-415.

McLachlan M.S., Hinkel M., Reissinger M., Hippelein M. and Kaupp H. (1994) A study of the influence of sewage sludge fertilization on the concentrations of PCDD/F and PCB in soil and milk. *Environmental Pollution* 85, 337-343.

McPhail C.D. (1984) Use of Sewage Sludge as a Forest Fertilizer Montreathmont Forest Experiment - Application of Sludge to Pole-Stage Scots Pine. WRc Report No. 609-M. WRc Medmenham, Marlow.

Melanen M., Jaakkola A., Melkas M., Ahtiainen M. and Matinvesi J. (1985) Leaching Resulting from Land Application of Sewage Sludge and Slurry. Publications of the Water Research Institute, 61. Veshiallitus - National Board of Waters, Finland.

Mellor A. and McCartney C. (1994) The effects of lead shot deposition on soils and crops at a clay pigeon shooting site in northern England. *Soil Use and Management* 10, 124-129.

Mertz W. (ed) (1986) *Trace Elements in Human and Animal Nutrition*, Volume 2, 5th edition. Academic Press Inc. (London) Ltd. London, pp. 391-397.

Metzger L. and Yaron B. (1987) Influence of sludge organic matter on soil physical properties. *Advances in Soil Science* 7, 141-163.

Meyer L.D. and Mannering J.V. (1963) Crop residues as surface mulches for controlling erosion on sloping land under intensive cropping. *Transactions of the American Society of Agricultural Engineers* 6, 322.

MHPPE; Ministry of Housing, Physical Planning and Environment (no date). Highlights of the Dutch National Environmental Policy Plan. A Clean Environment: Choose it or lose it. Department for Information and International Relations, The Hague.

Miller R.H. (1974) Factors affecting the decomposition of an anaerobically digested sewage sludge in soil. *Journal of Environmental Quality* 3, 376-380.

Miller W.E., Peterson S.A., Greene J.C. and Callahan C.A. (1985) Comparative toxicology of laboratory organisms for assessing hazardous waste sites. *Journal of Environmental Quality* 14, 569-574.

Mills C.F. (1970) Trace element metabolism in animals. Proceedings of WAAP/IBP International Symposium, Aberdeen, July 1969. E & S Livingstone, Edinburgh.

Mills C.F. (1974) Trace element interactions: Effects of dietary composition on the developmnet of imbalance and toxicity. In: Hoekstra W.G., Suttie J.W., Ganther H.E. and Mertz W. (eds), *Trace Element Metabolism in Animals*-2. Butterworths, London, pp. 79-90.

Mills C.F. (1986) Interactions concerning inorganic nutrients. In: Taylor T.G. and Jenkins N.K. (eds), *Proceedings of the XIII International Congress of Nutrition*. John Libby and Co. Ltd, London, pp. 532-536.

Mills J.G. and Zwarich M.A. (1982) Movement and loss of nitrate following heavy applications of sewage sludge to a poorly drained soil. *Canadian Journal of Soil Science* 62, 249-257.

Mills C.F., Campbell J.K., Bremner I. and Quarterman J. (1980) The influence of dietary composition on the toxicity of cadmium, copper, zinc and lead to animals. In: *Inorganic Pollution and Agriculture*. MAFF Reference Book 326. HMSO, London, pp. 11-21.

Misselbrook T.H., Pain, B.F., Stone A.C. and Scholefield D. (1995) Nutrient run-off following applications of livestock wastes to grassland. *Environmental Pollution* 88, 51-56.

Mitchell G.A., Bingham F.T. and Page A.L. (1978a) Yield and metal composition of lettuce and wheat grown on soils amended with sewage sludge enriched with cadmium, copper, nickel and zinc. *Journal of Environmental Quality* 7, 165-171.

Mitchell M.J., Hartenstein R., Swift B.L., Neuhauser E.F., Abrams B.I., Mulligan R.H., Brown B.A., Craig D. and Kaplan D. (1978b) Effects of different sewage sludges on some chemical and biological characteristics of soil. *Journal of Environmental Quality* 7, 551-559.

Moore B. (1971) The health hazards of pollution. In: Sykes G. and Skinner F.A. (eds), *Microbial Aspects of Pollution.* Academic Press, New York, pp. 11-32.

Morgan H. and Simms D.L. (1988) Discussion and conclusions. The Shipham report: An investigation into cadmium and its implications for human health. *The Science of the Total Environment* 75, 135-143

Morse G.K., Lester J.N. and Perry R. (1993) The Economic and Environmental Impact of Phosphorus Removal from Wastewater in the European Community. Selper Publications, Chiswick.

Morsing M. (1994) The use of sludge in forestry and agriculture. A comparison of the legislation in different countries. Forskningsserien Nr. 5. Danish Forest and Landscape Research Institute, Lyngby, Denmark.

Mosier A.R., Hutchinson G.L., Sabey B.R. and Baxter J. (1982) Nitrous oxide emissions from barley plots treated with ammonium nitrate or sewage sludge. *Journal of Environmental Quality* 22, 78-81.

Mostaghimi S., Deizman M.M., Dillaha T.A. and Heatwole C.D. (1989) Impact of land application of sewage sludge on run-off water quality. *Transactions of the American Society of Agricultural Engineers* 32, 491-496.

Mott C.J.B. (1988) Surface chemistry of soil particles. In: Wild A. (ed), *Russell's Soil Conditions and Plant Growth,* 11th edition. Longman Scientific and Technical, Harlow, pp. 237-281.

Mounsey, A.D. (ed) (1994) *Handbook of Medicinal Feed Additives 1994-1995.* HGM Publications, Bakewell.

Moza P., Scheunert I., Klein W. and Korte F. (1979) Studies with 2,4',5-trichlorobiphenyl-^{14}C and 2,2',4,4',6-pentachlorobiphenyl-^{14}C in carrots, sugar beets and soil. *Journal of Agricultural and Food Chemistry* 27, 1120-1124.

Muir P.C.G. and Baker B.E. (1973) Pesticide residues in soil and foodstuff. I. Chlorinated pesticides in cattle feed and milk produced in orchard and non-orchard areas. *Pesticide Science* 4, 113-119.

Mullen M.D., Wolf D.C., Beveridge T.J. and Bailey G.W. (1992) Sorption of heavy metals by the soil fungi *Aspergillus niger* and *Mucor rouxii. Soil Biology and Biochemistry* 24, 129-135.

Mullins G.L. and Sommers L.E. (1986) Characterization of cadmium and zinc in four soils treated with sewage sludge. *Journal of Environmental Quality* 15, 382-387.

Mullins G.L., Sommers L.E. and Barber S.A. (1986) Modelling the plant uptake of cadmium and zinc from soils treated with sewage sludge. *Soil Science Society of America Journal* 50, 1245-1250.

Murphy B.L. (1989) Modelling the leaching and transport of 2,3,7,8-TCDD from incinerator ash from landfills. *Chemosphere* 19, 433-438.

Narwal R.P., Singh B.R. and Panhwar A.R. (1983) Plant availability of heavy metals in a sludge-treated soil: I. Effect of sewage sludge and soil pH on the yield and chemical composition of rape. *Journal of Environmental Quality* 12, 358-365.

Nathan M.V. and Malzer G.L. (1994) Dynamics of ammonia volatilization from turkey manure and urea applied to soil. *Soil Science Society of America Journal* 58, 985-990.

Neal R.H. (1995) Selenium. In: Alloway B.J. (ed), *Heavy Metals in Soils*, Second edition. Blackie Academic and Professional, Glasgow, pp. 260-283.

Nell J.H., Engelbrecht J.F.P., Smith L.S. and Nupen E.M. (1980) Health aspects of sludge disposal; South African experience. *Progress in Water Technology* 13, 153-170.

Newman J. (1988) Soil fauna other than protozoa. In: Wild A. (ed), *Russell's Soil Conditions and Plant Growth,* 11th edition. Longman Scientific and Technical, Harlow, pp. 500-525.

NFU; National Farmers Union (1987) The nitrate in water debate. In: *Insight*, December 1987.

Nicholls P.H. (1991) Organic contaminants in soils and groundwaters. In: Jones K.C. (ed), *Organic Contaminants in the Environment.* Elsevier Science Publishers Ltd, Barking, pp. 87-132.

Nicholson J.K., Kendall M.D. and Osborn D. (1983) Cadmium and nephrotoxicity. *Nature* 304, 633-635.

Nielsen V.C., Voorburg J.H. and L'Hermite P. (eds) (1986) *Odour Prevention and Control of Organic Sludge and Livestock Farming*. Elsevier Applied Science Publishers Ltd, Barking.

Nielsen V.C., Voorburg J.H. and L'Hermite P. (eds) (1988) *Volatile Emissions from Livestock Farming and Sewage Operations.* Elsevier Applied Science Publishers Ltd, Barking.

Nielsen V.C., Voorburg J.H. and L'Hermite P. (eds) (1991) *Odour and Ammonia Emissions from Livestock Farming*. Elsevier Applied Science Publishers Ltd, Barking.

Nielson R.L. (1951) Effect of soil minerals on earthworms. *New Zealand Journal of Agriculture* 83, 433-435.

NRA; National Rivers Authority (1992a) *The Influence of Agriculture on the Quality of National Waters in England and Wales.* Water Quality Series No. 6. NRA, Bristol.

NRA; National Rivers Authority (1992b) Policy and Practice for the Protection of Groundwater. NRA, Bristol.

NRC; National Research Council (1980) *Mineral Tolerance of Domestic Animals.* National Academic Press, Washington.

NSDO; National Seed Development Organisation (no date). *Wheat: A Guide to Varieties from the Plant Breeding Institute.* PBI, Cambridge.

Nutman P.S. and Ross G.J.S. (1970) *Rhizobium* in the soils of the Rothamsted and Woburn farms. In: *Rothamsted Experimental Station Report for 1969*, Part 2. Rothamsted Experimental Station, Harpenden, pp. 148-167.

Oake R.J., Booker C.S. and Davis R.D. (1984) Fractionation of heavy metals in sewage sludges. *Water Science and Technology* 17, 587-598.

Oakes D. (1989) The impact of agricultural practices on groundwater nitrate concentration. *World Water '89*, Thomas Telford Limited, London, pp. 45-49.

Oakes D.B. and Slade S. (1992) Pollution Risk Assessment. NRA R&D Note 118. NRA, Bristol.

Obbard J.P. and Jones K.C. (1993) The effect of heavy metals on dinitrogen fixation by *Rhizobium*-white clover in a range of long-term sewage sludge amended and metal-contaminated soils. *Environmental Pollution* 79, 105-112.

Obbard J.P., Jones K.C. and Smith S.R. (1992a) Microbial Effects of Heavy Metals in Sewage Sludge Amended Soils - Effects on Symbiotic Nitrogen Fixation. Report No. FR 0308. Foundation for Water Research, Marlow.

Obbard J.P., Jones K.C. and Smith S.R. (1992b) Nitrogen fixation in sewage sludge amended soils: A review of recent studies. In: Hall J.E., Sauerbeck D.R. and L'Hermite P. (eds), *Effects of Organic Contaminants in Sewage Sludge on Soil Fertility, Plants and Animals.* Commission of the European Communities, Luxembourg, pp. 193-206.

Obbard J.P., Sauerbeck D.R. and Jones K.C. (1993a) *Rhizobium leguminosarum* bv. *trifolii* in soils amended with heavy metal contaminated sewage sludges. *Soil Biology and Biochemistry* 25, 227-231.

Obbard J.P., Jones K.C. and Smith S.R. (1993b) Microbial Effects of Heavy Metals in Sewage Sludge Amended Soils - Effects on Vesicular-Arbuscular Mycorrhizal Infection of Host Plant Root Tissue. WRc Unpublished Report. WRc Medmenham, Marlow.

O'Brien P. and Mitsch W.J. (1980) Root zone nitrogen simulation model for land application of sewage sludges. *Ecological Modelling* 8, 233-257.

Obrist W. (1979) Research and practice of sludge pasteurization in Switzerland. In: *Proceedings of a Conference on Utilization of Sewage Sludge on Land.* Water Research Centre, Stevenage, pp. 503-511.

O'Connor G.A., Fairbanks B.C. and Doyle E.A. (1981) Effects of sewage sludge on phenoxy herbicide adsorption and degradation in soils. *Journal of Environmental Quality* 10, 510-515.

O'Connor G.A., Kiehl D., Eiceman G.A. and Ryan J.A. (1990a) Plant uptake of sludge-borne PCBs. *Journal of Environmental Quality* 19, 113-118.

O'Connor G.A., Lujan J.R. and Jin Y. (1990b) Adsorption, degradation and plant availability of 2,4-dinitrophenol in sludge-amended calcareous soils. *Journal of Environmental Quality* 19, 587-593.

O'Connor G.A., Chaney R.L. and Ryan J.A. (1991) Bioavailability to plants of sludge-borne toxic organics. *Reviews of Environmental Contamination and Toxicology* 121, 129-155.

Odum, E. (1985) Trends expected in stressed ecosystems. *Bioscience* 35, 419-422.

OECD; Organization for Economic Cooperation and Development (1984) Guidelines for the testing of chemicals No. 207. Earthworm acute toxicity tests. Adopted 4 April 1984.

Offenbacher G. (1992) The PCB transfer from soil into plants depending on supply and degree of chlorination. In: Hall J.E., Sauerbeck D.R. and L'Hermite P. (eds), *Effects of Organic Contaminants in Sewage Sludge on Soil Fertility, Plants and Animals.* Commission of the European Communities, Luxembourg, pp. 90-102.

O'Keefe B.E., Axley J. and Meisinger J.J. (1986) Evaluation of nitrogen availability indexes for a sludge compost amended soil. *Journal of Environmental Quality* 15, 121-128.

Olson B.H. and Thornton I. (1982) The resistance patterns to metals of bacterial populations in contaminated land. *Journal of Soil Science* 33, 271-277.

Orazio C.E., Kapila S., Puri R.K. and Yanders A.F. (1992) Persistence of chlorinated dioxins and furans in the soil environment. *Chemosphere* 25, 1469-1474.

Overcash M.R. (1983) Land treatment of municipal effluent and sludge: specific organic compounds. In: Page A.L., Gleason III T.L., Smith Jr J.E., Iskander I.K. and Sommers L.E. (eds), *Proceedings of the 1983 Workshop, on Utilization of Municipal Wastewater and Sludge on Land.* University, of California, Riverside, pp. 199-231.

Page A.L. (1974) Fate and Effects of Trace Elements in Sewage Sludge when Applied to Agricultural Lands. A Literature Review. US EPA Report No. EPA-670/2-74-005. National Technical Information Service, Springfield, Virginia.

Page A.L., Gleason III T.L., Smith Jr J.E., Iskander, I.K. and Sommers L.E. (eds) (1983) *Proceedings of the 1983 Workshop on Utilization of Municipal Wastewater and Sludge on Land.* University of California, Riverside.

Page A.L., Logan T.J. and Ryan J.A. (eds) (1987) *Land Application of Sludge: Food Chain Implications.* Lewis Publishers Inc., Chelsea, Michigan.

Page A.L. and Logan T.J. (1989) Peer Review: Standards for the Disposal of Sewage Sludge US EPA Proposed Rule 40 CFR Parts-257 and 503 (February 6, 1989 Federal Register pp. 5746-5902). Organized by Cooperative State Research Service (CSRS) Technical Committee W-170. University of California, Riverside.

Pagliai M. and Sequi P. (1981) The influence of applications of slurries on soil properties related to run-off - experimental approach in Italy. In: Brogan J.C. (ed), *Nitrogen Losses and Surface Run-off from Landspreading of Manures.* Martinus Nijhoff/Dr W. Junk Publishers, The Hague, pp. 44-65.

Pagliai M., Guidi G., La Marca M., Giachetti M. and Lucamente G. (1981) Effects of sewage sludges and composts on soil porosity and aggregation. *Journal of Environmental Quality* 10, 556-561.

Påhlsson A-M.B. (1989) Toxicity of heavy metals (Zn, Cu, Cd, Pb) to vascular plants: A literature review. *Water, Air, and Soil Pollution* 47, 287-319.

Pahren H.R., Lucas J.B., Ryan J.A. and Dotson G.K. (1977) An appraisal of the relative health risks associated with land application of municipal sludge. 50th Annual Conference of the Water Pollution Control Federation. Philadelphia, October 1977.

Pain B.F. and Thompson R.B. (1989) Ammonia volatilization from livestock slurries applied to land. In: Hansen J.A. and Henriksen K. (eds), *Nitrogen in Organic Wastes Applied to Soils.* Academic Press Limited, London, pp. 202-212.

Pain B.F., Thompson R.B., Rees Y.J. and Skinner J.H. (1990) Reducing gaseous losses of nitrogen from cattle slurry applied to grassland by the use of additives. *Journal of the Science of Food and Agriculture* 50, 141-153.

Palzenberger M. (1995) Earthworms as bioindicators of soil copper - what might be indicated? Contaminated soils: Third International Conference on the Biogeochemistry of Trace Elements, 15-19 May, Paris.

Parker C.F. and Sommers L.E. (1983) Mineralization of nitrogen in sewage sludges. *Journal of Environmental Quality* 12, 150-156.

Parkin T.B. and Berry E.C. (1994) Nitrogen transformations associated with earthworm casts. *Soil Biology and Biochemistry* 26, 1233-1238.

Pawlowski Z.S. and Schultzberg K. (1986) Ascariasis and sewage in Europe. In: Block J.C. Havelaar A.H. and L'Hermite P. (eds), *Epidemiological Studies of Risks Associated with the Agricultural Use of Sewage Sludge: Knowledge and Needs.* Elsevier Applied Science Publishers Ltd, Barking, pp. 83-93.

Payne G.G., Martens D.C., Winarko C. and Perera N.F. (1988) Form and availability of copper and zinc following long-term copper sulfate and zinc sulfate applications. *Journal of Environmental Quality* 17, 707-711.

Pepper I.L., Bezdicek D.F., Baker A.S. and Sims J.M. (1983) Silage corn uptake of sludge-applied zinc and cadmium as affected by soil pH. *Journal of Environmental Quality* 12, 270-275.

Pera A., Giovannetti M., Vallini G. and de Bertoldi M. (1982) Land application of sludge: Effects on soil microflora. In: Catroux G., L'Hermite P. and Suess E. (eds), *The Influence of Sewage Sludge Application on Physical and Biological Properties of Soils.* D. Reidel Publishing Company, Dordrecht, pp. 141-166.

Pereira Neto J.T., Stentiford E.I. and Mara D.D. (1987) Comparative survival of pathogenic indicators in windrow and static pile. In: de Bertoldi M., Ferranti M.P., L'Hermite P. and Zucconi F. (eds), *Compost: Production, Quality and Use.* Elsevier Applied Science Publishers Ltd, Barking, pp. 276-295.

Persson T., Lundkvist H., Wirén A., Hyvönen R. and Wessén B. (1989) Effects of acidification and liming on carbon and nitrogen mineralization and soil organisms in mor humus. *Water, Air, and Soil Pollution* 45, 77-96.

Petruzzelli G., Guidi G. and Lubrano L. (1978) Cadmium occurrence in soil organic matter and its availability to wheat seedlings. *Water, Air, and Soil Pollution* 9, 263-269.

Pierzynski G.M. and Jacobs L.W. (1986) Extractability and plant availability of molybdenum from inorganic and sewage sludge sources. *Journal of Environmental Quality* 15, 323-326.

Pike E.B. (1975) Aerobic bacteria. In: Curds C.R. and Hawkes H.A. (eds), *Ecological Aspects of Used-Water Treatment,* Volume 1. Academic Press Limited, London, pp. 1-63.

Pike E.B. (1981) The control of salmonellosis in the use of sewage sludge on agricultural land. In: L'Hermite P. and Ott, H. (eds), *Proceedings of the Second European Symposium on Characterisation, Treatment and Use of Sewage Sludge.* D. Reidel Publishing Company, Dordrecht, pp. 404-407.

Pike E.B. (1986a) Pathogens in sewage sludge: (1) Agricultural use of sewage sludge and the control of disease. *Water Pollution Control* 85, 472-475.

Pike E.B. (1986b) Recent UK research on incidence, transmission and control of salmonella and parasitic ova in sludge. In: Block J.C., Havelaar A.H. and L'Hermite P. (eds), *Epidemiological Studies of Risks Associated with the Agricultural Use of Sewage Sludge: Knowledge and Needs.* Elsevier Applied Science Publishers Ltd. London, pp. 50-59.

Pike E.B. (1987) Aids: Implications for the Water Industry. WRc Report No. PRU 1529-M. WRc Medmenham, Marlow.

Pike E.B. and Carrington E.G. (1986) Stabilization of sludge by conventional and novel processes - a healthy future. In: *Proceeding of Symposium: The Agricultural Use of Sewage Sludge - Is There a Future?* Doncaster, 12 November. The Institute of Water Pollution Control, Maidstone, pp. 1-43.

Pike E.R., Graham L.C. and Fogden M.W. (1975) An appraisal of toxic metal residue in the soils of a disused sewage farm. *Journal of the Association of Public Analysts* 13, 48-63.

Pike E.B., Morris D.L. and Carrington E.G. (1983) Inactivation of ova of the parasites *Taenia saginata* and *Ascaris suum* during heated anaerobic digestion. *Water Pollution Control* 82, 501-509.

Pike E.B., Carrington E.G. and Harman S.A. (1988) Destruction of salmonellas, enteroviruses and ova of parasites by pasteurization and anaerobic digestion. *Water Science and Technology* 20, 337-343.

Pollard G. (1991) Sites With a Long History of Sludge Disposal: Phase II (ET 9500). Progress Report to the Department of the Environment: October 1990 to March 1991. WRc Report No. DoE 2769-M. WRc Medmenham, Marlow.

Poole D.B.R. (1981) Implications of applying copper rich pig slurry to grassland - effects on the health of grazing sheep. In: L'Hermite P. and Dehandtschutter J. (eds), *Copper in Animal Wastes and Sewage Sludge.* D. Reidel Publishing Company, Dordrecht, pp. 273-282.

Powlesland C. and Frost R. (1990) A Methodology for Undertaking BPEO Studies of Sewage Sludge Treatment and Disposal. WRc Report No. 2305-M/1. WRc Medmenahm, Marlow.

Powlson D.S., Poulton P.R., Addiscott T.M. and McCann D.S. (1989) Leaching of nitrate from soils receiving organic or inorganic fertilizers continuously for 135 years. In: Hansen J.A. and Henriksen K. (eds), *Nitrogen in Organic Wastes Applied to Soils.* Academic Press Limited, London, pp. 334-345.

Premi P.R. and Cornfield A.H. (1969) Incubation study of nitrification of digested sewage sludge added to soil. *Soil Biology and Biochemistry* 1, 1-4.

Premi P.R. and Cornfield A.H. (1971) Incubation study of nitrogen mineralization in soil treated with dried sewage sludge. *Environmental Pollution* 2, 1-5.

Prins W.H., Dilz K. and Neeteson J.J. (1988) Current recommendations for nitrogen fertilization within the EEC in relation to nitrate leaching. *Proceedings of the Fertilizer Society,* 276.

Quensen III J.F., Tiedje J.M. and Boyd S.A. (1988) Reductive dechlorination of polychlorinated biphenyls by anaerobic micro-organisms from sediments. *Science,* 242, 752-754.

Ram N. and Verloo M. (1985) Effect of various organic materials on the mobility of heavy metals in soils. *Environmental Pollution* 10, 241-248.

Ramel C. (1983) Advantages of and problems with short-term mutagenicity tests for the assessment of mutagenic and carcinogenic risk. *Environmental Health Perspectives* 47, 153-159.

Ransome M.E. and Carrington E.G. (1993) Environmentally Stressed Oocysts. Report No. FR 0388. Foundation for Water Research, Marlow.

Rappaport B.D., Martens D.C., Reneau Jr R.B. and Simpson T.W. (1988) Metal availability in sludge-amended soils with elevated metal levels. *Journal of Environmental Quality* 17, 42-47.

Reasoner D.J. (1976) Microbiology - detection of bacterial pathogens and their occurrence. *Journal of the Water Pollution Control Federation* 48, 1397-1410.

Reaves G.A. and Berrow M.L. (1984) Total copper contents of Scottish soils. *Journal of Soil Science* 35, 583-592.

Reddy M.R. and Dunn S.J. (1986) Distribution coefficients for nickel and zinc in soils. *Environmental Pollution (Series B)* 11, 303-313.

Reddy M.V. and Reddy V.R. (1992) Effects of organochlorine, organophosphorus and carbonate insecticides on the population structure and biomass of earthworms in a semi-arid tropical grassland. *Soil Biology and Biochemistry* 24, 1733-1738.

Reddy G.B., Cheng C.N. and Dunn S.J. (1983) Survival of *Rhizobium japonicum* in soil-sludge environment. *Soil Biology and Biochemistry* 15, 343-345.

Reddy K.R., Khaleel R. and Overcash M.R. (1981) Behaviour and transport of microbial pathogens and indicator organisms in soils treated with organic wastes. *Journal of Environmental Quality* 10, 255-266.

Reid R.L. and Horrath D.J. (1980) Soil chemistry and mineral problems in farm livestock. *Animal Feed Science and Technology* 5, 95-167.

Reinecke A.J. and Nash R.G. (1984) Toxicity of 2,3,7,8-TCDD and short-term bioaccumulation by earthworms (oligochaeta). *Soil Biology and Biochemistry* 16, 45-49.

Reischl A., Reissinger M., Thoma H. and Hutzinger O. (1989) Uptake and accumulation of PCDD/F in terrestrial plants: Basic considerations. *Chemosphere* 19, 467-474.

Riffaldi R., Levi-Minzi R., Saviozzi A. and Tropea M. (1983) Sorption and release of cadmium by some sewage sludges. *Journal of Environmental Quality* 12, 253-256.

Ritter W.F. and Chirnside A.E.M. (1984) Impact of land use on groundwater quality in southern Delaware. *Ground Water* 22, 39-47.

Roberts G. (1987) Nitrogen inputs and outputs in a small agricultural catchment in the eastern part of the United Kingdom. *Soil Use and Management* 3, 148-154.

Roberts R.D. and Johnson M.S. (1978) Dispersal of heavy metals from abandoned mine workings and their transferance through terrestrial food chains. *Environmental Pollution* 16, 293-310.

Robertson W.K., Lutrick M.C. and Yuan T.L. (1982) Heavy applications of liquid-digested sludge on three Utisols: I. Effects on soil chemistry. *Journal of Environmental Quality* 11, 278-282.

Robson M.J., Parsons A.J. and Williams T.E. (1989) Herbage production: Grasses and legumes. In: Holmes W. (ed), *Grass: Its Production and Utilisation*, Second edition. Blackwell Scientific Publications, Oxford, pp. 7-88.

Roche (1985-87) VITEC-2. Roche Products Ltd, Welwyn Garden City.

Roelofs J.G.M. and Houdijk A.L.F.M. (1991) Ecological effects of ammonia. In: Nielsen V.C., Voorburg J.H. and L'Hermite P. (eds), *Odour and Ammonia Emissions from Livestock Farming*. Elsevier Science Publishers Ltd, Barking, pp. 10-16.

Rogers H.R. (1987) Organic Contaminants in Sewage Sludge (EC 9322 SLD): Occurrence and Fate of Synthetic Organic Compounds in Sewage Sludge - A Review. WRc Report No. PRD 1539-M. WRc Medmenham, Marlow.

Romero J.C. (1970) The movement of bacteria and viruses through porous media. *Ground Water* 8, 37-48.

Roslycky E.B. (1982) Glyphosate and the response of the soil microbiota. *Soil Biology and Biochemistry* 14, 87-92.

Ross, I.J., Sizemore S., Bowden J.P. and Haan L.T. (1979) Quality of run-off from land receiving surface application and injection of liquid dairy manure. *Transactions of the Amercian Society of Agricultural Engineers* 22, 1058-1062.

Rother J.A., Millbank J.W. and Thornton I. (1982a) Effects of heavy-metal additions on ammonification and nitrification in soils contaminated with cadmium, lead and zinc. *Plant and Soil* 69, 239-258.

Rother J.A., Millbank J.W. and Thornton I. (1982b) Seasonal fluctuations in nitrogen fixation (acetylene reduction) by free-living bacteria in soils contaminated with cadmium, lead and zinc. *Journal of Soil Science* 33, 101-113.

Rowlands C.L. (1992) Sewage sludge in agriculture: A UK perspective. Water Environment Federation 65th Annual Conference and Exposition. New Orleans, 20-24 September, pp. 305-315.

Royle S.M., Chandrasekar N.C. and Unwin R.J. (1989) The effect of zinc, copper and nickel applied to soil in sewage sludge on the growth of white clover. In: Hall, J.E. (ed), *Alternative Uses for Sewage Sludge*, Poster papers presented at a conference organized by WRc, University of York, 5-7 September. WRc Report No. CP 596. WRc Medmenham, Marlow, pp. 57-61.

Rudolphs W., Falk L.L. and Rogotzke R.A. (1950) Survival of enteric, pathogenic and related organisms in soil, water, sewage and sludges, and on vegetation. II Animal parasites. *Sewage and Industrial Wastes* 22, 1417-1427.

Rudolphs W., Falk L.L. and Rogotzke R.A. (1951) Contamination of vegetables grown in polluted soil. *Sewage and Industrial Wastes* 23, (1) 253-268, (2) 478-485, (3) 656-660.

Rulkens W.H., van Voorneburg F. and Joziasse J. (1989) Removal of heavy metals from sewage sludges: State of the art and perspectives. In: Dirkzwager A.H. and L'Hermite P. (eds), *Sewage Sludge Treatment and Use: New Developments, Technological Aspects and Environmental Effects*. Elsevier Science Publishers Ltd, Barking, pp. 145-159.

Ryan J.A. (1993) Utilization of risk assessment in the development of limits for land application of municipal sewage sludge. In: *Sewage Sludge: Land Utilization and the Environment*, August 11-13, Bloomington, Minnesota.

Ryan J.A. and Chaney R.L. (1994) Development of limits for land application of municipal sewage sludge: Risk assessment. In: *Proceedings of the 15th International Congress of Soil Science*, Acapulco, Mexico, pp. 534-553.

Ryan J.A. and Keeney D.R. (1975) Ammonia volatilization from surface applied wastewater sludge. *Journal of the Water Pollution Control Federation* 47, 386-393.

Ryan J.A., Pahren H.R. and Lucas J.B. (1982) Controlling cadmium in the human food chain: A review and rationale based on health effects. *Environmental Research* 28, 251-302.

SAC; Scottish Agricultural Colleges (1986) Disposal of Sewage Sludge on Land-Hazards and Value. Scottish Agricultural Colleges Publication No. 170.

Sacchi G.A., Vigariò P., Fortunati G. and Cocucci S.M. (1986) Accumulation of 2,3,7,8-tetrachlorodibenzo-*p*-dioxin from soil and nutrient solution by bean and maize plants. *Experientia* 42, 586-588.

Sanders J.R. and Adams T. McM. (1987) The effects of pH and soil type on concentrations of zinc, copper and nickel extracted by calcium chloride from sewage sludge-treated soils. *Environmental Pollution* 43, 219-228.

Sanders J.R. and El Kherbawy M.I. (1987) The effect of pH on zinc adsorption equilibria and exchangeable zinc pools in soils. *Environmental Pollution* 44, 165-176.

Sanders D.A., Malina J.F., Moore B.E. Sagik B.P. and Sorber C.A. (1979) Fate of poliovirus during anaerobic digestion. *Journal of the Water Pollution Control Federation* 51, 333-343.

Sanders J.R., McGrath S.P. and Adams T. McM. (1986) Zinc, copper and nickel concentrations in ryegrass grown on sewage sludge-contaminated soils of different pH. *Journal of the Science of Food and Agriculture* 37, 961-968.

Sanders J.R., McGrath S.P. and Adams T. McM. (1987) Zinc, copper and nickel concentrations in soil extracts and crops grown on four soils treated with metal-loaded sewage sludge. *Environmental Pollution* 44, 193-210.

Sanson D.W., Hallford P.M. and Smith G.S. (1984) Effects of long-term consumption of sewage solids on blood, milk and tissue elemental composition of breeding ewes. *Journal of Animal Science* 59, 416-424.

Satchell J.E., Martin K. and Krishnamoorthy, R.V. (1984) Stimulation of microbial phosphate production by earthworm activity. *Soil Biology and Biochemistry* 16, 195.

Sattar S.A., Ramia S. and Westwood J.C. (1976) Calcium hydroxide and the elimination of human pathogenic viruses from sewage. Studies with experimentally contaminated and pilot plant samples. *Canadian Journal of Public Health* 67, 221-226.

Sauerbeck D.R. and Leschber R. (1992) German proposals for acceptable contents of inorganic and organic pollutants in sewage sludges and sludge-amended soils. In: Hall J.E., Sauerbeck D.R. and L'Hermite P. (eds), *Effects of Organic Contaminants in Sewage Sludge on Soil Fertility, Plants and Animals.* Commission of the European Communities, Luxembourg, pp. 3-13.

Sauerbeck D.R. and Rietz E. (1983) Soil-chemical evaluation of different extractants for heavy metals in soils. In: Davis R.D., Hucker G. and L'Hermite P. (eds), *Environmental Effects of Organic and Inorganic Contaminants in Sewage Sludge.* D. Reidel Publishing Company, Dordrecht, pp. 147-160.

Sauerbeck D.R. and Styperek P. (1986) Long-term effects of contaminants. In: L'Hermite P. (ed), *Processing and Use of Organic Sludge and Liquid Agricultural Wastes.* D. Reidel Publishing Company, Dordrecht, pp. 318-335.

Sawhney B.L. (1978) Leaching of phosphate from agricultural soil to groundwater. *Water, Air, and Soil Pollution* 9, 499-505.

Scaife M.A. (1975) Field measurements of nitrate and ammonium nitrogen under growing crops. In: *National Vegetable Research Station Annual Report 1974*. National Vegetable Research Station, Wellesbourne, pp. 49-50.

Schauer P.S., Wright W.R. and Pelchat J. (1980) Sludge-borne heavy metal availability and uptake by vegetable crops under field conditions. *Journal of Environmental Quality* 9, 69-73.

Schauss A. and Costin C. (1989) *Zinc and Eating Disorders*. Keats Publishing Inc., New Canaan, Connecticut.

Schmitzer J.L., Scheunert I. and Korte F. (1988) Fate of bis(2-ethylhexyl)[^{14}C] phthalate in laboratory and outdoor soil-plant systems. *Journal of Agricultural and Food Chemistry* 36, 210-215.

Schnitzer M. and Skinner S.I.M. (1966) Organo-metallic interactions in soils 5. Stability constants of Cu^{2+}, Fe^{2+} and Zn^{2+} fulvic acid complexes. *Soil Science* 102, 361-365.

Schnitzer M. and Skinner S.I.M. (1967) Organo-metallic interactions in soils 7. Stability constants of Pb^{2+}, Ni^{2+}, Mn^{2+}, Ca^{2+} and Mg^{2+} fulvic acid complexes. *Soil Science* 103, 247-252.

Schwartzbrod J., Mathieu C., Thevenot M.T., Baradel J.M. and Schwartzbood L. (1987) Watewater sludge: Parasitological and virological contamination. *Water Science and Technology* 19, 33-40.

Seip H.M., Alstad J., Carlberg G.E., Martinsen K. and Skaane R. (1986) Measurement of mobility of organic compounds in soils. *The Science of the Total Environment* 50, 87-101.

Senapati B.K., Biswal J., Pani S.C. and Sahu S.K. (1992) Ecotoxicological effects of malathion on earthworms. *Soil Biology and Biochemistry* 24, 1719-1722.

Senesi N., Sposito G., Holtzclaw K.M. and Bradford G.R. (1989) Chemical properties of metal-humic acid fractions of a sewage sludge-amended aridisol. *Journal of Environmental Quality* 18, 186-194.

Sewart A., Harrad S.J., McLachlan M.S., McGrath S.P. and Jones K.C. (1995) PCDD/Fs and non-*o*-PCBs in digested UK sewage sludges. *Chemosphere* 30, 51-67.

Sharpley A.N. and Menzel R.G. (1987) The impact of soil and fertilizer phosphorus on the environment. *Advances in Agronomy* 41, 297-324.

Sharpley A.N., Smith S.J., Jones O.R., Berg W.A. and Coleman G.A. (1992) The transport of bioavailable phosphorus in agricultural runoff. *Journal of Environmental Quality* 21, 30-35.

Sheaffer C.C., Decker A.M., Chaney R.L. and Douglass L.W. (1979) Soil temperature and sewage sludge effects on metals in crop tissue and soils. *Journal of Environmental Quality* 8, 455-459.

Shen S.M., Hart P.B.S., Powlson D.S. and Jenkinson D.S. (1989) The nitrogen cycle in the Broadbalk wheat experiment: ^{15}N-labelled fertilizer residues in the soil and in the soil microbial biomass. *Soil Biology and Biochemistry* 21, 529-533.

Shepherd M.A. (1990) Gleadthorpe update No. 4 - Effect of crop rotation and husbandry on nitrate leaching losses: Sandland. ADAS Gleadthorpe, Mansfield.

Shepherd M.A. (1993) Nitrate Losses from Application of Sewage Sludge to Farmland (PECD 7/7/389). ADAS Gleadthorpe, Mansfield.

Sherlock J.C. (1983) The intake by man of cadmium from sludged land. In: Davis R.D., Hucker G. and L'Hermite P. (eds), *Environmental Effects of Organic and Inorganic Contaminants in Sewage Sludge.* D. Reidel Publishing Company, Dordrecht, pp. 113-120.

Sherlock J.C., Smart G.A. and Walters B. (1983) Dietary surveys on a population at Shipham, Somerset, United Kingdom. *The Science of the Total Environment* 29, 121-142.

Sherwood M. and Fanning A. (1981) Nutrient content of surface run-off water from land treated with animal wastes. In: Brogan J.C. (ed), *Nitrogen Losses and Surface Run-off from Landspreading of Manures.* Martinus Nijhoff/Dr W. Junk Publishers, The Hague, pp. 5-17.

Shuval H.I. (1970) Detection and control of enteroviruses in the water environment. In: *Developments in Water Quality Research.* Ann Arbor Science Publications, Michigan, pp. 47-71.

SI; UK Statutory Instrument (1985) The Lead in Food (Amendment) Regulations No. 912. HMSO, London.

SI; UK Statutory Instrument (1989a) The Sludge (Use in Agriculture) Regulations 1989. Statutory Instrument No. 1263. HMSO, London.

SI; UK Statutory Instrument (1989b) The Water Supply (Water Quality) Regulations 1989. Statutory Instrument No. 1147. HMSO, London.

Siepel H. and Maaskamp F. (1994) Mites of different feeding guilds affect decomposition of organic matter. *Soil Biology and Biochemistry* 26, 1389-1394.

Silver S., Laddaga R.A. and Misra T.K. (1989) Plasmid-determined resistance to metal ions. In: Poole R.K. and Gadd G.M. (eds), *Metal-Microbe Interactions*, Special Publications of the Society for General Microbiology, Volume 26. IRL Press, Oxford, pp. 49-63.

Silverman P.H. (1955) Bovine cysticercosis on Great Britain from July 1950 to Dec. 1953, with some notes on meat inspection and the incidence of *Taenia saginata* in man. *Annals of Tropical Medicine and Parasitology* 49, 429-437.

Silverman P.H. and Guiver K. (1960) Survival of eggs of Taenia saginata (the human-beef tapeworm) after mesophyllic anaerobic digestion. *Journal of the Institution of Sewage Purification* for 1960, 345-347.

Simms D.L. and Beckett M.J. (1987) Contaminated land: Setting trigger concentrations. *The Science of the Total Environment* 65, 121-134.

Simpson J.R. and Steele K.W. (1983) Gaseous nitrogen exchanges in grazed pastures. In: Freney J.R. and Simpson J.R. (eds), *Gaseous Loss of Nitrogen from Plant-soil Systems.* Martinus Nijhoff/Dr W. Junk Publishers, The

Hague, pp. 215-236.

Sims J.T. (1986) Soil pH effects on the distribution and plant availability of manganese, copper and zinc. *Soil Science Society of America Journal* 50, 367-373.

Sims, J.T. and Boswell F.C. (1980) The influence of organic wastes and inorganic nitrogen sources on soil nitrogen, yield and elemental composition of corn. *Journal of Environmental Quality* 9, 512-517.

Sims J.T. and Wolf D.L. (1994) Poultry waste management: Agricultural and environmental issues. *Advances in Agronomy* 52, 1-83.

Singleton P.W., El Swaify S.A. and Bohlool B.B. (1982) Effect of salinity on *Rhizobium* growth and survival. *Applied and Environmental Microbiology* 44, 884-890.

Skinner R.J., Church B.M. and Kershaw C.D. (1992) Recent trends in soil pH and nutrient status in England and Wales. *Soil Use and Management* 8, 16-20.

Slangen J.H.G. and Kerkhoff P. (1984) Nitrification inhibitors in agriculture and horticulture: A literature review. *Fertilizer Research* 5, 1-76.

Sleeman P.J. (1984) Determination of Pollutants in Effluents (MPC 4332 C). Detailed Analysis of the Trace Element Contents of UK Sewage Sludges. WRc Report No. 280-S. WRc Medmenham, Marlow.

Smith D.W. and Wischmeier W.H. (1962) Rainfall erosion. *Advances in Agronomy* 14, 109-148.

Smith G.S., Hallford D.M. and Watkins J.B. (1985) Toxicological effects of gamma irradiated sewage solids fed as seven percent of diet to sheep for four years. *Journal of Animal Science* 61, 931-941.

Smith K.A. and Chambers B.J. (1992) Improved utilization of slurry nitrogen for arable cropping. In: *Nitrate and Farming Systems. Aspects of Applied Biology* 30, 127-134.

Smith K.A., Crichton I.J., McTaggart I.P. and Long R.W. (1989) Inhibition of nitrification by dicyandiamide in cool temperate conditions. In: Hansen J.A. and Henriksen K. (eds), *Nitrogen in Organic Wastes Applied to Soils.* Academic Press Limited, London, pp. 289-303.

Smith S.R. (1991) Effects of sewage sludge application on soil microbial processes and soil fertility. *Advances in Soil Science* 16, 191-212.

Smith S.R. (1992) Sites with a Long History of Sludge Disposal: Phase II. The Occurrence of Rhizobium in Soils with a Long History of Sludge Disposal. Interim Report to the Department of the Environment. WRc Report No. DoE 2752(P). WRc Medmenham, Marlow.

Smith S.R. (1994a) Effect of soil pH on availability to crops of metals in sewage sludge-treated soils. I. Nickel, copper and zinc uptake and toxicity to ryegrass. *Environmental Pollution* 85, 321-327.

Smith S.R. (1994b) Effect of soil pH on availability to crops of metals in sewage sludge-treated soils. II Cadmium uptake by crops and implications for human dietary intake. *Environmental Pollution* 86, 5-13.

Smith S.R. (1994c) Effects of Heavy Metals on the Size and Activity of the Soil Microbial Biomass After Long-term Treatment with Sewage Sludge. Report No. FR 0469. Foundation for Water Research, Marlow.

Smith S.R. (1994d) Effects of Heavy Metals in Sewage Sludge-treated Soils on Nitrogen Turnover. II. Field Observations on Nitrification. WRc Report No. UM 1427/1. WRc Medmenham, Marlow.

Smith S.R. and Giller K.E. (1992) Effective *Rhizobium leguminosarum* biovar *trifolii* present in five soils contaminated with heavy metals from long-term applications of sewage sludge or metal mine spoil. *Soil Biology and Biochemistry* 24, 781-788.

Smith S.R. and Hadley P. (1988) A comparison of the effects of organic and inorganic nitrogen fertilizers on the growth response of summer cabbage (*Brassica oleracea* var. *capitata* cv Hispi F_1). *Journal of Horticultural Science* 63, 615-620.

Smith S.R. and Hadley P. (1990) Carbon and nitrogen mineralization characteristics of organic nitrogen fertilizers in a soil-less incubation system. *Fertilizer Research* 23, 97-103.

Smith S.R. and Hadley P. (1992) Nitrogen fertilizer value of activated sewage derived protein: Effect of environment and nitrification inhibitor on NO_3^- release, soil microbial activity and yield of summer cabbage. *Fertilizer Research* 33, 47-57.

Smith S.R. and Powlesland C.B. (1990) Impact of Nitrate Sensitive Areas on Agricultural Utilization of Sewage Sludge. Report No. FR 0146. Foundation for Water Research, Marlow.

Smith S.R. and Woods V. (1994) Patterns of Nitrate Accumulation in Sewage Sludge-Treated Agricultural Soils. WRc Report No. UC 2453. WRc Medmenham, Marlow.

Smith S.R., Obbard J.P., Kwan K.H.M. and Jones K.C. (1990) Symbiotic N_2-fixation and Microbial Activity in Soils Contaminated with Heavy Metals Resulting from Long-Term Sewage Sludge Application. Report No. FR 0128. Foundation for Water Research, Marlow.

Smith S.R., Hall J.E. and Hadley P. (1992a) Composting sewage sludge wastes in relation to their suitability for use as fertilizer materials for vegetable crop production. *Acta Horticulturae* 302, 203-215.

Smith S.R., Sweet N.R., Davies G.K. and Hallett J.E. (1992b) Uptake of chromium and mercury by crops. Sites with a Long History of Sludge Disposal: Phase II (EHA 9019). Final Report to the Department of the Environment. WRc Report No. DoE 3023. WRc Medmenham, Marlow, pp. 53-73.

Smith S.R., Sweet N.R., Davies G.K. and Hallett J.E. (1992c) Effect of soil pH on availability to crops of cadmium and other heavy metals in sewage sludge-treated soils. Sites with a Long History of Sludge Disposal: Phase II (EHA 9019). Final Report to the Department of the Environment. WRc Report No. DoE 3023. WRc Medmenham, Marlow, pp. 7-53.

Smith S.R., Sweet N.R., Davies G.K. and Hallett J.E. (1992d) Concentrations of PTEs in ryegrass relative to their depth in the soil profile. Sites with a Long History of Sludge Disposal: Phase II (EHA 9019). Final Report to the Department of the Environment. WRc Report No. DoE 3023. WRc Medmenham, Marlow, pp. 73-101.

Smith S.R., Sweet N.R., Davies G.K. and Hallett J.E. (1992e) Effects of PTEs in sludge-treated soils on *Rhizobium leguminosarum* biovar *trifolii*. Sites with a Long History of Sludge Disposal: Phase II (EHA 9019). Final Report to the Department of the Environment. WRc Report No. DoE 3023. WRc Medmenham, Marlow, pp. 101-143.

Smith S.R., Tibbett M. and Evans T.D. (1992f) Nitrate accumulation potential of sewage sludge applied to soil. In: *Nitrate and Farming Systems. Aspects of Applied Biology* 30, 157-161.

Smith S.R., Hallett J.E., Reynolds S.E., Brookman S.J., Carlton-Smith C.H., Woods V. and Sweet N. (1994a) Nitrate Leaching Losses from Sewage Sludge-Treated Agricultural Land. WRc Report No. UM 1448. WRc Medmenham, Marlow.

Smith S.R., Reynolds S.E., Hallett J.E., Carlton-Smith C.H., Woods V. and Brookman S.J. (1994b) Studies on Management Strategies to Minimize Nitrate Leaching Losses from Sewage Sludge-treated Agricultural Land. WRc Report No. UC 2458. WRc Medmenham, Marlow.

Sneath R.W. and Phillips V.R. (1992) New Techniques and Equipment for Applying Sewage Sludge to Grassland and Growing Crops. WRc Report No. UM 1327. WRc Medmenham, Marlow.

SO; Danish Statutory Order (1989) Statutory Order No. 736 of October 26, 1989, on application of sludge, sewage and compost etc. for agricultural purposes. Ministry of the Environment, National Agency of Environmental Protection, Copenhagen.

SOAFD; Scottish Office Agriculture and Fisheries Department (1991) Prevention of Environmental Pollution from Agricultural Activity Code of Practice. SOAFD, Edinburgh.

Sommer S.G. and Christensen B.T. (1991) Effect of dry matter content on ammonia loss from surface applied cattle slurry. In: Nielsen V.C., Voorburg J.H. and L'Hermite P. (eds), *Odour and Ammonia Emissions from Livestock Farming*. Elsevier Science Publishers Ltd, Barking, pp. 141-147.

Sommers L., van Volk V., Giordano P.M., Sopper W.E. and Bastian R. (1987) Effects of soil properties on accumulation of trace elements by crops. In: Page A.L., Logan T.J. and Ryan J.A. (eds), *Land Application of Sludge Food Chain Implications*. Lewis Publishers Inc., Chelsea, Michigan, pp. 5-24.

Sommers L.E. and Nelson D.W. (1981) Nitrogen as a limiting factor in land application of sewage sludges. In: *Proceedings Fourth Annual Madison Conference of Applied Research and Practice in Municipal and Industrial Waste*. University of Wisconsin - Extension, Madison, pp. 425-448.

Sommers L.E., Nelson D.W. and Silviera D.J. (1979) Transformations of carbon, nitrogen and metals in soils treated with waste materials. *Journal of Environmental Quality* 8, 287-294.

Soni R. and Abbasi S.A. (1981) Mortality and reproduction in earthworm *Pheretima posthuma* exposed to chromium, VI. *International Journal of Environmental Studies* 17, 147-149.

Soon Y.K., Bates T.E. and Moyer J.R. (1978a) Land application of chemically treated sewage sludge: II. Effects on plant and soil phosphorus, potassium, calcium and magnesium and soil pH. *Journal of Environmental Quality* 7, 269-273.

Soon Y.K., Bates T.E., Beauchamp E.G. and Moyer J.R. (1978b) Land application of chemically treated sewage sludge: I. Effects on crop yield and nitrogen availability. *Journal of Environmental Quality* 7, 264-269.

Soon Y.K., Bates T.E., and Moyer J.R. (1980) Land application of chemically treated sewage sludge: III. Effects on soil and plant heavy metal content. *Journal of Environmental Quality* 9, 497-504.

Sorber C.A. and Moore B.E. (1987) Survival and Transport of Pathogens in Sludge-Amended Soil. A Critical Literature Review. US EPA Report No. EPA/600/2-87/028. National Technical Information Service, Springfield, Virginia.

Southey J.F. and Glendinning K.R. (1966) The effect of mesophilic anaerobic digestion of sludge on potato root eelworm cysts. *Journal of the Proceedings of the Institute of Sewage Purification* for 1966, 186-189.

Spaull A.M. and McCormack D.M. (1989) The incidence and survival of potato cyst nematode (*Globodera* spp.) in various sewage sludge treatments and processes. *Nematologica* 34, 452-461.

Speirs R.B. and Frost C.A. (1987) The enhanced acidification of a field soil by very low concentrations of atmospheric ammonia. *Research and Development in Agriculture* 4, 83-86.

Springett J.A. and Gray R.A.J. (1992) Effect of repeated low doses of biocides on the earthworm *Aporrectodea caliginosa* in laboratory culture. *Soil Biology and Biochemistry* 24, 1739-1744.

Spurgeon D.J., Hopkins S.P. and Jones D.J. (1994) Effects of cadmium, copper, lead and zinc on growth, reproduction and survival of the earthworm *Eisenia foetida* (Savigny): Assessing the environmental impact of point-source metal contamination in terrestrial ecosystems. *Environmental Pollution* 84, 123-130.

Stahl R.S. and James B.R. (1991) Zinc sorption by B horizon soils as a function of pH. *Soil Science Society of America Journal* 55, 1592-1597.

Stark B.A. (1988) Effects on Grazing Animals of Ingestion of Inorganic and Organic Materials Contained in Sewage Sludge. WRc Report No. PRU 1691-M. WRc Medmenham, Marlow.

Stark B.A. (1989) Sites with a Long History of Sludge Disposal: Phase II. Possible Implications of Soil Ingestion by Grazing Animals. WRc Report

No. DoE 2123-M. WRc Medmenham, Marlow.

Stark J.H. and Carlton-Smith C.H. (1985) Sites with a Long History of Sludge Depositon (SO 4166C). Progress Report to the Department of the Environment: April 1984-March 1985. WRc Report No. 942-M. WRc Medmenham, Marlow.

Stark S.A. and Clapp C.E. (1980) Residual nitrogen availability from soils treated with sewage sludge in a field experiment. *Journal of Environmental Quality* 9, 505-512.

Stark B.A. and Hall J.E. (1992) Implications of sewage sludge application to pasture on the intake of contaminants by grazing animals. In: Hall J.E., Sauerbeck D.R. and L'Hermite P. (eds), *Effects of Organic Contaminants in Sewage Sludge on Soil Fertility, Plants and Animals*. Commission of the European Communities, Luxembourg, pp. 134-157.

Stark J.H. and Lee D.H. (1988) Sites with a History of Sludge Deposition. Final Report on Rehabilitation Field Trials and Studies Relating to Soil Microbial Biomass (LDS 9166 SLD). Final Report to the Department of the Environment. WRc Report No. DoE 1768-M. WRc Medmenham, Marlow.

Stark B.A. and Wilkinson J.M. (1994) Accumulation of Potentially Toxic Elements by Sheep Given Diets Containing Sewage Sludge (OC 8910, CSA 1826). Final Report to the Ministry of Agriculture, Fisheries and Food. Report No. 7. Chalcombe Agricultural Resources, Canterbury.

Stark J.H. and Whitelaw K. (1986) Crop Uptake of Tungsten, Cerium, Niobium, Lanthanum, Silver, Antimony, Bismuth and Cobalt from Sludge-treated Soils and their Effect on Yields. WRc Report No. ER 1154-M. WRc Medmenham, Marlow.

Stark J.H., Carlton-Smith C.H. and Campbell C.A. (1986) Sites with a Long History of Sludge Deposition (EI 9166 SLD). Progress Report to the Department of the Environment: April 1985 - March 1986. WRc Report No. DoE 1192-M. WRc Medmenham, Marlow.

Stark B., Suttle N., Sweet N. and Brebner J. (1995) Accumulation of PTEs in Animals Fed Dried Grass Containing Sewage Sludge. Final Report to the Department of the Environment. WRc Report No. DoE 3753/1. WRc, Medmenham, Marlow.

Steenhuis T.S. and Naylor L.M. (1985) Relative risk to groundwater from chemicals in land applicated sludge. American Society of Agricultural Engineers, Paper No. 85-5527.

Steenvoorden J.H.A.M. (1981) Landspreading of animal manure and run-off: Comments on the draft guidelines. In: Brogan J.C. (ed), *Nitrogen Losses and Surface Run-off from Landspreading of Manures*. Martinus Nijhoff/Dr W. Junk Publishers, The Hague, pp. 26-33.

Steenvoorden J.H.A.M. (1986) Nutrient leaching losses following application of farm slurry and water quality concentrations in the Netherlands. In: Dam Kofoed A., Williams J.H. and L'Hermite P. (eds), *Efficient Land Use of Sludge and Manure*. Elsevier Applied Science Publishers Ltd, Barking, pp. 168-176.

Steenvoorden J.H.A.M. (1989) Agricultural practices to reduce nitrogen losses via leaching and surface run-off. In: German J.C. (ed), *Management Systems to Reduce Impact of Nitrates.* Elsevier Applied Science Publishers Ltd, Barking, pp. 72-84.

Steinhilber P.M. (1981) Fate of sewage sludge derived Zn relative to soil factors and plant utilization. PhD Thesis, University of Georgia. University Microfilms, Ann Arbor, Michigan.

Steinhilber P. and Boswell F.C. (1983) Fractionation and characterization of two aerobic sewage sludges. *Journal of Environmental Quality* 12, 529-534.

Stenersen J., Brekke E. and Engelstad F. (1992) Earthworms for toxicity testing; species differences in response towards cholinesterase inhibiting insecticides. *Soil Biology and Biochemistry* 24, 1761-1764.

Stevenson F.J. (1977) Nature of divalent transition metal complexes of humic acids as revealed by a modified potentiometric titration method. *Soil Science* 123, 10-17.

Stevenson F.J. and Ardakani M.S. (1972) Organic matter reactions involving micronutrients in soils. In: *Micro-nutrients in Agriculture.* Soil Science Society of America Inc., Madison, Wisconsin.

Stevenson F.J. and Chen Y. (1991) Stability constants of copper(II)-humate complexes determined by modified potentiometric titration. *Soil Science Society of America Journal* 55, 1586-1591.

Stewart N.E., Beauchamp E.G., Corke C.T. and Webber L.R. (1975) Nitrate nitrogen distribution in cornland following applications of digested sewage sludge. *Canadian Journal of Soil Science* 55, 287-294.

Storey G.W. and Phillips R.A. (1985) The survival of parasite eggs throughout the soil profile. *Parasitology* 91, 585-590.

Stover R.C., Sommers L.E. and Silviera D.J. (1976) Evaluation of metals in waste water sludges. *Journal of the Water Pollution Control Federation* 48, 2165-2175.

Street J.J., Lindsay W.L. and Sabey B.R. (1977) Solubility and plant uptake of cadmium in soils amended with cadmium and sewage sludge. *Journal of Environmental Quality* 6, 72-77.

Strehlow C.D. and Barltrop D. (1988) Health studies. The Shipham report: An investigation into cadmium contamination and its implications for human health. *The Science of the Total Environment* 75, 101-133.

Strutt N. (1970) Modern Farming and the Soil. Report of the Agricultural Advisory Council on Soil Structure and Soil Fertility. HMSO, London.

Stukenberg J.R., Shimp G., Sandino J., Clark J.H. and Crosse J.T. (1994) Compliance outlook: Meeting 40CFR part 503, class B pathogen reduction criteria with anaerobic digestion. *Water Environment Research* 66(3), 255-263.

Suttle N.F. (1975) Trace element interactions in animals. In: Nicholas D.J.D.

and Egeen A.R. (eds), *Trace Elements in Soil-Plant-Animal Systems.* Academic Press Inc. (London) Ltd, London, pp. 271-289.

Suttle N.F., Alloway B.J. and Thornton I. (1975) An effect of soil ingestion on the utilization of dietary Cu by sheep. *Journal of Agricultural Science, Cambridge* 84, 249-254.

Suttle N.F., Abrahams P. and Thornton I. (1984) The role of a soil x dietary sulphur interaction in the impairment of copper absorption by ingested soil in sheep. *Journal of Agricultural Science, Cambridge* 103, 81-86.

Sweet N., Reynolds S. and Hallett J. (1994) The Application of Sewage Sludge to Pasture - An Investigation of Potential Sward Contamination. WRc Report No. UM 1449. WRc Medmenham, Marlow.

Sweetman A.J. (1991) Review of Dioxins in Sludge and Their Significance to Sludge Utilization in Agriculture. WRc Report No. UM 1199. WRc Medmenham, Marlow.

Sweetman A.J., Rogers H.R., Harms H. and Mosbaek H. (1992a) Organic Contaminants in Sewage Sludge and their Effects on Soil and Crops. Final Report to the Department of the Environment. WRc Report No. DoE 2745(P). WRc Medmenham, Marlow.

Sweetman A.J., Rogers H.R., Fullman J.R., Alcock R. and Jones K.C. (1992b) Organic Contaminants in Sewage Sludge: Phase III (Env 9031). Progress Report to the Department of the Environment: July to December. WRc Report No. DoE 3277. WRc Medmenham, Marlow.

Sweetman A., Rogers H.R., Watts C.D., Alco R. and Jones K.C. (1994) Organic Contaminants in Sewage Sludge: Phase III (ENV 9031). Final Report to the Department of the Environment. WRc Report No. DoE 3625/1. WRc Medmenham, Marlow.

Telford J.N., Thonney M.L., Hogue D.E., Stouffer J.R., Bache C.A., Gutenmann W.H. and Lisk D.J. (1982) Toxicologic studies in growing sheep fed silage corn cultured on municipal sludge-amended acid soil. *Journal of Toxicology and Environmental Health* 10, 73-85.

Telford J.N., Babish J.G., Johnson B.E., Thonney M.L., Currie B.W., Bache C.A., Gutenmann W.H. and Lisk D.J. (1984) Toxicologic studies with pregnant goats fed grass-legume silage grown on municipal sludge-amended subsoil. *Archives of Environmental Contamination and Toxicology* 13, 635-640.

Tena M., Garrido R. and Magallanes M. (1984) Sugar beet herbicides and soil nitrification. *Soil Biology and Biochemistry* 16, 223-226.

Tester C.F., Sikora L.J., Taylor J.M. and Parr J.F. (1977) Decomposition of sewage sludge compost in soil. I. Carbon and nitrogen transformation. *Journal of Environmental Quality* 6, 459-463.

Theis J.H., Bolton V. and Storm D.R. (1978) Helminth ova in soil and sludge from twelve US urban areas. *Journal of the Water Pollution Control Federation* 50, 2485-2493.

Thiel D.A., Martin S.G., Duncan J.W. and Lance W.R. (1989) The effects of a sludge containing dioxin on wildlife in pine plantations. *Tappi Journal, January 1989*, 94-99.

Thompson R.B. and Pain B.F. (1989) Denitrification from cattle slurry applied to grassland. In: Hansen J.A. and Henriksen K. (eds), *Nitrogen in Organic Wastes Applied to Soils*. Academic Press Limited, London, pp. 247-260.

Thornton I. (1974) Biogeochemical and soil ingestion studies in relation to the trace element nutrition of livestock. In: Hoekstra W.G., Suttie J.W., Ganther H.E. and Mertz W. (eds), *Trace Element Metabolism in Animals*, Second edition. Butterworths, London, pp. 451-454.

Thornton I. and Abrahams P. (1983) Soil ingestion - a major pathway of heavy metals into livestock grazing contaminated land. *The Science of the Total Environment* 28, 287-294.

Tierney J.J., Sullivan R. and Larkin E. (1977) Persistence of poliovirus I in soil and on vegetables grown in soil previously flooded with inoculated sewage sludge or effluent. *Applied Environmental Microbiology* 33, 109-113.

Tipping E., Thompson D.W., Ohnstad M. and Hetherington N.B. (1986) Effects of pH on the release of metals from naturally-occurring oxides of Mn and Fe. *Environmental Technology Letters* 7, 109-114.

Titchen N.M. and Scholefield D. (1992) The potential of a rapid test for soil mineral nitrogen to determine tactical applications of fertilizer nitrogen to grassland. In: *Nitrate and Farming Systems. Aspects of Applied Biology* 30, 223-229.

Todd J.R. (1978) The copper status of ruminant animals in Northern Ireland in relation to the usage of copper compounds in agriculture. In: Kirchgessner M. (ed), *Proceedings of the 3rd International Symposium on Trace Element Metabolism in Man and Animals*. Freising-Weihenstephan, pp. 486-489.

Tritt, W.P. (1994) Problems concerning the acceptability of sewage sludges. *Korrespondenz Abwasser* 41(8), 1306-1316.

Turner A.P., Giller K.E. and McGrath S.P. (1993) Long term effects on *Rhizobium leguminosarum* bv. *trifolii* of heavy metal contamination of land from the application of sewage sludge. In: Allan R.J. and Nriagu J.O. (eds), *International Conference Heavy Metals in the Environment*, Volume 1. CEP Consultants Ltd, Edinburgh, pp. 442-445.

Turner J., Stafford D.A., Hughes D.E. and Clarkson J. (1983) The reduction of three plant pathogens (*Fusarium, Corynebacterium* and *Globodera*) in anaerobic digesters. *Agricultural Wastes* 6, 1-11.

Turpin P.E., Dhir V.K., Maycroft K.A., Rowlands C. and Wellington E.M.H. (1992) The effect of *Streptomyces* species on the survival of *Salmonella* in soil. *FEMS Microbiology Ecology* 101, 271-280.

Turpin P.E., Maycroft K.A., Rowlands C.L. and Wellington E.M.H. (1993) Viable but non-culturable salmonellas in soil. *Journal of Applied Bacteriology* 74, 421-427.

Twigg G.I., Hughes D.M. and McDiarmid A. (1973) The low incidence of leptospirosis in British deer. *Veterinary Record* 93, 98-100.

Tyler G. (1981) Heavy metals in soil biology and biochemistry. In: Paul E.A. and Ladd J.N. (eds), *Soil Biochemistry*, Volume 5. Marcel Dekker, Inc., New York, pp. 371-414.

Tyson K.C., Roberts D.H., Clement C.R. and Garwood E.A. (1990) Comparison of crop yields and soil conditions during 30 years under annual tillage or grazed pasture. *Journal of Agricultural Science, Cambridge* 115, 29-40.

Uhlen G. (1981) Surface run-off and the use of farm manure. In: Brogan J.C. (ed), *Nitrogen Losses and Surface Run-off from Landspreading of Manures.* Martinus Nijhoff/Dr W. Junk Publishers, The Hague, pp. 34-43.

Underwood E.J. (1971) The history and philosphy of trace element research. In: Mertz W. and Corratzer W.E. (eds), *Newer Trace Elements in Nutrition.* Marcel Dekker Inc., New York, pp. 1-18.

Underwood E.J. (1977) *Trace Elements in Human and Animal Nutrition*, Fourth edition. Academic Press Inc. (London) Ltd, London.

Underwood E.J. (1981) *The Mineral Nutrition of Livestock*, Second edition. Commonwealth Agricultural Bureaux, Farnham Royal.

Unwin, R.J., Royle S.M. and Chandrasekhar N.C. (1989) Apparent recovery of zinc, copper and nickel applied to soil in sewage sludge. In: Hall J.E. (ed), *Alternative Uses for Sewage Sludge*, Poster papers presented at a conference organized by WRc, University of York, 5-7 September. WRc Report No. CP 596. WRc Medmenham, Marlow, pp. 63-69.

Unwin R.J., Shepherd M.A. and Smith K.A. (1991) Controls on manure and sludge applications to limit nitrate leaching, does the evidence justify the restrictions which are being proposed? In: L'Hermite P. (ed), *Treatment and Use of Sewage Sludge and Liquid Agricultural Wastes.* Elsevier Science Publishers Ltd, London, pp. 261-270.

Ure A.M. and Berrow M.L. (1982) The elemental constituents of soils. In: *Environmental Chemistry*, Volume 2. The Royal Society of Chemistry, London, pp. 94-204.

USDA; United States Department of Agriculture (1978) *Improving Soils with Organic Wastes.* Report to the Congress in response to Sec. 1461 of the Food and Agriculture Act of 1977. US Govt. Printing Office, Washington D.C.

US EPA; US Environmental Protection Agency (1979) Criteria for classification of solid waste disposal facilities and practices. *Federal Register* 44, 53438-53468.

US EPA; US Environmental Protection Agency (1980) Sewage Sludge: Factors Affecting the Uptake of Cadmium by Food-Chain Crops Grown on Sludge-Amended Soils. W-124 SEA-CR Technical Research Committee. US EPA Report No. SW-882. National Technical Information Service, Springfield, Virginia.

US EPA; US Environmental Protection Agency (1987) Proceedings: Workshop on Effects of Sewage Sludge Quality and Soil Properties on Uptake of

Sludge-Applied Trace Constituents. EPA 600/9-87/002. National Technical Information Service, Springfield, Virginia.

US EPA; US Environmental Protection Agency (1989a) 40 CFR Parts 257 and 503 Standards for the Disposal of Sewage Sludge; Proposed Rule. *Federal Register* 54, 5745-5902.

US EPA; US Environmental Protection Agency (1989b) *Development of Risk Assessment Methodology for Land Application and Distribution and Marketing of Municipal Sludge.* EPA 600/6-89/001. National Technical Information Service, Springfield, Virginia.

US EPA; US Environmental Protection Agency (1992a) *Technical Support Document for Land Application of Sewage Sludge,* Volume I. Eastern Research Group, Lexington.

US EPA; US Environmental Protection Agency (1992b) *Technical Support Document for Land Application of Sewage Sludge,* Volume II, Appendices. Eastern Research Group, Lexington.

US EPA; US Environmental Protection Agency (1993) Part 503-Standards for the Use or Disposal of Sewage Sludge. *Federal Register* 58, 9387-9404.

US EPA; US Environmental Protection Agency (1994a) Health Assessment Document for 2,3,7,8-Tetrachlorodibenzo-*p*-dioxin (TCDD) and Related Compounds. EPA/600/BP-92/001a-c. US EPA, Cincinnati.

US EPA; US Environmental Protection Agency (1994b) Estimating Exposure to Dioxin-like Compounds. EPA/600/6-88/005Ca-c. US EPA, Cincinnati.

Van Breemen N. (1991) Ecological effects of ammonia deposition. In: L'Hermite P. (eds), *Treatment and Use of Sewage Sludge and Liquid Agricultural Wastes.* Elsevier Science Publishers Ltd, Barking, pp. 90-105.

Vance E.D., Brookes P.C. and Jenkinson D.S. (1987) An extraction method for measuring soil microbial biomass C. *Soil Biology and Biochemistry* 19, 703-707.

Van den Abbeel R., Claes A. and Vlassak K. (1989) Gaseous nitrogen losses from slurry-manured land. In: Hansen J.A. and Henriksen K. (eds), *Nitrogen in Organic Wastes Applied to Soils.* Academic Press Limited, London, pp. 213-224.

Van der Voet E., van Egmond L., Kleijn R. and Huppes G. (1994) Cadmium in the European Community: A policy-orientated analysis. *Waste Management and Research* 12, 507-526.

Van Gestel C.A.M., Dirven-van Breeman E.M. and Baerselman R. (1993) Accumulation and elimination of cadmium, chromium and zinc and effects on growth and reproduction in *Eisenia andrei* (Oligochaeta, Annelida). *The Science of the Total Environment,* Supplement, 585-597.

Van Gestel M., Ladd J.N. and Amato M. (1992) Microbial biomass responses to seasonal change and imposed drying regimes at increasing depths of undisturbed topsoil profiles. *Soil Biology and Biochemistry* 24, 103-111.

Van Loon J.C. (1974) Mercury contamination of vegetation due to the application of sewage sludge as a fertilizer. *Environmental Letters* 6,

211-218.

Van Rhee J.A. (1967) Development of earthworm populations in orchard soils. In: Graff O. and Satchell J. (eds), *Progress in Soil Biology.* North Holland Publishing Company, Amsterdam, pp. 360-371.

Van Rhee J.A. (1969) Effects of biocides and their residues on earthworms. *Mededelingen Rijksfaculteit Landbouwweten-schappen (Gent)* 34, 682-689.

Van Rhee J.A. (1975) Copper contamination effects on earthworms by disposal of pig wastes in pastures. In: Vanek J. (ed), *Proceedings of the 5th International Colloquium on Soil Zoology.* Dr Junk W./B.V. Publishers, The Hague, pp. 451-456.

Varma U.M., Christian B.A. and McKinstry D.W. (1974) Inactivation of Sabin oral poliomyelitis type I virus. *Journal of the Water Pollution Control Federation* 46, 987-992.

Verloo M. (1979) Influence of soil organic matter on the behaviour of heavy metals in soils and sediments. *Laboratoire de Analyst et Agrochemie,* R.U. Gent.

Verloo M. and Cottenie A. (1972) Stability and behaviour of complexes of Cu, Zn, Fe, Mn and Pb with humic substances of soils. *Pedologie* 22, 174-184.

Vetter H. and Steffens G. (1981) Surface run-off. In: Brogan J.C. (ed), *Nitrogen Losses and Surface Run-off from Landspreading of Manures.* Martinus Nijhoff/Dr W. Junk Publishers, The Hague, pp. 84-86.

Vigerust E. and Selmer-Olsen A.R. (1986) Basis for metal limits relevant to sludge utilisation. In: Davis R.D., Haeni H. and L'Hermite P. (eds), *Factors Influencing Sludge Utilization Practices in Europe.* Elsevier Applied Science Publishers Ltd, Barking, pp. 26-42.

Vincent J.M. (1970) *A Manual for the Practical Study of the Root-Nodule Bacteria.* IBP Handbook No. 15. Blackwell, Oxford.

Vivier F.S., Pieterse S.A. and Aucamp P.J. (1988) Guidelines for the use of sewage sludge. Paper presented at the Symposium on Sewage Sludge Handling, November 15. Division for Water Technology, CSIR, Pretoria.

Vlamis J., Williams D.E., Corey J.E., Page A.L. and Ganje T.J. (1985) Zinc and cadmium uptake by barley in field plots fertilized seven years with urban and suburban sludge. *Soil Science* 139, 81-87.

Voets J.P., Meerschman P. and Verstraete W. (1974) Soil microbiological and biochemical effects of long-term atrazine applications. *Soil Biology and Biochemistry* 6, 149-152.

Wade S.E., Bache C.A. and Lisk D.J. (1982) Cadmium accumulation by earthworms inhabiting municipal sludge-amended soil. *Bulletin of Environmental Contamination and Toxicology* 28, 557-560.

Wadman W.P. and Neeteson J.J. (1992) Nitrate leaching losses from organic manures - the Dutch experience. In: *Nitrate and Farming Systems. Aspects of Applied Biology* 30, 117-126.

Wadman W.P., Neeteson J.J. and Wijnen G. (1989) Effects of slurry with and without the nitrification inhibitor dicyandiamide on soil mineral and nitrogen response of potatoes. In: Hansen J.A. and Henriksen K. (eds) *Nitrogen in Organic Wastes Applied to Soils.* Academic Press Limited, London, pp. 304-314.

Wagener G.J. (1993) Accumulation of cadmium in crop plants and its consequences to human health. *Advances in Agronomy* 51,173-212.

Wainwright M. (1978) A review of the effects of pesticides on microbial activity in soils. *Journal of Soil Science* 29, 287-298.

Wainwright M. and Pugh G.J.F. (1973) The effect of three fungicides on nitrification and ammonification in soil. *Soil Biology and Biochemistry* 5, 577-584.

Walther W. (1989) The nitrate leaching out of soils and their significance for groundwater: Results of long-term tests. In: Hansen J.A. and Henriksen K., *Nitrogen in Organic Wastes Applied to Soils.* Academic Press Limited, London, pp. 346-356.

Wang M-J. and Jones K.C. (1994) Behaviour and fate of chlorobenzenes (CBs) introduced into soil-plant systems by sewage sludge application: A review. *Chemosphere* 28, 1325-1360.

Ward R.C. (1975) *Principles of Hydrology*, Second edition. McGraw-Hill Book Company (UK) Ltd, Maidenhead.

Ward R.C. and Ashley C.S. (1977) Identification of the virucidal agent in wastewater sludge. *Applied Environmental Microbiology* 33, 860-864.

Watson D.C., Stachwell M. and Jones C.E. (1980) A study of the prevalence of parasitic helminth eggs and cysts in sewage sludges disposed to agricultural land. Paper presented to IWPC, November 1980.

Webb J. and Archer J.R. (1994) Pollution of soils and water courses by wastes from livestock production systems. In: Ap Dewi I., Axford R.F.E., Marai I.F.M. and Omed H.M. (eds), *Pollution in Livestock Production Systems.* CAB International, Wallingford, pp. 189-204.

Webber J. (1980) Metals in sewage sludge applied to the land and their effects on crops. In: *Inorganic Pollution and Agriculture.* MAFF Reference Book 326. HMSO, London, pp. 222-234.

Webber J. (1981) Trace metals in agriculture. In: Lepp N.W. (ed), *Effect of Heavy Metal Pollution on Plants*, Volume 2. Applied Science Publishers Ltd., London, pp. 159-184.

Webber M.D. and Goodin J.D. (1992) Studies on the fate of organic contaminants in sludge treated soils. In: Hall J.E., Sauerbeck D.R. and L'Hermite P. (eds), *Effects of Organic Contaminants in Sewage Sludge on Soil Fertility, Plants and Animals.* Commission of the European Communities, Luxembourg, pp. 54-69.

Webber M.D. and Monks T.L. (1983) Cadmium concentrations in field and vegetable crops - A recommended maximum cadmium loading to agricultural soils. In: Davis, R.D., Hucker G. and L'Hermite P. (eds), *Environmental Effects of Organic and Inorganic Contaminants in Sewage Sludge.* D. Reidel Publishing Company, Dordrecht, pp. 130-136.

Webber M.D., Pietz R.I., Granato T.C. and Svoboda M.L. (1994) Plant uptake of PCBs and other organic contaminants from sludge-treated coal refuse. *Journal of Environmental Quality* 23, 1019-1026.

Wegener K.E., Alday R. and Meyer B. (1985) Soil algae: Effects of herbicides on growth and C_2H_2 reduction (nitrogenase) activity. *Soil Biology and Biochemistry* 17, 641-644.

Weiss B. and Larink O. (1991) Influence of sewage sludge and heavy metals on nematodes in an arable soil. *Biology and Fertility of Soils* 12, 5-9.

Wekerle J. (1986) Agricultural use of sewage sludge as a vector for transmission of viral disease. In: Block J.C., Havelaar A.H. and L'Hermite P. (eds), *Epidemiological Studies of Risks Associated with the Agricultural Use of Sewage Sludge: Knowledge and Needs.* Elsevier Applied Science Publishers Ltd, London, pp. 106-122.

Welch J.E. and Lund L.J. (1987) Soil properties, irrigation water quality and soil moisture level influences on the movement of nickel in sewage sludge-treated soils. *Journal of Environmental Quality* 16, 403-410.

Welp G. and Brümmer G.W. (1992) Toxicity of organic pollutants to soil micro-organisms. In: Hall J.E., Sauerbeck D.R. and L'Hermite P. (eds), *Effects of Organic Contaminants in Sewage Sludge on Soil Fertility, Plants and Animals.* Commission of the European Communities, Luxembourg, pp. 161-168.

White R.J. (1983) Nitrate in British waters. *Aqua* 2, 51-57.

Whitehead D.C. (1990) Atmospheric ammonia in relation to grassland agriculture and livestock production. *Soil Use and Management* 6, 63-65.

Whitmore T.N. (1993) Fate of *Cryptosporidium* During Sewage Sludge Treatment (EHA 9036). WRc Report No. DoE 3265/1. WRc Medmenham, Marlow.

WHO; World Health Organization (1981) Sewage Sludge to Land: Human Health Implications of Microbial Content. Report on a WHO Working Group, Stevenage 6-9 January 1981. EURO Reports and Studies No. 95-S. WHO, Copenhagen.

WHO; World Health Organization (1984) *Guidelines for Drinking Water Quality. Volume 1: Recommendations.* WHO, Geneva.

WHO; World Health Organization (1989) *DDT and its Derivatives - Environmental Aspects.* Environmental Health Criteria, 83. WHO, Geneva.

WHO/FAO; World Health Organization/Food and Agriculture Organization (1972) *Evaluation of Certain Food Additives and the Contaminants Mercury, Lead and Cadmium.* WHO Technical Report Series No. 505. WHO, Geneva.

Widdowson F.V., Penny A., Darby R.J., Bird E. and Hewitt M.V. (1987) Amounts of NO_3-N and NH_4-N in soil, from autumn to spring, under winter wheat and their relationship to soil type, sowing date, previous crop and N uptake at Rothamsted, Woburn and Saxmundham, 1979-85. *Journal of Agricultural Science, Cambridge* 108, 73-95.

Wild A. (1988a) Plant nutrients in soil: nitrogen. In: Wild A. (ed), *Russell's Soil Conditions and Plant Growth*, 11th edition. Longman Scientific and Technical, Harlow, pp. 652-694.

Wild A. (1988b) Plant nutrients in soil: Phosphate. In: Wild A. (ed), *Russell's Soil Conditions and Plant Growth*, 11th edition. Longman Scientific and Technical, Harlow, pp. 695-742.

Wild S.R. and Jones K.C. (1991) Organic contaminants in wastewaters and sewage sludges: Transfer to the environment following disposal. In: Jones K.C. (ed), *Organic Contaminants in the Environment.* Elsevier Science Publishers Ltd, Barking, pp. 133-158.

Wild S.R. and Jones K.C. (1992a) Organic chemicals entering agricultural soils in sewage sludges: Screening for their potential to transfer to crop plants and livestock. *The Science of the Total Environment* 119, 85-119.

Wild S.R. and Jones K.C. (1992b) The Fate and Behaviour of Polynuclear Aromatic Hydrocarbons in Sewage Sludge Amended Soils. WRc Unpublished Report. WRc Medmenham, Marlow.

Wild S.R., Berrow M.L. and Jones K.C. (1991) The persistence of polynuclear aromatic hydrocarbons (PAHs) in sewage sludge amended agricultural soils. *Environmental Pollution* 72, 141-157.

Wild S.R., Harrad S.J. and Jones K.C. (1994) The influence of sewage sludge applications to agricultural land on human exposure to polychlorinated dibenzo-*p*-dioxins (PCDDs) and -furans (PCDFs). *Environmental Pollution* 83, 357-369.

Williams D.E., Vlamis J., Pukite A.H. and Corey J.E. (1987) Metal movement in sludge-amended soils: A nine-year study. *Soil Science* 143, 124-131.

Williams J.H. (1980) Effect of soil pH on the toxicity of zinc and nickel to vegetable crops. In: *Inorganic Pollution and Agriculture.* MAFF Reference Book 326. HMSO, London, pp. 211-218.

Williams J.H. (1988) *Chromium in Sewage Sludge Applied to Agricultural Land.* Commission of the European Communities, Brussels.

Williams J.H. and Hall J.E. (1986) Efficiency of utilization of nitrogen in sludges and slurries. In: L'Hermite P. (ed), *Processing and Use of Organic Sludge and Liquid Agricultural Wastes.* D. Reidel Publishing Company, Dordrecht, pp. 258-289.

Williams P.H., Shenk J.S. and Baker D.E. (1978) Cadmium accumulation by meadow voles (*Microtus pennsylvanicus*) from crops grown on sludge-treated soil. *Journal of Environmental Quality* 7, 450-454.

Wilson B. and Jones B. (1994) The Phosphate Report: A Life Cycle Study to Evaluate the Environmental Impact of Phosphates and Zeolite A-PCA as Alternative Builders in UK Laundry Detergent Formulations. Landbank Environmental Research and Consulting, London.

Wilson S.C., Burnett V., Waterhouse K.S. and Jones K.C. (1994) Volatile organic compounds in digested United Kingdom sewage sludges. *Environmental Science and Technology* 28, 259-266.

Wischmeier W.H. and Mannering J.V. (1965) Effect of organic matter content of the soil on infiltration. *Journal of Soil and Water Conservation* 20, 150-151.

Wischmeier W.H. amd Mannering J.V. (1969) Relation of soil properties to its erodibility. *Soil Science Society of America Proceedings* 33, 131-137.

Withers P.J.A. and Sharpley A.N. (in press) Phosphorus management in sustainable agriculture. In: Cook H.F. and Lee H.C. (eds), *Soil Management in Sustainable Agriculture*, Proceedings of the 3rd Wye International Conference on Sustainable Agriculture, 31 August - 4 September, 1993.

Witte I.H. (1989) Investigations of the entry of selected organic pollutants into soils and plants by use of sewage sludge in agriculture. In: Quaghebeur D., Temmerman I. and Angeletti G. (eds), *Organic Contaminants in Waste Water, Sludge and Sediment Occurrence, Fate and Disposal.* Elsevier Science Publishers Ltd, Barking, 183-204.

Witter E. (1989) Agricultural use of sewage sludge-controlling metal contamination of soils. *Staten Naturvårdverket Rapport*, 3620, Sweden.

Witter E., Mårtensson A.M. and Garia F.V. (1993) Size of the soil microbial biomass in a long-term field experiment as affected by different N-fertilizers and organic manures. *Soil Biology and Biochemistry* 25, 659-669.

Witty, J.F. and Minchin, F.R. (1988) Measurement of nitrogen fixation by the acetylene reduction assay: Myths and mysteries. In: Beck D.P. and Materon L.A. (eds), *Nitrogen Fixation by Legumes in Mediterranean Agriculture.* ICARDA.

Wolstenholme R., Dutch J. and Riddell-Black D. (1991). The Use of Sewage Sludge in Forestry. Final Report. WRc Report No. UM 1253. WRc Medmenham, Marlow.

WRc; Water Research Centre (1979) Agricultural and environmental aspects of fluorides in sewage sludge. Notes on Water Research No. 21. WRc Medmenham, Marlow.

WRc; Water Research Centre (1985) *The Agricultural Value of Sewage Sludge - a Farmers' Guide.* WRc Medmenham, Marlow.

WRc; Water Research Centre (1989) *Soil Injection of Sewage Sludge. A Manual of Good Practice (Second Edition).* Reference No. FR 0008. WRc Medmenham, Marlow.

Yamaguchi T. and Aso S. (1977) Chromium from the stand-point of plant nutrition: I. Effect of chromium concentration on the germination and growth of several kinds of plants. *Journal of the Science of Soil and Manure, Japan* 48, 466-470.

Ye Q., Puri R.K., Kapila S. and Yanders A.F. (1992) Studies on the transport and transformation of PCBs in plants. *Chemosphere* 25, 1475-1479.

Zelles L., Bai Q.Y., Ma R.X., Rackwitz R., Winter K. and Beese F. (1994) Microbial biomass, metabolic activity and nutritional status determined from fatty acid patterns and poly-hydroxybutyrate in

agriculturally-managed soils. *Soil Biology and Biochemistry* 21, 211-221.
Zook E.G., Greene F.E. and Morris E.R. (1970) Nutrient composition of selected wheats and wheat products. VI. Distribution of manganese, copper, nickel, zinc, magnesium, lead, tin, cadmium, chromium and selenium as determined by atomic absorption spectroscopy and colorimetry. *Cereal Chemistry* 47, 720-731.

Index

Note: page numbers in *italics* refer to figures and tables